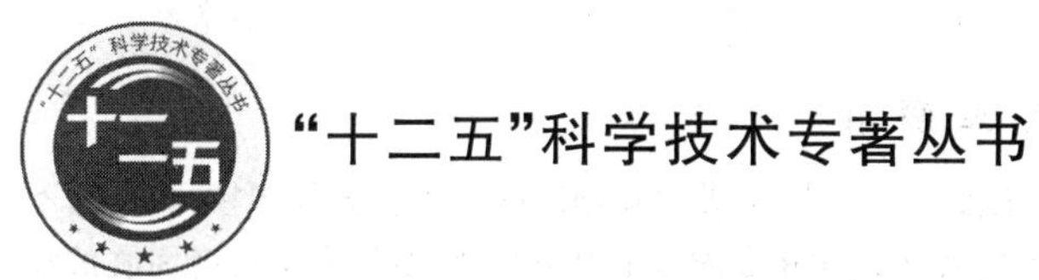

Java Web 核心技术

毋建军 著

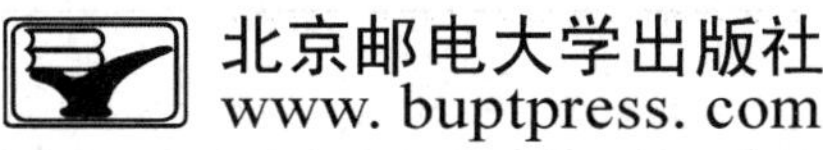

内容简介

Java Web应用已经成为当前互联网主流的应用发展方向之一，尤以移动互联网为代表的领域，更是离不开Java Web技术的支撑，如何快速地从繁多、杂乱的Web技术中学习核心基础内容，是许多学习者面临的技术门槛，本书作者希望通过理论和实际应用相结合的方式，让初入者能够理解其关注的技术内容。

本书作为Java Web原理与技术应用的入门书籍，内容全面且通俗易懂，对Java Web应用所涉及的关键核心技术进行了全面的详解，除了对Web基础技术HTML、JavaScript、Servlet、JSP解析之外，还详细讲述了核心技术应用开发中的设计模式、框架模式、Web应用服务器、JDBC、XML、Struts、Hibernate、Spring等技术及实践应用，以帮助读者学习和深入理解相关技术。

本书适用于对Java Web应用技术感兴趣的初学者、技术人员，可作为大中专院校软件开发、移动应用开发相关专业的教材，也可作为Java Web开发人员的参考书。

图书在版编目(CIP)数据

Java Web核心技术 / 毋建军著. -- 北京 ：北京邮电大学出版社，2015.5（2019.1重印）

ISBN 978-7-5635-4301-4

Ⅰ. ①J… Ⅱ. ①毋… Ⅲ. ①JAVA语言－程序设计 Ⅳ. ①TP312

中国版本图书馆CIP数据核字（2015）第033713号

书　　名：Java Web核心技术
著作责任者：毋建军　著
责 任 编 辑：徐振华　孙宏颖
出 版 发 行：北京邮电大学出版社
社　　址：北京市海淀区西土城路10号（邮编：100876）
发 行 部：电话：010-62282185　传真：010-62283578
E-mail：publish@bupt.edu.cn
经　　销：各地新华书店
印　　刷：保定市中画美凯印刷有限公司
开　　本：787 mm×1 092 mm　1/16
印　　张：19.5
字　　数：507千字
版　　次：2015年5月第1版　2019年1月第2次印刷

ISBN 978-7-5635-4301-4　　定　价：42.00元

· 如有印装质量问题，请与北京邮电大学出版社发行部联系 ·

前　言

近年来，随着互联网行业技术的不断成熟和发展，传统的软件开发基础技术已经远远不能满足当前社会的需求，无论是以淘宝、京东为代表的电子商务网站应用，还是以移动互联网为基础的移动应用，都不再是仅靠原有的C和Java基础语言就能实现Web方面的应用，以C系列(C、C++、C#)和Java系列为主的两大技术阵营，各自都推出了基于Web的应用开发技术，围绕Visual Studio Net集成开发平台为代表的C系列Web开发和MyEclipse集成开发平台为代表的Java系列Web开发，催生了许多关键技术，特别是Java Web开发技术，在开源代码和开源框架的有力推动下，得到了快速的发展，影响并改变着整个互联网技术生态链条，也深刻地影响着以Android为代表的移动互联网生态群落的当前和未来的发展趋势。

现在，Java Web技术在企业项目开发中的应用越来越广泛，围绕Java衍生的Web核心技术、Java Web开发框架、设计模式等已经成为技术开发研究者的深入研究领域。同时，其相关的一些核心技术也已经成为院校软件开发相关专业学生未来就业和企业Java开发人员快速提升的必备技术，也被许多开发人员当作一项专项技能来学习和掌握。因而，了解Java Web应用后面的核心技术、技术原理及应用对很多人而言非常重要。

诚然，目前市面涉及Java Web核心技术的书籍门类繁多，但通常都为厚重的实践项目案例集成，非常繁琐且没有把理论分析和实践技术进行结合，更没有对整个Java Web开发涉及的核心技术全面、整体、由浅入深进行介绍的书籍。另外，由于本人身处教育行业，从近年来软件工程技术领域研究和教育研究来看，关于Java Web核心技术的介绍比较通俗易懂，适合没有技术背景的人员阅读。另外，比较全面的Java Web开发方面的书籍较少，能够应用于专业教学、符合专业人才培养、实践能力培养的更少，如何有效地解决这些问题，也是本书撰写的基本动因。

随着Java Web及其应用的不断丰富、充实，Java Web技术涵盖的范围也不再局限于传统的Java技术和人们常说的Web技术，它是一个不断扩展的技术范畴，一直以来，本人想就其与Java Web开发框架和Java Web技术的前生今世、渊源及发展现状，做一个统领性的解析和梳理，以便读者和学生对此有个初步的认识和了解。但由于个人认知的局限和Java Web技术门类的琐细、繁多，本书只对Java Web核心的一些关键技术及应用做了详述，其涵盖了软件架构模式、开发模式、设计模式和应用服务器等顶层设计内容，这部分内容通常被大多数人用来在设计模式的书籍中进行讲解，并没有结合具体应用来进行分析详述。实践中，通常大多数人在学习设计模式后，在具体工程、案例中并不能很好地应用，而另一方面，初级入门的代码开发者，关于设计模式在工程中如何贯彻应用，并不能有一个清晰的思路。无论哪一种情况，最后结果都是不能对它们实现有效的结合，这也是本书为什么把设计模式、架构模式设计加入Java Web核心技术的初衷，希望能帮助初学者在走进、学习和使用Java Web技术来盖大楼之前，知道如何建设大楼的框架，而不要拘泥于某一细节技术，而放弃全局的视野，迷失在为学技术而学技术的浩瀚的技术之海，无法自拔。

基于此，本书的 Java Web 核心技术包含了初级、中级和高级三方面的核心技术。初级技术包含了页面设计、创建的初级技术和 JavaScript 技术；中级技术包含了 JSP 技术、JDBC 常用及高级技术和 XML 技术；高级技术包含了 Struts 初级技术、Struts 高级技术、Hibernetate 技术、Spring 初级技术、Spring 高级技术以及 Spring 和 Struts 的结合集成 Java Web 项目开发技术框架 SSH(Struts-Spring-Hibernate)。其目的是为了使初学者和读者对整个 Java Web 技术从设计模式到初、中、高级技术有个了解和认识性的循序渐进的学习，学习者在阅读中会发现，技术的讲解是一个方面，更为重要的是符合人们认知规律的螺旋式渐进技术体系安排，有利于读者培养理论和技术应用有效结合的学习模式，通过项目技术引导可以使读者明白为什么而学技术(学习的目标性)，技术核心要点、原理之间的关系及衔接(学哪些内容及学习内容之间先后次序关系)，同时，也可以通过技术应用了解自己的学习深入程度及效果(学的效果如何)。

此外，由于移动互联网方面应用的快速发展，Java Web 核心技术的一个分支在移动互联网领域也得到了快速的发展，但限于篇幅，本书并没有对此进行介绍，本书只是在其基础、中级、高级技术的基础上，对 Java Web 核心技术中的一个具有代表性的开源框架系统 SSH 及其在项目中的应用，进行了深入的技术解析和详解，此方面包含了 4 个部分的内容。

第一部分包含了 Struts-Spring-Hibernate 框架概述、Struts 基础、Struts 2 标签、Struts 设计模式、工作流程、表达式、Struts 高级技术、Struts 拦截器和文件上传等。

第二部分包含了 Hibernate 基础、Hibernate 实体关系映射、Hibernate 查询语言和表的设计等。

第三部分包含了 Spring 技术、IoC 模式、Bean 的应用和 Spring Bean 的开发技术等。

第四部分包含了 Spring 持久层、Spring AOP、面向方面编程和事务处理、Struts-Spring 框架技术集成、Struts-Spring-Hibernate 框架集成等。

总而言之，本书内容技术体系及应用初步实现了自己的初衷，是对整个 Java Web 技术体系过往的深入导引，但由于自己水平和认识的局限，难免有遗漏和错误之处，希望阅读者能对此进行反馈，共同推动 Java Web 技术在国内的推广和应用，促进开源技术的快速发展。

致谢

感谢学院领导一直以来的支持和帮助，感谢创新团队项目对我的支持和帮助，是你们促使了这项工作的进展和人才培养教育的落地。

特别感谢我的家人，书稿的写作是一项耗时耗力的工作，没有你们的支持，这项工作基本无法完成，你们的鼓励促使我不断前行和进步。

感谢阅读本书的读者，您的建议和反馈，将是本书完善的基础，希望与你们一起促进国内开源技术的普及和推广。

毋建军

2015 年 1 月

目　　录

第1章　Java Web 技术概述

随着技术的快速发展和广泛应用，Java Web 技术已经成为当前 Web 开发领域应用最为广泛和流行的技术，广泛地应用于各种电子商务网站、政府电子政务及其他软件的开发，如京东、当当、Apache 平台下的开源的项目 Nutch 等。Java Web 技术是以 Java 为核心基础的 Web 应用开发技术的统称，其涵盖了客户端开发技术（HTML 及 CSS 样式、JSP 语言、XML 技术、JavaScript 技术、Servlet、HTTP 协议、JQuery 技术、JavaFX、Ajax 技术（Ajax 框架 Prototype、JQuery））、服务器端开发技术（Java 核心技术、UML 建模开发、JDBC、Java Web 开发框架（SSH、RIA））、数据库开发技术和 SQL 语言、Java Web 项目管理技术及工具（ANT、CVS、SVN）、Java Web 测试技术及工具（JMeter、Junit、Log4j、LoadRunner）等。因其涉及的技术非常广泛，本书只对其中 Java Web 部分核心技术进行深入的解析和应用，使其对 Web 开发有兴趣的程序员和学习者有深入的帮助和了解。

本章将就 Java Web 开发的基础、相关的基本概念、关键技术的当前进展、开发环境的搭建、Web 开发中常用的设计模式（C/S、B/S 模式）和 Java Web 开发技术架构进行阐述，并就其运行的原理及其技术模式进行比较和分析。

1.1　Java Web 开发模式

1.1.1　软件架构模式（C/S、B/S）

随着全球网络开发、互联、信息共享技术要求的不断提高，以前软件开发的 C/S（客户服务端）模式（单机服务软件），已经远远不能满足人们的需要和信息共享发展的要求，其中，以 B/S（浏览器/服务器）模式为代表的新型应用模式被广泛应用。在此模式下用户可以通过 WWW 浏览器去访问 Internet 上的文本、数据、图像、动画、视频点播和声音信息，这些信息都是由许许多多的 Web 服务器产生的，而每一个 Web 服务器又可以通过各种方式与数据库服务器连接，大量的数据实际存放在数据库服务器中。客户端除了 WWW 创览器，一般无须任何用户程序，只需从 Web 服务器上下载程序到本地来执行，在下载过程中若遇到与数据库有关的指令，由 Web 服务器交给数据库服务器来解释执行，并返回给 Web 服务器，Web 服务器又返回给用户。

C/S 模式主要由客户应用程序（Client）、服务器管理程序（Server）和中间件（Middleware）3 个部件组成。客户端应用程序是 Web 系统中被用户用来与数据进行交互的前端。服务器程序负责有效地管理系统资源，其主要工作是当多个客户并发地请求服务器上的相同资源时，对这些资源进行最优化管理。中间件负责联结客户端应用程序与服务器管理程序，协同完成一个作业，以满足用户查询管理数据的需求。

浏览器/服务器(Browser/Server,B/S)模式是一种以 Web 技术为基础的新型的系统平台模式。把传统 C/S 模式中的服务器部分分解为一个数据服务器与一个或多个应用服务器(Web 服务器)，从而构成一个三层结构的客户服务器体系。

第一层客户端是用户与整个系统的接口，客户的应用程序简化为一个通用的浏览器软件，如 IE、Chrome、火狐等，浏览器将 HTML 代码转化成图文并茂的网页。网页具备一定的交互功能，允许用户在网页提供的申请表上输入信息提交给后台第二层的 Web 服务器，并提出处理请求。

第二层 Web 服务器将启动相应的进程来响应这一请求，并动态生成一串 HTML 代码，其中嵌入处理的结果，返回给客户端的浏览器。如果客户端提交的请求包括数据的存取，Web 服务器还需与数据库服务器协同完成这一处理工作。

第三层数据库服务器的任务类似于 C/S 模式，负责协调不同的 Web 服务器发出的 SQ 请求，管理数据库。

在实践应用软件开发中，应用软件开发的层次，通常分为 3 层，分别是表现层(P)、逻辑层(B)和数据层(D)，它们各自的作用主要表现为：

- 表现层——向访问客户端展现 UI；
- 逻辑层——完成客户需求的程序功能；
- 数据层——保存业务数据的持久化。

无论基于 CS/BS，都需要有与客户产生交互作用的层面，因而，C/S 模式开发通常为用来访问服务程序的客户端程序，需要独立开发，则称为 CS 模式的开发；如利用已有的 Browser，并在此基础上开发 Browser 可以运行的程序，称为 B/S 模式。基于 B/S 结构的软件开发层次通常为：

- P：浏览器→HTML
- B：应用程序服务器→ASP、ASP. NET、JSP、Servlet、Python、CGI、…
- D：关系型数据库→Table

在此层次中，HTML 通常只负责显示用户需要呈现的数据或者收集客户提交的数据信息。

(1) B/S 模式的优势主要有以下几方面。

① 简化了客户端，它无须像 C/S 模式那样在不同的客户机上安装不同的客户应用程序，而只需安装通用的浏览器软件，这样不但可以节省客户机的硬盘空间与内存，而且使安装过程更加简便、网络结构更加灵活。

② 简化了系统的开发和维护，系统的开发者无须再为不同级别的用户设计开发不同的客户应用程序，只需把所有的功能都实现在 Web 服务器上，并就不同的功能为各个组别的用户设置权限就可以了。各个用户通过 HTTP 请求在权限范围内调用 Web 服务器上不同处理程序，从而完成对数据的查询或修改。相对于 C/S，B/S 的维护具有更大的灵活性，当其变化时，它无须再为每一个现有的客户应用程序升级，而只需对 Web 服务器上的服务处理程序进行修订。

③ 使用户的操作变得更简单，对于 C/S 模式，客户端应用程序有自己特定的使用流程，使用者需要接受专门培训。而采用 B/S 模式时，客户端只是一个简单易用的浏览器软件，无论是决策层还是操作层的人员都无须培训，就可以直接使用。B/S 模式的这种特性，还使 MIS

系统维护的限制因素更少。B/S 特别适用于网上信息发布，这是 C/S 所无法实现的。

(2) C/S 模式的优势主要有以下几方面。

① 交互性强是 C/S 自身的优势。在 C/S 中，客户端有一套完整的应用程序，在出错提示、在线帮助等方面都有强大的功能，并且可以在子程序间自由切换。B/S 虽然由 JavaScript、VBScript 提供了一定的交互能力，但与 C/S 的一整套客户应用相比太有限了。

② C/S 模式提供了更安全的存取模式。由于 C/S 是配对的点对点的结构模式，采用适用于局域网、安全性比较好的网络协议，安全性可以得到较好的保证。而 B/S 采用点对多点、多点对多点这种开放的结构模式，并采用 TCP/IP 这一类运用于 Internet 的开放性协议，其安全性只能靠数据服务器上管理密码的数据库来保证。

③ 采用 C/S 模式将降低网络通信量。B/S 采用了逻辑上的 3 层结构，而在物理上的网络结构仍然是原来的以太网或环形网，这样，第一层与第二层结构之间的通信、第二层与第三层结构之间的通信都需占用同一条网络线路。而 C/S 只有两层结构，网络通信量只包括 Client 与 Server 之间的通信量。所以，C/S 处理大量信息的能力是 B/S 所无法比拟的。

④ 此外，由于 C/S 在逻辑结构上比 B/S 少一层，对于相同的任务，C/S 完成的速度总比 B/S 快，使得 C/S 更利于处理大量数据。

另外，在实践开发中，有些需要采用 C/S 模式与 B/S 模式相结合的方案，开发者根据一定的原则，将系统的所有子功能分类，决定哪些子功能适合采用 C/S，哪些适合采用 B/S，针对不同的模式进行开发、部署。在软件维护阶段，针对不同模式的子功能应采取不同维护方式。

1.1.2 软件初期设计模式

在早期的 Web 开发中，因为业务比较简单，软件开发中开发者通常将大部分 Web 应用程序用像 ASP、PHP 等过程化语言来创建。它们将像数据库查询语句这样的数据层代码和像 HTML 这样的表示层代码混在一起，并没有上述软件开发层次 3 层的划分。用户数据的呈现及输入的接收、封装、验证、处理，以及对数据库的操作，都放在 JSP 页面中。此时的 Web 软件开发代码混杂在一起，即双层设计模式：Browser → JSP(Web Server，DataBase Server)，JSP 端充当 Web 服务器和数据库服务器的角色。

随着业务越来越复杂，人们开始考虑更好地利用 OOP 来解决上述存在的混杂问题。实践中有人发现把业务逻辑抽取出来，并形成与显示和持久化无关的一层，能够让业务逻辑清晰，产品更便于维护，即 SUN 开始倡导并推行的 JSP 初期模型开发模式：多层设计模式。

多层设计模式，具体可分为以下几种：

Browser→JSP(Web Server)→DataBase Server

Browser→JSP(Web Server)→beans→DataBase Server

Browser→Servlet→JSP→beans→DataBase Server

下面就它们进行详细介绍。

1. 传统的 JSP 初期模型(JSP Model 1)

传统的 JSP 初期开发模型中，JSP 是独立的，自主完成所有的任务，其模型如图 1-1 所示。

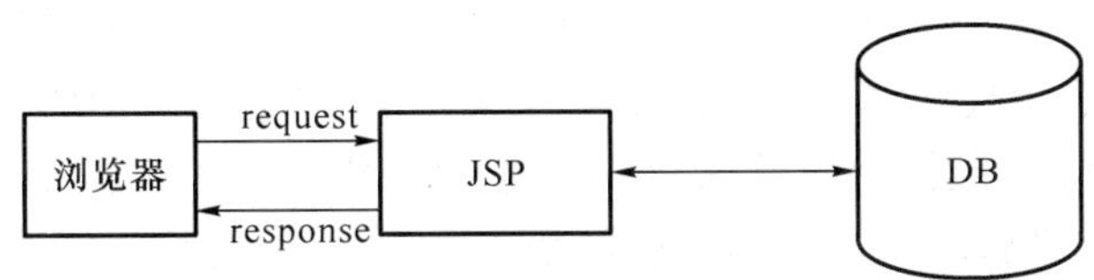

图 1-1 JSP 初期模型

JSP 初期模型的开发方式中，并没有对数据如何持久化给出建议。如图 1-1 所示，在许多产品开发中，产品是以数据库为中心进行架构和设计的。在其产品设计中，虽然也有 DAO 层，但是职责不清，其对 DAO 层的职责简单定位为增删改查。尤其随着业务逻辑变得越来越复杂，复杂的对象关系很难理清，如何把数据存储起来(通常的情况下是存到关系型数据库中)是一个棘手的问题。在此过程中，软件设计中引入面向对象设计的思想，即数据持久化。即在自己的应用中添加一个新的层，持久化层来专门负责对象状态的持久化保存及同步。持久化意味着对关系型数据库的依赖减少。通常，有经验的软件开发者会将数据应用从表示层分离出来，后续讲到的设计模式 MVC 从根本上强制性的将其分开。在此过程中，有人提出了改进后的 JSP 模型——JSP Model 改进模型。

2. 改进的 JSP Model 初期模型(JSP Model 1 改进)

针对 JSP 初期模型数据应用和代码混杂的缺陷，有人提出了 JSP 页面与 JavaBean 协作响应浏览器页面请求，并 response 回复完成任务的 JSP Model 改进模型，其工作模式如图 1-2 所示。

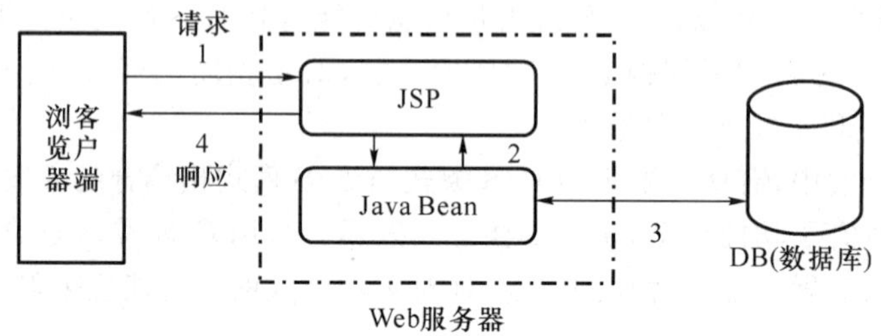

图 1-2　JSP Model 1 改进模型工作模式

从上述的 JSP Model 初期模型可以看出，其实现比较简单，适用于快速开发小规模项目。但就项目产品工程化来分析，其缺陷显而易见，JSP 页面充当了 View 和 Controller 两种角色，将控制逻辑和表现逻辑混杂在一起，从而导致其代码的重用性较低，对应用的后期扩展和维护都增加了难度。而改进后的 JSP Model 模型，增加了 JavaBean，把数据存储及数据库打交道的功能，分解给 JavaBean 进行完成。

早期有大量采集 JSP 技术开发出来的 Web 应用，都采用了 JSP Model 初期架构。

3. JSP Model 2

随着业务变得越来越复杂，在 JSP Model 1 改进的基础上，针对上述的缺陷，出现了 JSP Model 2 架构，JSP Model 2 在 JSP、JavaBeans 的基础上，增加了 Servlet 角色，在此结构中：JSP 负责生成动态网页，只用作显示页面；Servlet 负责流程控制，用来处理各种请求的响应及分派；JavaBeans 负责业务逻辑，实现对数据库的各种操作。

JSP Model 2 的交互过程：用户通过浏览器向 Web 应用中的 Servlet 发送请求，Servlet 接收到请求后实例化 JavaBeans 对象，调用 JavaBeans 对象的方法，JavaBeans 对象返回从数据库中读取的数据，Servlet 选择合适 JSP，并且把从数据库中读取的数据通过这个 JSP 进行显示，最后 JSP 页面把最终的结果返回给浏览器。其交互过程工作模式如图 1-3 所示。

JSPModel 2 已经是 MVC 设计思想下的架构，由此引入了 MVC 模式，使 JSP Model 2 具有组件化的特点，更适用于大规模应用的开发，但也增加了应用开发的复杂程度。从 JSP Model 2 模型理论上看，其已经具备了 Web 三层架构的模式。但在实际开发产品中，我们发现在上述的 UI 层和业务层之间有交叉逻辑。这些交叉逻辑即不属于业务逻辑层管理的范畴，也不属于 UI 层管理的范畴，如页面跳转、表单数据的验证及封装、页面的国际化(在 JSP 页面根据用户的配置或请求信息判断应该为该用户提供哪一种语言的页面信息)等，这些交叉逻辑如何分配，交给业务逻辑层还是 UI 层，如何实现业务逻辑层和 UI 层的分离呢？下述的 MVC 模式便为解决上述模式中存在的问题，而得到广泛的应用。

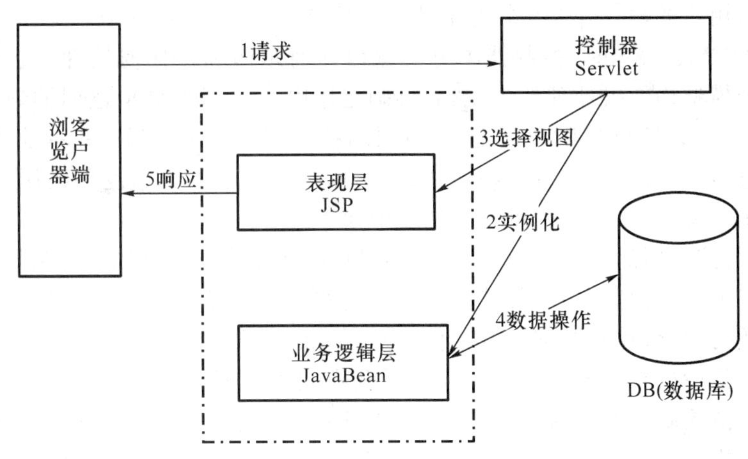

图 1-3 JSP Model 2 交互工作模式

1.1.3 MVC 模式

MVC 模式(Model-View-Controller)，即“模型-视图-控制器”。MVC 是 Xerox PARC 在 20 世纪 80 年代为编程语言 Smalltalk-80 发明的一种软件设计模式，后来被推荐为 Oracle 下 Sun 公司 Java EE 平台的设计模式。它开始是存在于桌面程序中的，M 是指业务模型，V 是指用户界面，C 则是控制器，使用 MVC 的目的是将 M 和 V 的实现代码分离，从而使同一个程序可以使用不同的表现形式，如用户表格数据的提交，统计数据可以分别用柱状图、饼图等表示。C 存在的目的则是确保 M 和 V 的同步，一旦 M 改变，V 应该同步更新。MVC 是一个框架模式，它强制性地使应用程序的输入、处理和输出分开。使 MVC 应用程序被分成 3 个核心部件：模型、视图和控制器。它们各自处理自己的任务。最典型的 MVC 就是上述的 JSP+Servlet+JavaBean 的模式。下面就其核心组件的作用和功能进行详细解析。

1. 视图

视图(View)是用户看到并与之交互的界面。对老式的 Web 应用程序而言，视图就是由 HTML 元素组成的界面，在新式的 Web 应用程序中，HTML 依旧在视图中扮演着重要的角色，但如 Adobe Flash、XHTML、XML/XSL、WML 等一些标识语言及 Web Services 等新技术已经广泛应用。一个 Web 应用可能有很多不同的视图，MVC 设计模式对于视图的处理仅限于视图上数据的采集和处理，以及用户的请求，但不包括在视图上的业务流程的处理。业务流程的处理交给模型(Model)处理，如一个订单的视图只接受来自模型的数据并显示给用户，以及将用户界面的输入数据和请求传递给控制和模型。

2. 模型

模型即业务流程/状态的处理以及业务规则的制定。业务流程的处理过程对其他层而言是黑箱操作，模型接受视图请求的数据，并返回最终的处理结果。业务模型的设计是 MVC 最主要的核心。以前流行的 EJB 模型就是一个典型的应用例子，模型拥有最多的处理任务，被模型返回的数据是中立的，就是说模型与数据格式无关，这样一个模型能为多个视图提供数据。此外，业务模型还有一个很重要的模型是数据模型。数据模型主要指实体对象的数据保存(持续化)，例如，将一张订单保存到数据库，从数据库获取订单，即可以将此模型单独列出，所有有关数据库的操作都限制在该模型中。

3. 控制器

控制器接收用户的输入并调用模型和视图去完成用户的需求，所以当单击 Web 页面中的超链接和发送 HTML 表单时，控制器本身不输出任何东西和作任何处理。它只是接收请求并决定调用哪个模型构件去处理请求，然后再确定用哪个视图来显示返回的数据。控制器起着一个分发器的作用，所以当用户点击一个链接，控制层接受请求后，并不处理业务信息，它只把用户的信息传递给模型，告诉模型做什么，选择符合要求的视图返回给用户。因此，一个模型可能对应多个视图，一个视图可能对应多个模型。

模型、视图与控制器的分离，使得一个模型可以具有多个显示视图。如果用户通过某个视图的控制器改变了模型的数据，所有其他依赖于这些数据的视图都会发生变化。因此，无论何时发生了何种数据变化，控制器都会将变化通知所有的视图，导致显示的更新，实际上是一种模型的变化——传播机制。模型、视图、控制器及数据库之间的交互关系，如图 1-4 所示。视图中用户的输入被控制器解析后，控制器改变状态激活模型，模型根据业务逻辑维护数据，从数据库提取数据或写入数据，并通知视图，其对应的数据已经发生变化，视图得到消息通知后，从模型中获取新的数据并进行更新显示。

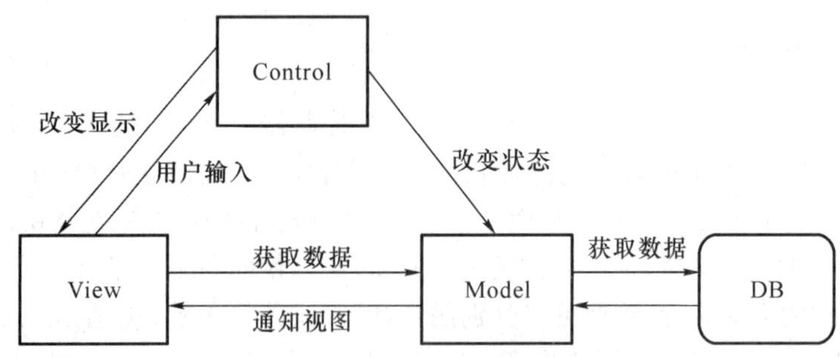

图 1-4　模型、视图、控制器及数据库的交互关系

如上所述 MVC 模式的关键是实现了视图和模型的分离。其实现原理为：MVC 模式通过建立一个“发布-订阅”(Publish-Subscribe)的机制来分离视图和模型。发布-订阅机制的目标是发布者，其发出通知时并不知道，也不需知道谁是它的观察者，可以有任意数目的观察者订阅消息并接收通知。其优点主要有：

- 多视图表示，一个模型提供不同的多个视图表现形式，也能够为一个模型创建新的视图而无须重写模型，一旦模型的数据发生变化，模型将通知有关的视图，每个视图相应地刷新自己；
- 模型可复用，因为模型是独立于视图的，所以可以把一个模型独立地移植到新的平台工作；
- 提高开发效率，在开发界面显示部分时，仅需要考虑的是如何布局一个好的用户界面，开发模型时，只需考虑业务逻辑和数据维护，从而能使开发者专注于某一方面的开发，提高开发效率。

MVC 模式中关键点利用了观察者模式(Observer)，而观察者模式实现了发布-订阅(Publish-Subscribe)机制，并能完成视图和模型的分离。其设计模式关系还涉及了组合模式(Composite)、策略模式(Strategy)，关于这 3 种模式由于篇幅的关系，本书不再详细介绍。

1.1.4　框架模式与设计模式

MVC 是一种框架模式，而不是一种设计模式。框架模式与设计模式，实际上完全是不同的概念。框架通常是指代码重用,而设计模式是指设计重用,架构则介于两者之间,部分代码重用,部分设计重用,有时分析也可重用。在软件生产中有 3 种级别的重用:内部重用,即在同一应用中能公共使用的抽象块;代码重用,即将通用模块组合成库或工具集,以便在多个应用和领域都能使用;应用框架的重用,即为专用领域提供通用的或现成的基础结构,以获得最高级别的重用性。

框架模式与设计模式虽然相似,但却有着根本的不同。设计模式是对在某种环境中反复出现的问题以及解决该问题的方案的描述,它比框架更抽象;框架可以用代码表示,也能直接执行或复用,而对模式而言只有实例才能用代码表示;设计模式是比框架更小的元素,一个框架中往往含有一个或多个设计模式,框架总是针对某一特定应用领域,但同一模式却可适用于各种应用。可以说,框架是软件,而设计模式是软件的知识。

常见的框架模式有:MVC、MTV、ORM、MVP 等。框架有:基于 PHP 语言的 smarty 框架,基于 C++语言的 QT、MFC、gtk 等,基于 Java 语言的 SSH、SSI 等。本书后续将对 SSH 框架进行深入的解析。

常用的设计模式有:工厂模式、适配器模式、策略模式、观察者模式等。

1.2　Java Web 应用服务器

JavaWeb 服务器是运行及发布 Java Web 应用的容器,只有将开发的 Web 项目放置到该容器中,才能使网络中的所有用户通过浏览器进行访问。开发 Java Web 应用所采用的服务器主要是与 JSP/Servlet 兼容的 Web 服务器,比较常用的 Java Web 服务器有 Apache、Tomcat、Resin、JBoss、WebSphere 和 WebLogic 等,下面将分别进行介绍。

1.2.1　Apache 服务器

Apache HTTP Server(Apache)是 Apache 软件基金会的一个开放源码的网页服务器,可以在大多数计算机操作系统中运行,由于其多平台和安全性被广泛使用,是当前最流行的 Web 服务器端软件之一。它快速、可靠并且可通过简单的 API 扩展,将 Java Web 及 Perl/Python 等解释器编译到服务器中。目前的版本是 2014 年 7 月发布的 Apache 2.4.10 ,其下载网址为:http://httpd.apache.org/,如图 1-5 所示。

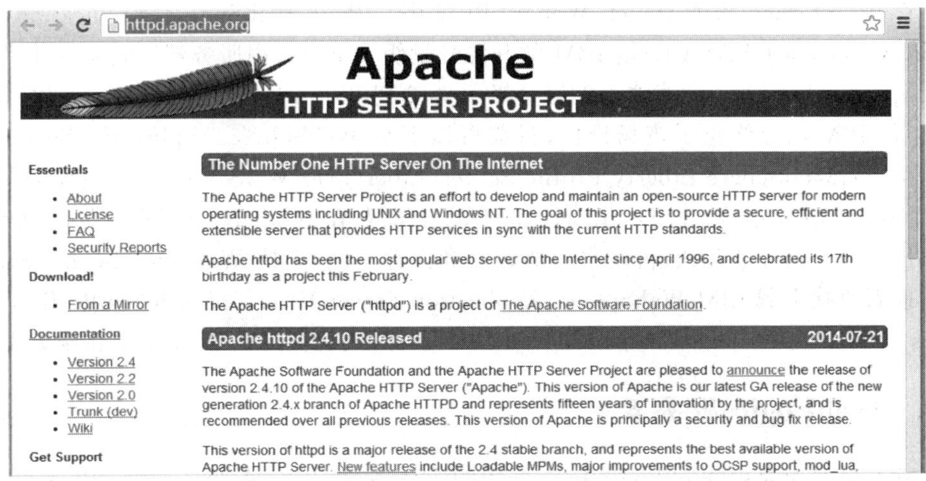

图 1-5　Apache 服务器下载网站

1.2.2 Tomcat 服务器

Tomcat 是 Apache 软件基金会(Apache Software Foundation)的 Jakarta 项目中的一个核心项目,由 Apache、Sun 和其他一些公司及个人共同开发而成。在 Sun 的参与和支持下,最新的 Servlet 和 JSP 规范在 Tomcat 中得到实现,目前最新版本是 8.0 版本,支持最新的 Servlet 3.1、JSP 2.3 规范。Tomcat 成为目前比较流行的 Web 应用服务器。它是一个小型、轻量级的支持 JSP 和 Servlet 技术的 Web 服务器,也是初学者学习开发 JSP 应用的首选 Web 服务器,其下载网站:http://tomcat.apache.org/,如图 1-6 所示。

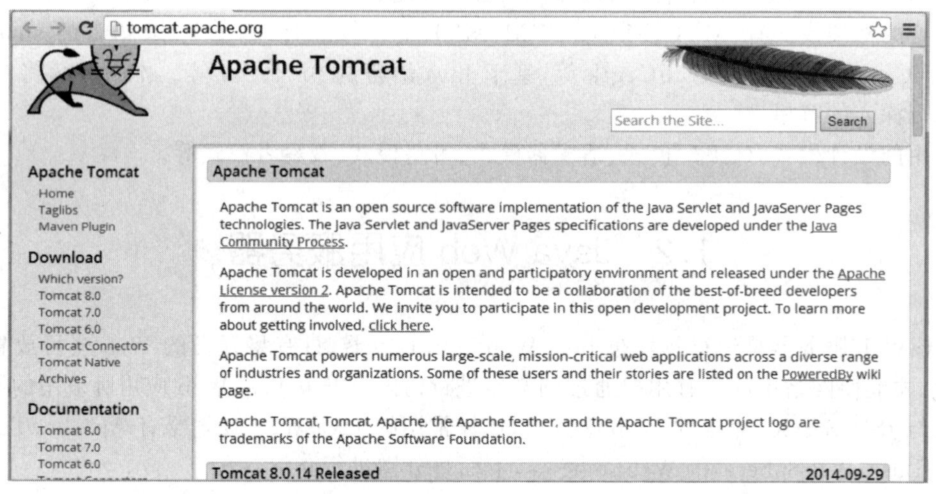

图 1-6　Tomcat 下载网站

1.2.3 WebSphere 服务器

WebSphere 是 IBM 公司的产品,可进一步细分为 WebSphere Performance Pack、Cache Manager 和 WebSphere Application Server 等系列,其中 WebSphere Application Server 是基于 Java 的应用环境,可以运行于 Sun Solaris、Windows NT 等多种操作系统平台,用于建立、部署和管理 Internet 和 Intranet Web 应用程序。其中 WebSphere Application Server Community Edition (WAS CE)是 IBM 的开源轻量级 J2EE 应用服务器。它是一个免费的、构建在 Apache Geronimo 技术之上的轻量级 Java 2 Platform Enterprise Edition(J2EE)应用服务器。但 IBM 已经终止了市场推广,目前可以下载 WebSphere Application Server V8.5 (WAS V8.5),WebSphere Liberty Profile Server(Liberty)是 WAS V8.5中最主要的新特性,它是一个基于 OSGi 内核、高模块化、高动态性的轻量级 WebSphere 应用服务器,其安装极为简单(解压即可)、启动非常快、占用很少的磁盘和内存空间,支持 Web、mobile 和 OSGi 应用的开发,或者直接下载 IBM WebSphere Application Server V8.5.5.3 Liberty Profile,下载网址:https://developer.ibm.com//wasdev/downloads/,或者直接在 Eclipse 中在线集成。

1.2.4 WebLogic 服务器

WebLogic 最早是由 WebLogic Inc. 开发的产品,后并入 BEA 公司,目前 BEA 公司又并入 Oracle 公司。WebLogic 细分为 WebLogic Server、WebLogic Enterprise 和 WebLogic

Portal 等系列，其中 WebLogic Server 的功能特别强大。WebLogic 支持企业级的、多层次的和完全分布式的 Web 应用，并且服务器的配置简单、界面友好。WebLogic 常用于开发、集成、部署和管理大型分布式 Web 应用、网络应用和数据库应用的 Java 应用服务器。目前 WebLogic 最新版本为 Oracle Weblogic Server 12c(12.1.3)，其下载网址：http://www.oracle.com/technetwork/middleware/weblogic/overview/index.html，如图 1-7 所示。

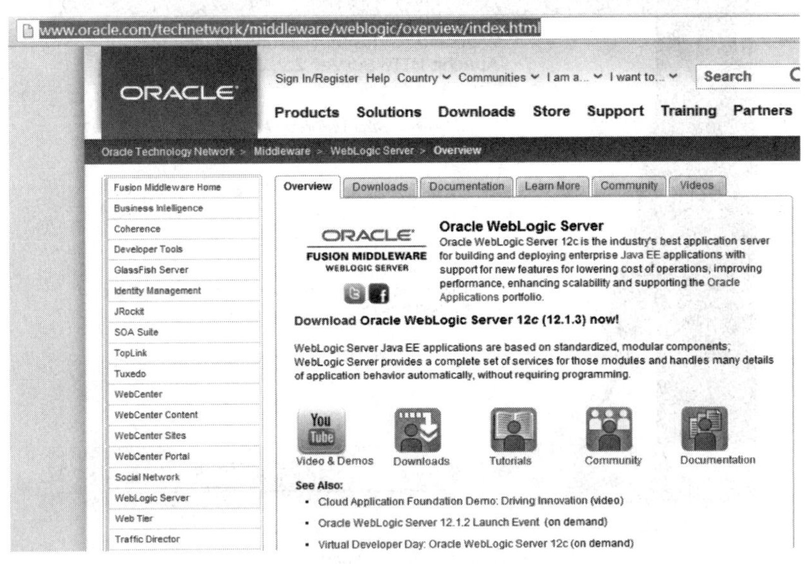

图 1-7　WebLogic 下载网址

1.2.5　Resin 服务器和 JBoss 服务器

Resin 是 Caucho 公司的产品，是一个非常流行的支持 Servlet 和 JSP 的服务器，速度非常快。Resin 本身包含了一个支持 HTML 的 Web 服务器，使它不仅可以显示动态内容，而且显示静态内容的能力也毫不逊色，Resin 也可以和许多其他的 Web 服务器一起工作，如 Apache Server 和 IIS 等。Resin 支持 Servlets 2.3 标准和 JSP 1.2 标准，支持负载平衡，因而许多使用 JSP 的网站用 Resin 服务器进行构建。

JBoss 是一个遵从 JavaEE 规范的、开放源代码的、纯 Java 的开放源代码的 EJB 服务器和应用服务器。因为 JBoss 代码遵循 LGPL 许可，可以在任何商业应用中免费使用它，而不用支付费用，对于 J2EE 有很好的支持。JBoss 采用 JML API 实现软件模块的集成与管理，是一个管理 EJB 的容器和服务器，支持 EJB 1.1、EJB 2.0 和 EJB 3.0 的规范，但 JBoss 核心服务不包括支持 Servlet/JSP 的 Web 容器，一般与 Tomcat 或 Jetty 绑定使用。

1.3　Java Web 服务器安装、测试

1.3.1　Apache 服务器

在 Windows 平台下，可以下载"Win32 Source"或者"Win32 Binary(MSI Installer)"，其中"Win32 Source"是源代码，需要自己编译执行，而"Win32 Binary(MSI Installer)"是已经做好

的安装程序,可以直接安装。

1. Apache 服务器安装

(1) 选择"apache_2.2.17-win32-x86-no_ssl.msi"安装程序进行双击,打开安装界面,如图 1-8 所示。

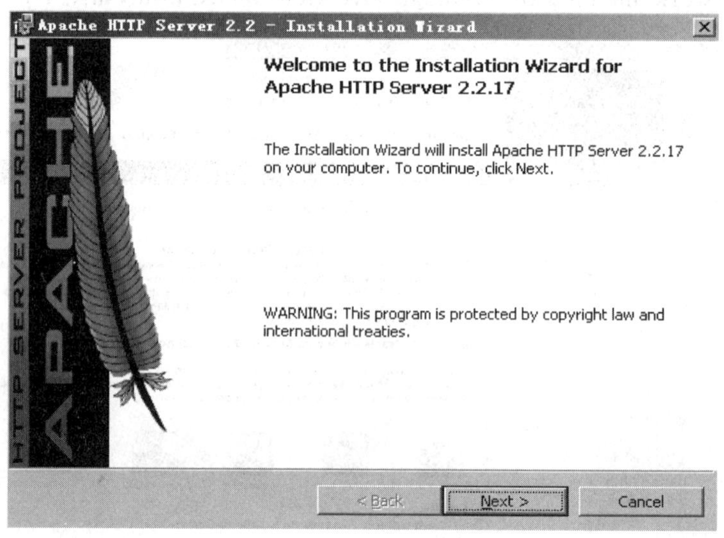

图 1-8 安装 Apache

(2) 单击"Next",出现"Iicense Agreement"对话框,选择"I accept the terms in the license agreement",如图 1-9 所示。

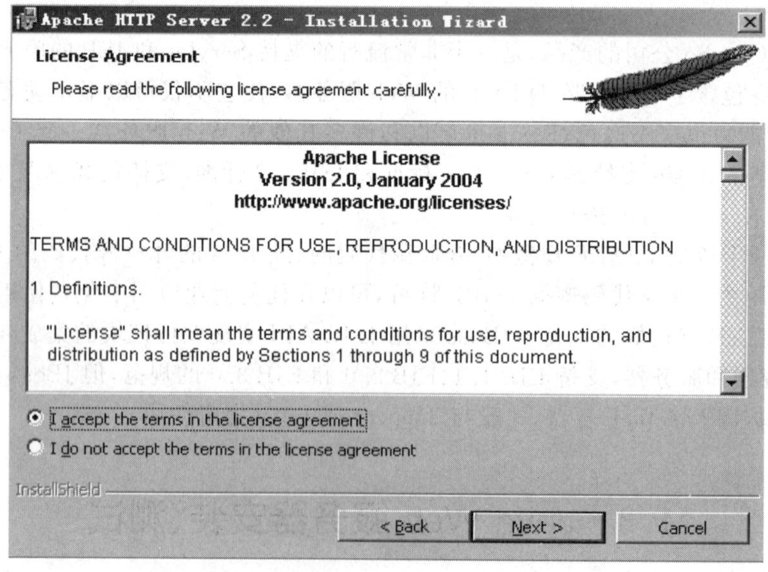

图 1-9 接受协议

(3) 单击"Next",出现"Read The First"对话框,然后再单击"Next",出现"Server Information"对话框,如图 1-10 所示,在该界面中,需要输入服务器的信息,包括网络域名、服务器名和管理员邮箱,这里可以根据自己的情况进行填写。在该界面下方的单选按钮组中,选择第一项表示任何用户都可以连接或使用服务器，同时设置服务器的侦听端口为“80”，选择第二项表示只有本地用户可以连接和使用。

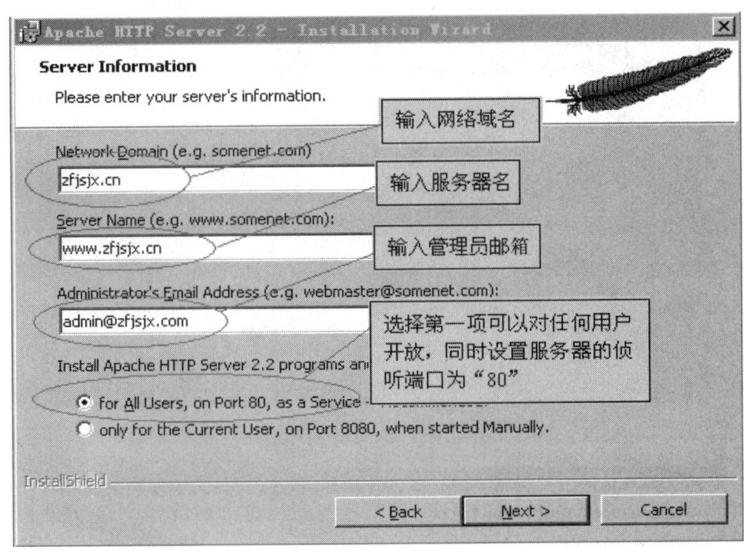

图 1-10　输入配置信息

需要注意的是，如果选择的是第一项，同时又安装了 IIS，那就必须修改 IIS 的默认端口，否则将导致 Apache 服务无法启动，不能正常工作。可以在 IIS 的管理中更改 IIS 的侦听端口，或者停止 IIS 的服务。

（4）单击“Next”，出现“Setup Type”界面，默认是“Typical”安装，如图 1-11 所示。

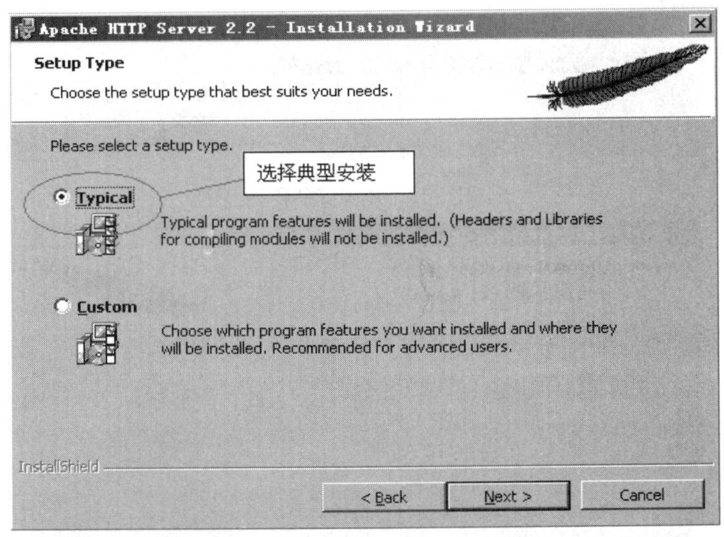

图 1-11　选择典型安装

（5）单击“Next”，出现如图 1-12 所示的默认安装界面，如果需要改变安装的路径，单击“Change”按钮，出现如图 1-13 所示的安装界面，修改程序的安装路径。

（6）单击“OK”，进入“Ready to Install the Program”对话框，如图 1-14 所示。

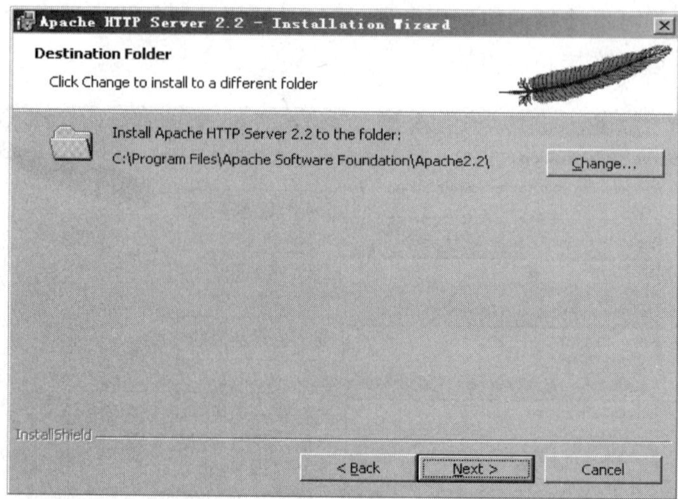

图 1-12　修改安装路径

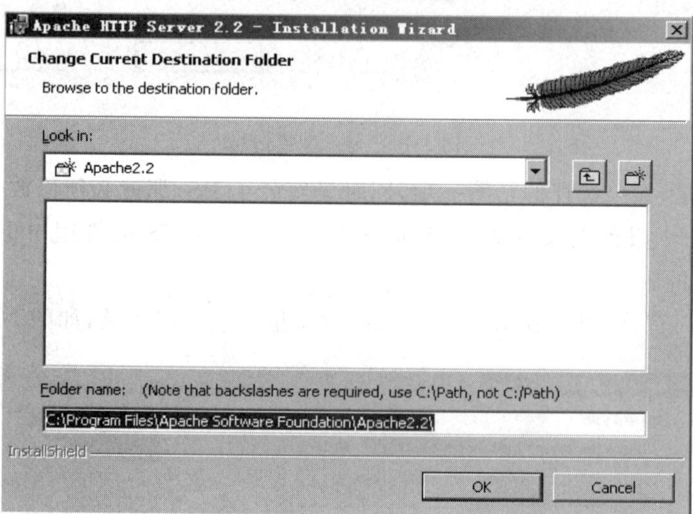

图 1-13　添加安装路径

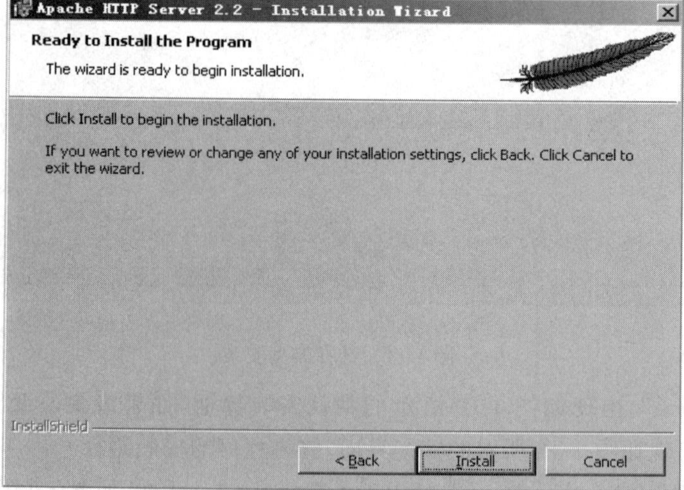

图 1-14　单击安装

(7) 单击“Install”按钮,开始正式安装,如图 1-15 所示,安装程序复制文件到系统中。

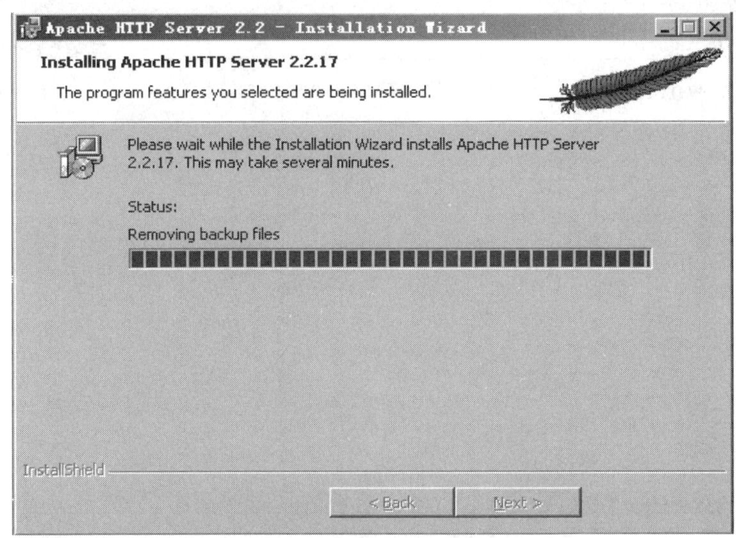

图 1-15　安装

(8) 单击“Next”,出现如图 1-16 所示的对话框,再单击“Finish”完成安装。

图 1-16　安装完成

(9) 在浏览器中输入 http://127.0.0.1/,如果能够看到如图 1-17 所示的对话框,则说明安装成功。

(10) 安装成功后,单击“开始”→“所有程序”→“Apache HTTP Server 2.2”→“Control Apache Server”→“Start”,启动 Apache 服务器,同时在桌面右下面会出现 图标,表示 Apache服务器已经启动,单击“Restart”可以重新启动服务器,如图 1-18 所示。

2. 利用 Apache 服务器创建网站

Apache 服务器安装成功后,接下来需对它进行配置,配置 Apache 服务器主要是在“C:\Program Files\Apache Software Foundation\Apache2.2\conf”目录下的 httpd.conf 文件中进行的。

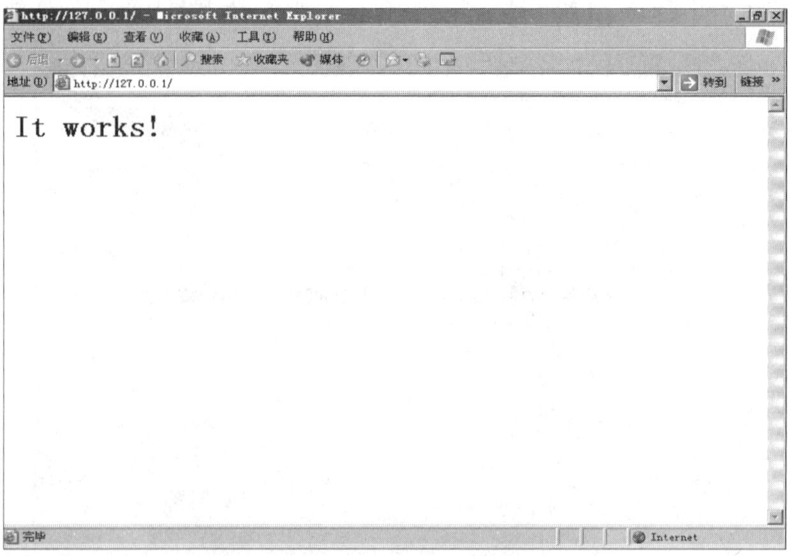

图 1-17　测试安装

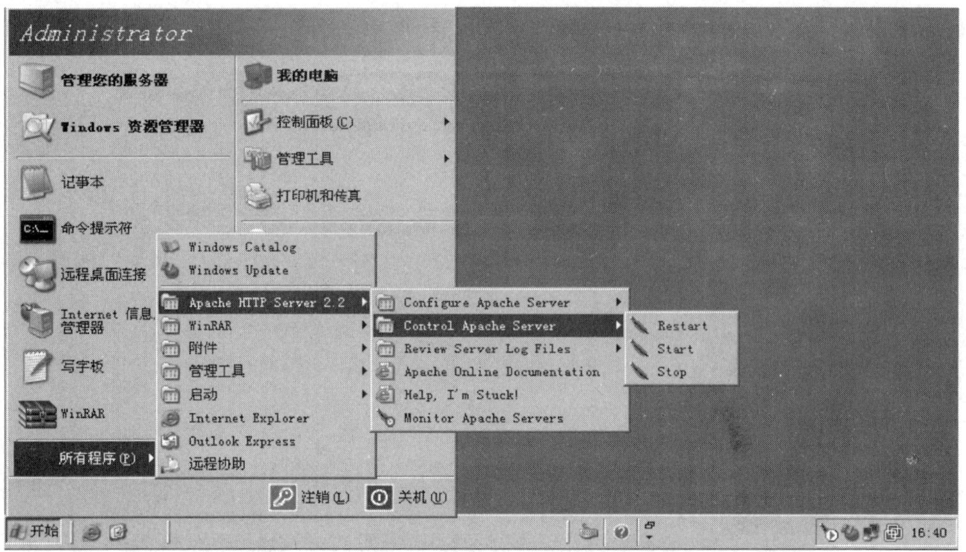

图 1-18　启动服务器

(1) 在 httpd. conf 文件中定位到“DocumentRoot”一行，可以将路径修改为你认为合适的路径，这里设置为“DocumentRoot ‘C:/ ascent’”，如图 1-19 所示。

(2) 定位到“DirectoryIndex index. htm defaut. php”行，在其后添加一个默认页 index. html，通常是 index. html，添加代码为：DirectoryIndex index. htm defaut. php index. html，如图 1-20 所示。

(3) 设置服务器网站目录的访问权限为指定的目录路径“C:/ascent”，如图 1-21 所示。

(4) 在对 httpd. conf 文件进行修改配置后，重新启动 Apache 服务器，才能生效。在 Apache 服务器配置生效后，可以根据配置在指定的路径下面放置一个网站系统，如本案例放置的是 AscentSys 医药商务系统，然后在浏览器中输入 http://127.0.0.1/或者在浏览器地址

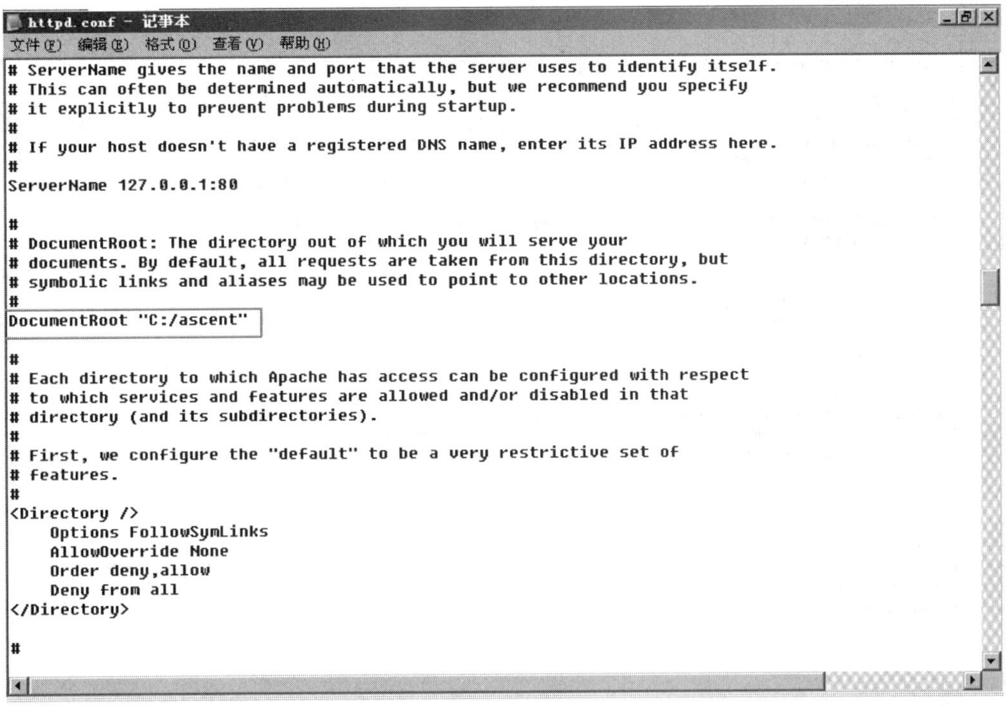

```
httpd.conf - 记事本
文件(F) 编辑(E) 格式(O) 查看(V) 帮助(H)
# ServerName gives the name and port that the server uses to identify itself.
# This can often be determined automatically, but we recommend you specify
# it explicitly to prevent problems during startup.
#
# If your host doesn't have a registered DNS name, enter its IP address here.
#
ServerName 127.0.0.1:80

#
# DocumentRoot: The directory out of which you will serve your
# documents. By default, all requests are taken from this directory, but
# symbolic links and aliases may be used to point to other locations.
#
DocumentRoot "C:/ascent"

#
# Each directory to which Apache has access can be configured with respect
# to which services and features are allowed and/or disabled in that
# directory (and its subdirectories).
#
# First, we configure the "default" to be a very restrictive set of
# features.
#
<Directory />
    Options FollowSymLinks
    AllowOverride None
    Order deny,allow
    Deny from all
</Directory>

#
```

图 1-19　修改根目录

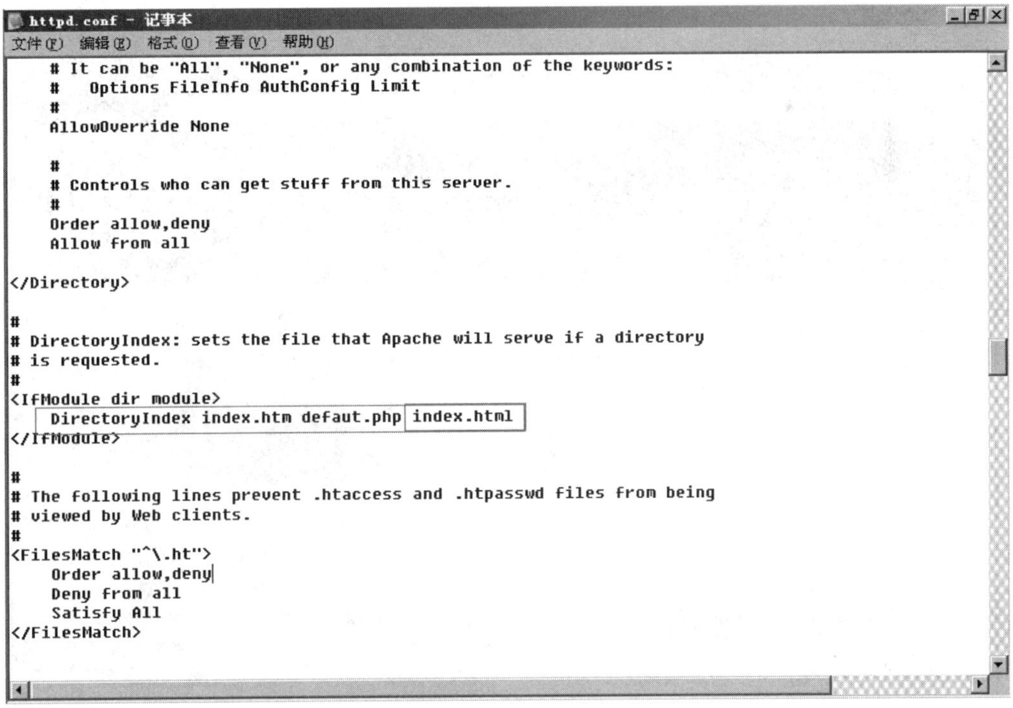

```
httpd.conf - 记事本
文件(F) 编辑(E) 格式(O) 查看(V) 帮助(H)
    # It can be "All", "None", or any combination of the keywords:
    #   Options FileInfo AuthConfig Limit
    #
    AllowOverride None

    #
    # Controls who can get stuff from this server.
    #
    Order allow,deny
    Allow from all

</Directory>

#
# DirectoryIndex: sets the file that Apache will serve if a directory
# is requested.
#
<IfModule dir_module>
    DirectoryIndex index.htm defaut.php index.html
</IfModule>

#
# The following lines prevent .htaccess and .htpasswd files from being
# viewed by Web clients.
#
<FilesMatch "^\.ht">
    Order allow,deny
    Deny from all
    Satisfy All
</FilesMatch>
```

图 1-20　添加默认路径

栏中输入服务器 IP 地址 http://192.168.3.22，就可以打开 AscentSys 医药系统网站，如图 1-22 所示。

```
httpd.conf - 记事本
文件(F) 编辑(E) 格式(O) 查看(V) 帮助(H)
# This should be changed to whatever you set DocumentRoot to.
#
<Directory "C:/ascent">
    #
    # Possible values for the Options directive are "None", "All",
    # or any combination of:
    #   Indexes Includes FollowSymLinks SymLinksifOwnerMatch ExecCGI MultiViews
    #
    # Note that "MultiViews" must be named *explicitly* --- "Options All"
    # doesn't give it to you.
    #
    # The Options directive is both complicated and important.  Please see
    # http://httpd.apache.org/docs/2.2/mod/core.html#options
    # for more information.
    #
    Options Indexes FollowSymLinks

    #
    # AllowOverride controls what directives may be placed in .htaccess files.
    # It can be "All", "None", or any combination of the keywords:
    #   Options FileInfo AuthConfig Limit
    #
    AllowOverride None

    #
    # Controls who can get stuff from this server.
    #
    Order allow,deny
    Allow from all

</Directory>
```

图 1-21　添加服务器目录路径

图 1-22　测试发布的网站

1.3.2　Tomcat 服务器

在开始安装之前，需下载 J2SE 和 Tomcat 软件，如果已经安装了 J2SE，就只需安装 Tomcat 即可。下载版本为 Tomcat 安装文件 apache-tomcat-7.0.6.exe，下载地址为 http://tomcat.apache.org/download-55.cgi，J2SDK(JDK)v1.6 版本和 JRE6，下载地址为 http://java.sun.com/javase/downloads/index.jsp。

1. JDK 安装

(1) 安装 JDK 和 JRE

双击 jdk-6u10-windows-i586-p.exe 文件进行安装，使用默认配置进行安装，JDK 的默认安装目录为“C:\Program Files\Java\jdk1.6.0_10\”，JRE 的默认安装目录为“C:\Program Files\Java\jre6”。

(2) 安装好 JDK 和 JRE 之后，需要配置一下环境变量，打开“我的电脑”→“属性”→“高级”→“环境变量”→“系统变量”，新建添加以下系统变量(假设 JDK 安装在 C:\Program Files\Java\jdk1.6.0_10\)：

```
JAVA_HOME = C:\Program Files\Java\jdk1.6.0_10\
classpath = .; %JAVA_HOME%\lib\dt.jar; %JAVA_HOME%\lib\tools.jar(.;一定不能少，因为它代表当前路径)
path = %JAVA_HOME%\bin
```

这一步做好了以后可以在字符窗口中用 DOS 命令来测试一下：

```
echo %JAVA_HOME%
echo %classpath%
echo %path%
```

(3) 接着可以写一个简单的 Java 程序来测试 JDK 是否已安装成功：

```
public class Test
 {  public static void main(String args[])
    { System.out.println("This is a test program."); }
 }
```

将上面的这段程序保存为文件名为 Test.java 的文件。然后打开命令提示符窗口，cd 到 Test.java 所在目录，然后键入下面的命令：

```
javac Test.java
java Test
```

此时如果看到打印出来“This is a test program.”的话说明安装成功了，如果没有打印出这句话，你需要仔细检查一下配置情况。

2. Tomcat 服务器安装

(1) 双击图标“apache-tomcat-7.0.6.exe”，进行 Tomcat-7.0.6 安装，如图 1-23 所示。

(2) 单击“Next”→“I Agree”，选择默认设置，如图 1-24 所示。然后选择需要安装的类型，本例选择是 Normal 安装 Tomcat 软件组件，如图 1-25 所示。

图 1-23　Tomcat 安装界面

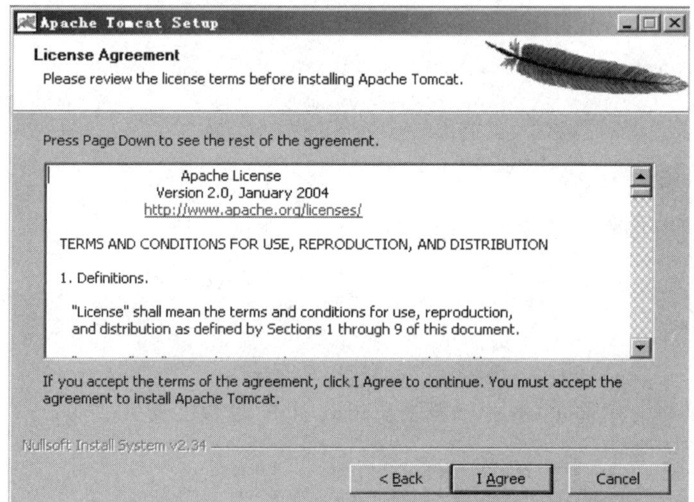

图 1-24　协议接受界面

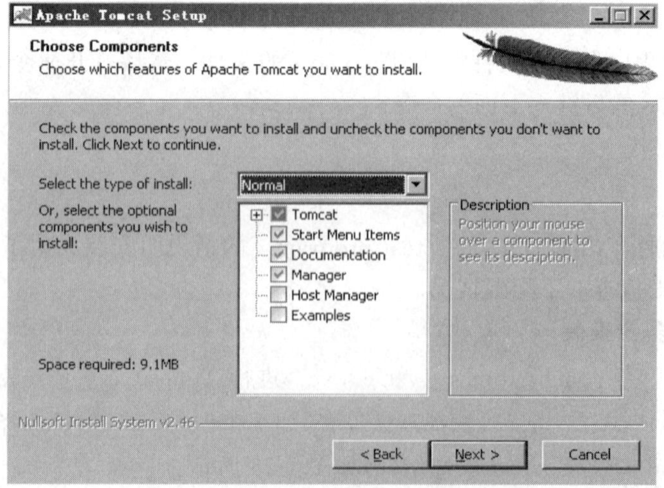

图 1-25　选择安装

（3）设置连接端口（默认 8080 端口）和管理员登录的用户名（默认是 admin）、密码及角色（默认的角色是 manager-gui），如图 1-26 所示。

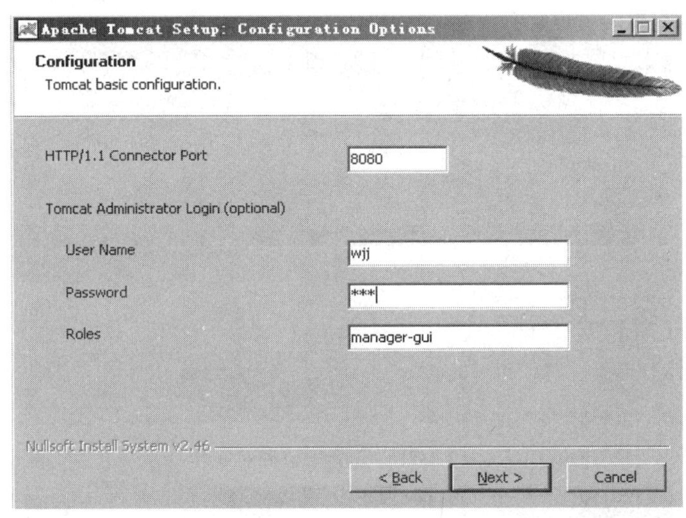

图 1-26　配置安装参数

（4）单击“Next”，选择 Java 虚拟机（JVM）的安装路径，一般采用默认选择的即可，如图 1-27 所示。

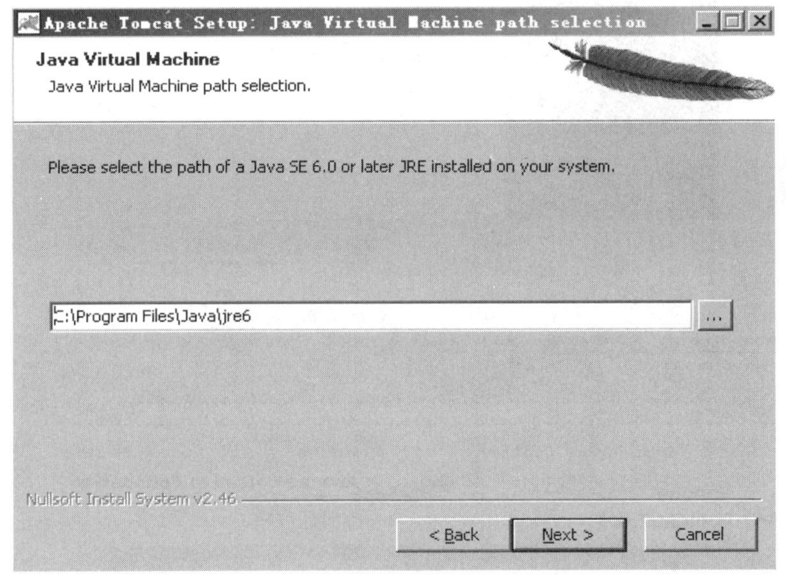

图 1-27　选择 Java 安装路径

（5）单击“Next”，进入安装路径界面，如果修改安装路径，单击“Browse”按钮进行选择，否则，直接单击“Next”，如图 1-28 所示。

（6）单击“Install”进行安装，然后单击“Finish”完成安装。默认选择运行 Tomcat 和显示 Readme 的内容，如果不想运行和显示内容，可以把它们复选框的“√”去掉，如图 1-29 所示。

（7）配置完成后，启动 Tomcat，在 IE 浏览器中访问 http://localhost:8080，如果看到 Tomcat 的欢迎页面的话说明安装成功了，如图 1-30 所示。

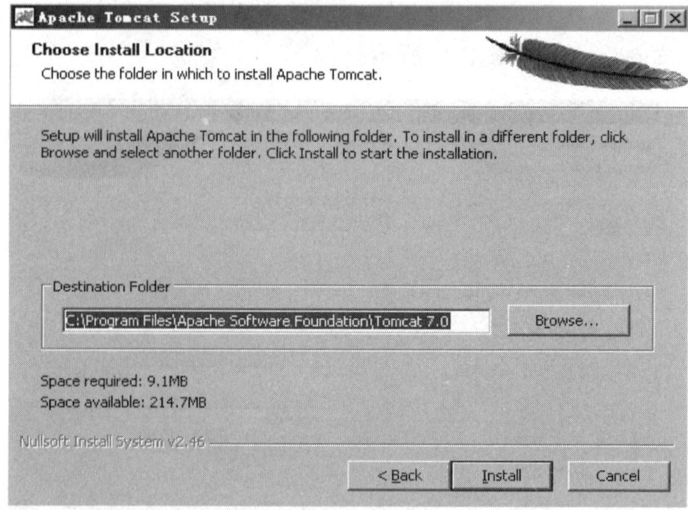

图 1-28　选择安装路径

图 1-29　安装完成

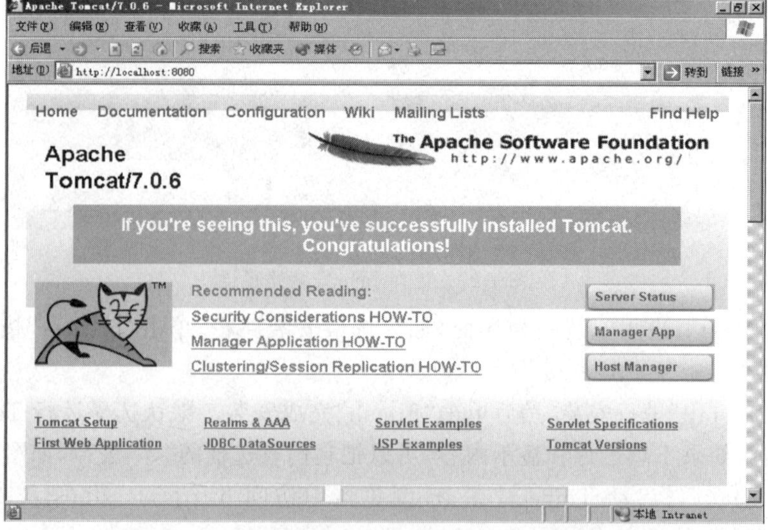

图 1-30　测试安装

3. 利用 Tomcat 服务器创建网站

(1) 在 Tomcat 的安装目录下的 webapps 目录中,可以看到 Root、tomcat-docs 等 Tomcat 自带的目录。

(2) webapps 目录是 Tomcat 服务器 Web 应用目录,最简单的方法是把所要应用的网站系统复制到 webapps 目录下,如本案例是把 AscentSys 医药商务系统根目录文件夹 ascent 复制到 webapps 目录下。

(3) 在 IE 地址栏输入 http://localhost:8080/ascent,如图 1-31 所示。

图 1-31　发布测试网站

同时,也可以使用另外一种方法创建 Web 站点,即把原来的 Root 目录改为别的名字,然后把 AscentSys 医药商务系统根目录文件夹 ascent 复制到 webapps 目录下,改名为 Root 目录,再在 IE 地址栏直接访问浏览网站系统。

1.3.3 WebSphere 服务器

1. WebSphere 下载

在网址 https://developer. ibm. com//wasdev/downloads 上下载 IBM WebSphere Application Server V8.5.5.3 Liberty Profile,并在 cmd 中使用命令 java-jar wlp-developers-runtime-8.5.5.3.jar,解压文件,解压后的文件夹目录如图 1-32 所示。

bin	2014/10/4 11:17	文件夹	
clients	2014/10/4 11:17	文件夹	
dev	2014/10/4 11:17	文件夹	
lafiles	2014/10/4 11:17	文件夹	
lib	2014/10/4 11:17	文件夹	
templates	2014/10/4 11:17	文件夹	
usr	2014/10/4 11:17	文件夹	
Copyright.txt	2014/10/4 11:17	文本文档	1 KB
README.TXT	2014/10/4 11:17	文本文档	12 KB

图 1-32　WebSphere V8.5.5.3 Liberty Profile 目录

2. WebSphere 服务器创建和启动

使用命令 bin\server. bat create 服务器名，创建 WebSphere 服务器，如图 1-33 所示。

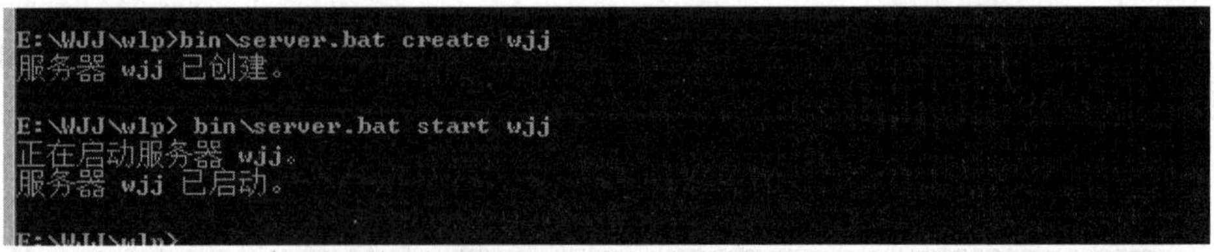
E:\WJJ\wlp>bin\server.bat create wjj
服务器 wjj 已创建。
E:\WJJ\wlp> bin\server.bat start wjj
正在启动服务器 wjj。
服务器 wjj 已启动。

图 1-33　创建和启动服务器

在创建好的服务器目录路径 usr\servers\服务器名\apps 下，放入需要发布的网站或服务，如图 1-34 所示。

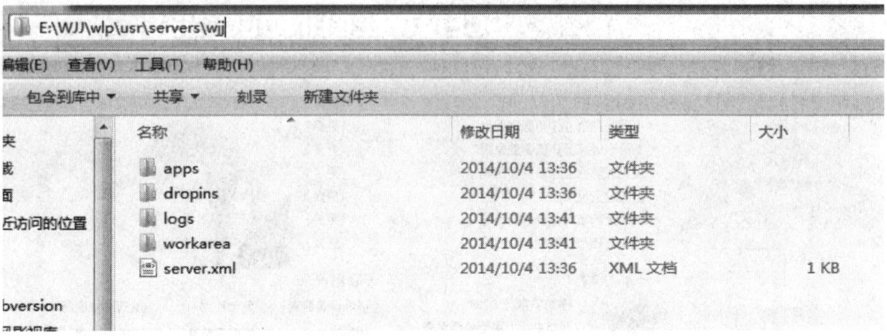
E:\WJJ\wlp\usr\servers\wjj
编辑(E)　查看(V)　工具(T)　帮助(H)
包含到库中　共享　刻录　新建文件夹
名称　修改日期　类型　大小
apps　2014/10/4 13:36　文件夹
dropins　2014/10/4 13:36　文件夹
logs　2014/10/4 13:41　文件夹
workarea　2014/10/4 13:41　文件夹
server.xml　2014/10/4 13:36　XML 文档　1 KB

图 1-34　放置发布的网站

1.4　Java Web 开发环境搭建

1.4.1　开发工具与环境

1. 集成开发工具：MyEclipse 10

MyEclipse 企业级工作平台（MyEclipse Enterprise Workbench，MyEclipse）是对 EclipseIDE 的扩展，利用它我们可以在数据库和 JavaEE 的开发、发布以及应用程序服务器的整合方面极大地提高工作效率。它是功能丰富的 JavaEE 集成开发环境，包括了完备的编码、调试、测试和发布功能，完整支持 HTML、Struts、JSP、CSS、Javascript、Spring、SQL、Hibernate、Java Servlet、Ajax、JSF、EJB3、JDBC 数据库链接工具等多项功能。

2. Web 服务器：apache-tomcat-7. 0. 42

Tomcat 是 Apache 软件基金会（Apache Software Foundation）的 Jakarta 项目中的一个核心项目，由 Apache、Sun 和其他一些公司及个人共同开发而成。在 Sun 的参与和支持下，最新的 Servlet 和 JSP 规范在 Tomcat 中得到实现，目前最新版本是 8. 0 版本，支持最新的 Servlet 3. 1、JSP 2. 3 规范。Tomcat 成为目前比较流行的 Web 应用服务器。它是一个小型、轻量级的支持 JSP 和 Servlet 技术的 Web 服务器，也是初学者学习开发 JSP 应用的首选 Web 服务器。

Tomcat 提供了各种平台的版本供下载，可以从本书前述的 Tomcat 网址下载其源代码版或者二进制版。由于 Java 的跨平台特性，基于 Java 的 Tomcat 也具有跨平台性。

3. 数据库：MySQL 5.5

MySQL 是一个关系型数据库管理系统，由瑞典 MySQL AB 公司开发，目前属于 Oracle 公司，它分为社区版和商业版，是一个多用户、多线程的 SQL 关系型数据库，是一个客户机/服务器结构的应用，它由一个服务器守护程序 mysqld、很多不同的客户程序和库组成。它是目前市场上运行最快的 结构化查询语言（Structured Query Language，SQL）数据库之一，其下载网站为 http://www.mysql.com，由于其体积小、速度快、总体拥有成本低，尤其是开放源码的特点，使得一般中小型网站的开发选择 MySQL 作为网站数据库。从应用架构上，MySQL 分为单点（Single，适合小规模）、复制（Replication，适合中小规模）和集群（Cluster，适合大规模）。

1.4.2　开发工具集成

1. MyEclipse 中集成 Tomcat 服务器

（1）在前面讲述的基础上，安装好 JDK 和 Tomcat，并设置好它们的环境变量。

（2）打开 MyEclipse，选择工具栏上的"Window"→"Preferences"，弹出对话框，并选择 MyEclipse 下的"Servers"→"Tomcat"，根据安装的"Tomcat"的版本号，选择"Servers"下的"Tomcat"，并指定其安装路径，并使其状态为"Enable"，然后单击"Apply"，如图 1-35 所示。

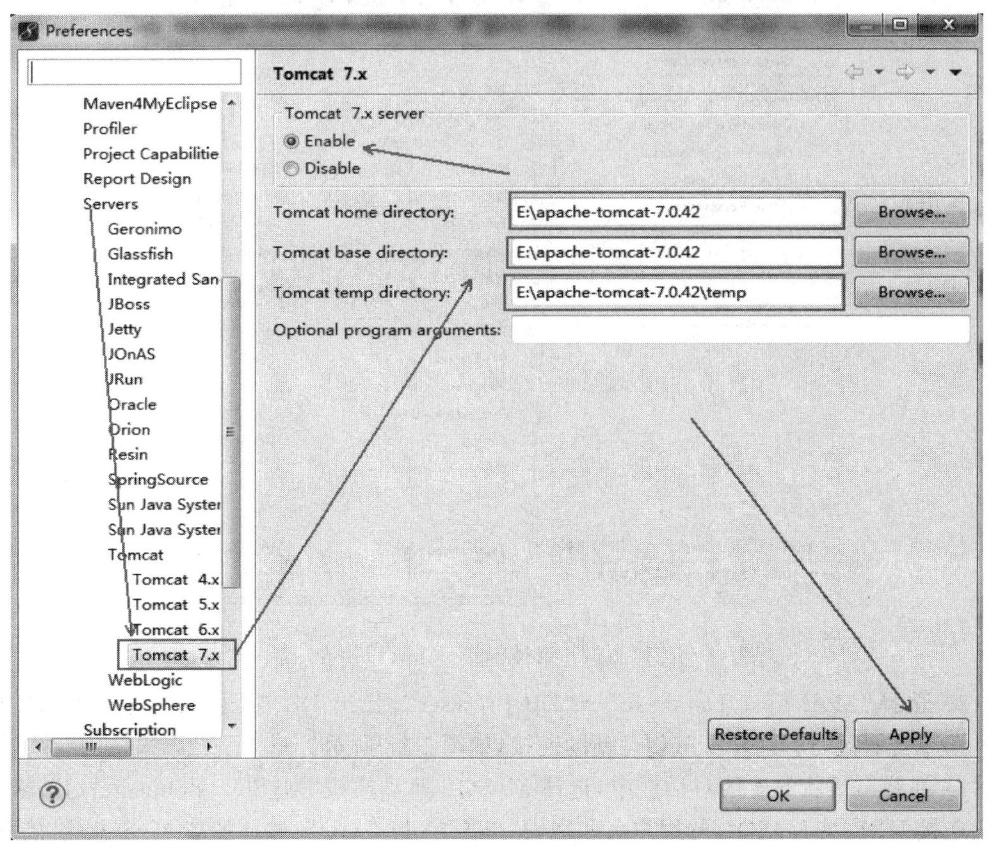

图 1-35　MyEclipse 中设置 Tomcat 安装路径

（3）配置成功后，即可在 MyEclipse 中发现，服务器图标下，已经出现 Tomcat 7.x 的图标，然后就可以在 MyEclipse 中，单击配置的“Tomcat 7.x”→“start”，启动 Tomcat 服务器，如图 1-36 所示。

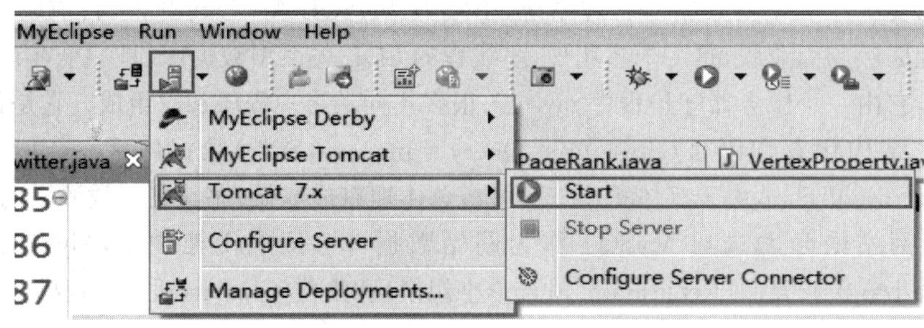

图 1-36　启动 Tomcat 服务器

(4) 启动 Tomcat 服务器后，然后即可创建 Java Web 工程，并部署新创建的工程到 Tomcat 服务器。

2. MyEclipse 中集成连接 MySQL 数据库

(1) 选择“Window”→“Show View”→“Other”，弹出 Show View 对话框，如图 1-37 所示。

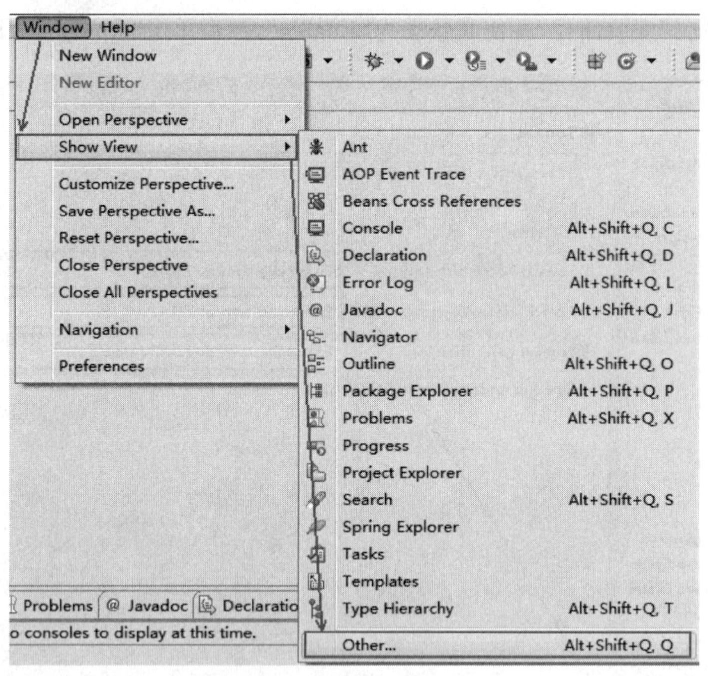

图 1-37　选择 Show View 路径

(2) 选择“MyEclipse Database”→“DB Browser”，弹出 DB 视图，然后选择“MyEclipse Derby”，单击右键，选择“New”，创建新的连接，如图 1-38 所示。

(3) 在弹出的新的连接对话框中，选择 MySQL 驱动模板“MySQL Connector/J”，输入驱动名、连接 URL 及 MySQL 的用户名和密码，单击“Add JARs”，选择加载 MySQL 连接驱动，然后单击“Test Driver”，测试连接。如果显示数据库连接成功建立，表明配置连接成功，否则，需要重新配置连接，如图 1-39 所示。

(4) 选择打开连接，即可直接编辑数据库的表及表信息，如图 1-40 所示。

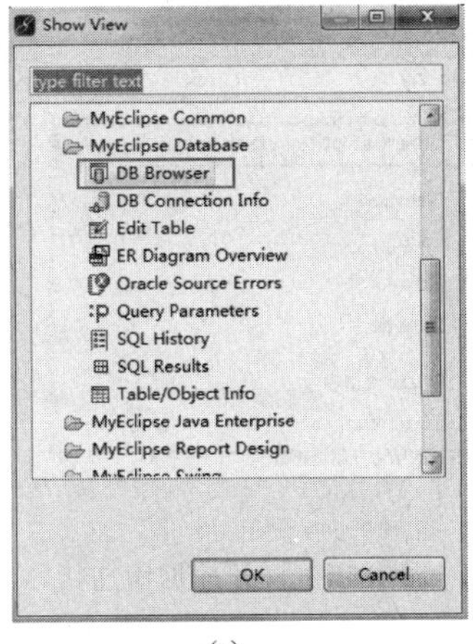

(a)

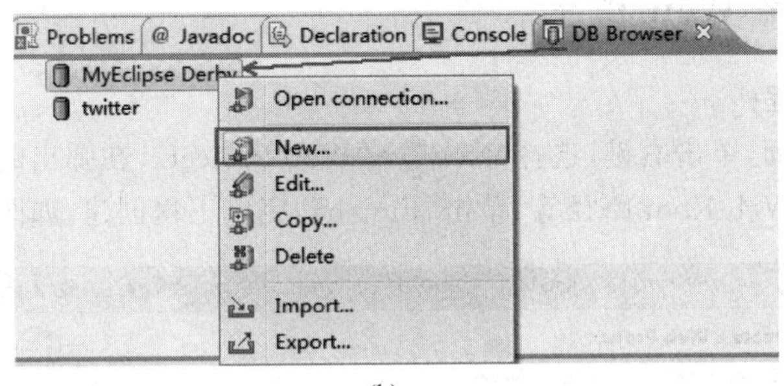

(b)

图 1-38　新的连接的创建

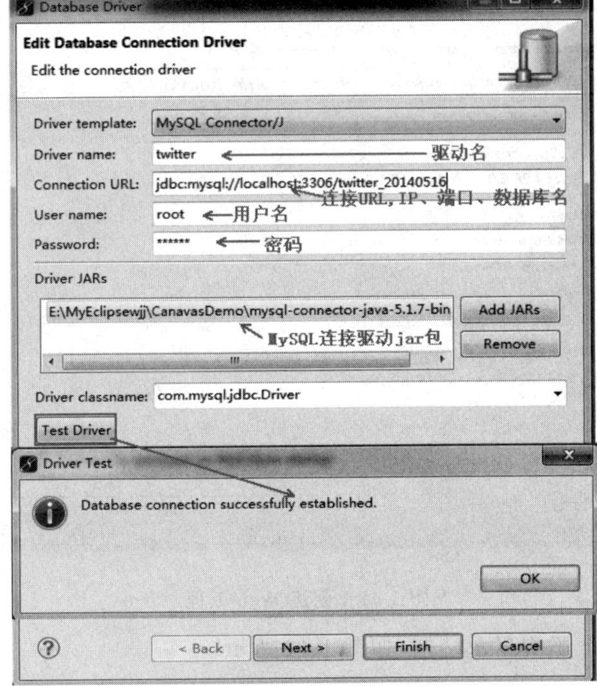

图 1-39　数据库连接信息配置及测试

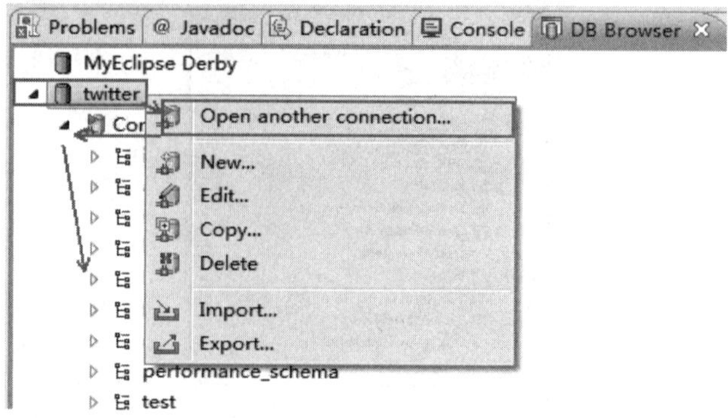

图 1-40　打开数据库连接

1.4.3　创建部署 Web 程序

1. 创建 Web 工程

(1) 在包浏览界面,单击右键,选择"New"→"Web Project",在弹出的对话框中,填写工程名、工程存储路径和 Web Root 路径等,单击"Finish",完成工程创建,如图 1-41 所示。

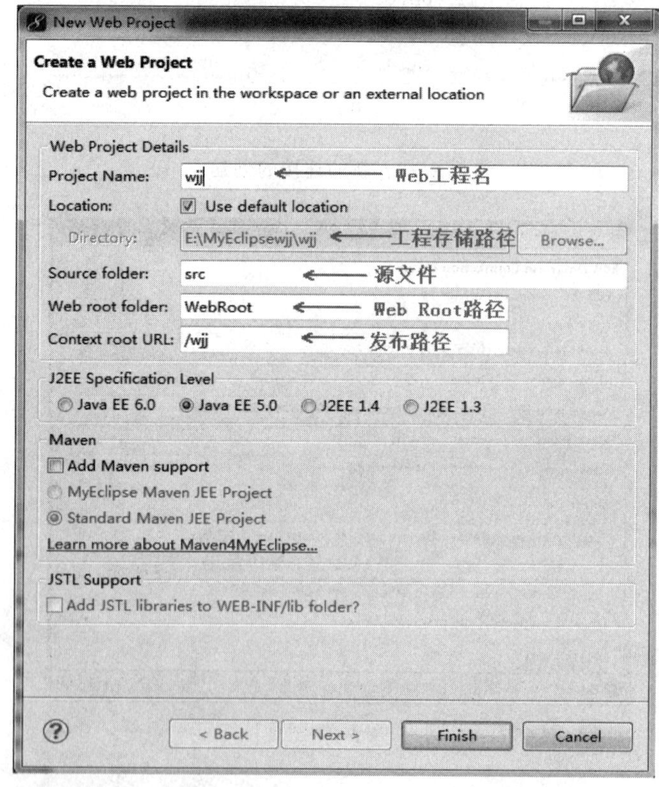

图 1-41　创建 Web 工程

(2) 创建完的 Web 工程目录结构,如图 1-42 所示。

2. 部署 Web 工程

(1) 选中新创建的 Web 工程,单击部署按钮,弹出 Project Deployments 对话框,默认部署的工程为刚才选中的工程名,然后单击"Add"按钮,弹出新的部署对话框,选中部署的服务器"Tomcat 7. x",以及默认的部署路径,如图 1-43 所示。

(2) 部署成功后,单击"OK"按钮,完成 Web 工程部署,如图 1-44 所示。

(3) 启动运行 Tomcat,并在浏览器地址栏中输入 http://localhost:8080/wjj/index. jsp,测试部署的工程,测试成功后的页面,如图 1-45 所示。

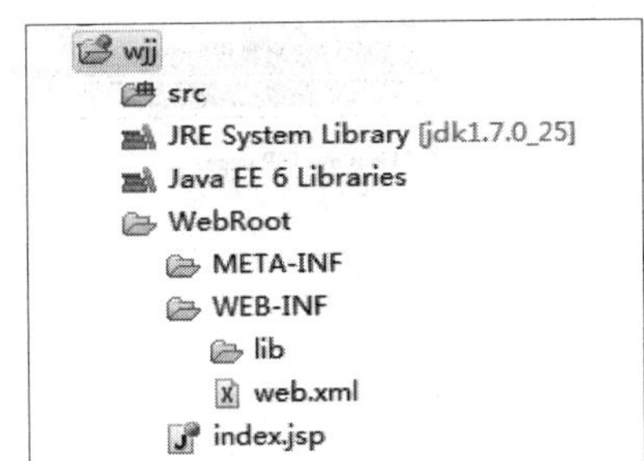

图 1-42　Web 工程目录结构

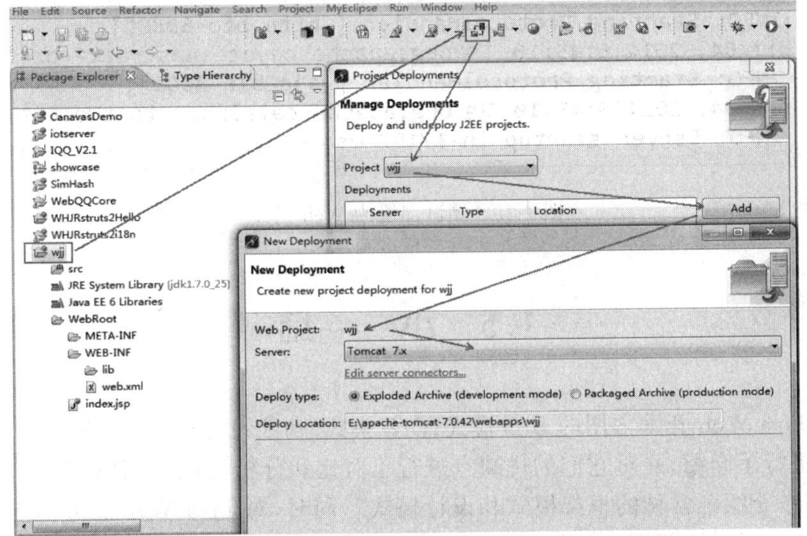

图 1-43　配置部署 Web 工程

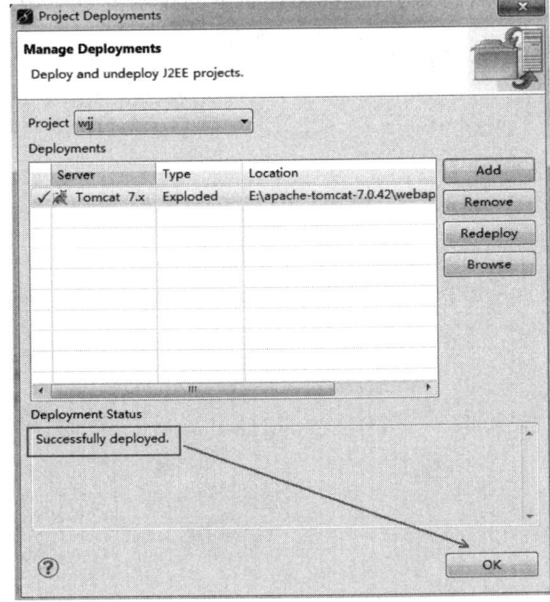

图 1-44　Web 工程部署成功

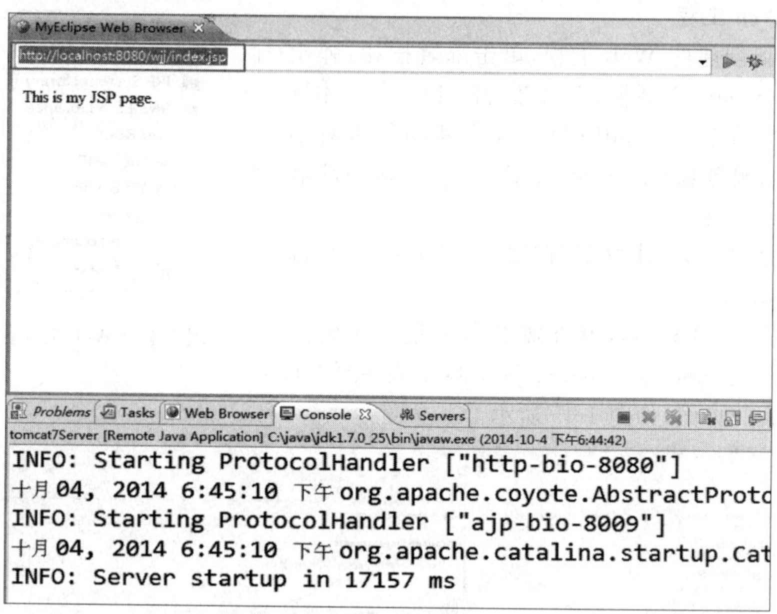

图 1-45　测试部署的 Web 工程

1.5　小　　结

本章就 Java Web 开发常用的 B/S 模式、C/S 模式、初期设计模式、经典的 MVC 模式和框架模式分别进行了介绍，并对它们的优缺点进行了对比和分析，澄清了框架模式和设计模式的概念混淆，简单介绍了常见的框架模式和设计模式。同时，对 Java Web 服务器的类型、应用、安装和测试进行了详细阐述，并就 Java Web 开发环境搭建、集成开发工具应用和创建部署 Web 应用程序进行了案例式入门导引，这些都是后续的技术要点应用的基础。

第2章　Java Web 基础

2.1　HTML 语言

2.1.1　HTML 简介

超文本链接标示语言(Hypertext Markup Language,HTML)最初的出现是为了世界各地的科学家们能够方便地进行合作研究,它不是程序语言,是由文字及标记组合而成,HTML不但可以用来结构化网页上的信息,如标题、段落和列表等,还可以用来将图片、链接、音乐和程序等非文字的元素添加到网页上。浏览器或者其他可以浏览网页的设备将这些 HTML 语言"翻译"过来,并按照定义的格式显示出来,转化成最终看到的网页。现在它已经成为国际标准,由万维网联盟(W3C)维护。

目前 HTML 编辑器有很多,可以是任何文本编辑器或者网页编辑制作工具,如 FrontPage、Dreamweave。生成的 HTML 文件最常用的扩展名是. html,htm 也很常用。

由于 HTML 的标准比较松散,很多不规范的 HTML 代码逐渐出现,使得页面体积越来越庞大,而且数据和表现混在一起,这样,XML 就被用来描述数据,而 HTML 则用来显示数据。另外,当前很多浏览器不是运行在计算机中,而是运行在移动电话和一些信息家电上,这些浏览器没有办法解析不规范的标记语言,因此,XHTML 应运而生,它是 HTML 和 XML 的结合,也是 HTML 向 XML 过渡的一个桥梁。它比 HTML 更加严密,代码也更加整洁,使我们能够编写出结构良好的文档,并且可以很好地工作于所有的浏览器、无线设备等。

2.1.2　HTML 基本结构

1. HTML 基本结构

HTML 文件是纯文本文件,可以用所有的文本编辑器进行编辑,如记事本等,也可以使用可视化编辑器,如 FrontPage、Dreamweaver 等。

在 HTML 中,由<>和</>括起来的文本称为"标签",<>表示开始标签,</>表示结束标签,开始标签和结束标签配对使用,它们之间的部分是该标签的作用域,如<html></html>等,HTML 就是以这些标签来控制内容的显示方式。

创建基本的 HTML 页面的方法如下所示。

(1) 在记事本中编辑、输入如下所示的代码,并保存为以. htm 或. html 为扩展名的文件,命名为 index. html。

```
<html>
<head>
    <title>北京政法信息技术</title>
</head>
<body>
      测试页面信息
</body>
</html>
```

(2) 双击 index. html 网页后，就可以在浏览器中显示网页了，如图 2-1 所示。

图 2-1　index. html 页面

上述代码是创建一个 HTML 文件的最基本结构，所有 HTML 文件都要包含这些基本部分。其中：

① <html>和</html>表示该文档是 HTML 文档；

② <head>和</head>标明文档的头部信息，一般包括标题和主题信息，该部分信息不会出现在页面正文中，也可以在其中嵌入其他标签，表示如文件标题、编码方式等属性；

③ <title>和</title>表示该文档的标题，标签间的文本显示在浏览器的标题栏中；

④ <body>和</body>是网页的主体信息，可以包括各种字符、表格、图像及各种嵌入对象等信息。

2. HTML 书写规范

在 HTML 中按照格式来划分标签可分为两类，大部分标签是成对出现的，需要开始标签和结束标签；也有一些标签不需要成对出现，单独出现一次就可以，这类标签通常不控制显示形态，如
表示换行。

标签是不区分大小写的。

2.1.3 HTML 常用标签

1. 常用文本标签

文本是网页中最基本也是最重要的元素之一，在网页上输入、编辑、格式化文本元素是制作网页的基本操作。文本的主要作用是帮助网页浏览者快速地了解网页的内容，它通常是网页内容的基础，是网页中必不可少的元素，常用的文本标签分为标题标签、段落标签和格式化标签 3 类。

(1) 标题标签<hi>

<hi>设置网页内容标题标签，通过<hi>…</hi>标签配对使用设置 HTML 网页内容标题，标题标签共分 6 种，分别表示不同字号的标题，i 可以取值为 1～6。同时，在<hi>中可以使用属性<align>来设置标题对齐方式，如果没有设置<align>属性，默认对齐方式是 left(左对齐)。

(2) 段落标签<p>

<p>用来标记段落的开始,用</p>可以标记一个段落的结束,也可以省略,到下一个<p>开始新的段落。换行标签
可以将文字强制换行,取消换行标签 nobr。右缩进标签 blockquote 可以用来使文字的段落缩进、居中对齐标签 center 等。

2. 超级链接

Web 上的网页都是互相连接的,通过超链接可以链接到其他页面,这里的超链接就是具有链接能力的文字或图片,可以链接文本、媒体等网络资源。

(1) 超级链接的基本格式

创建超级链接的标签为<a>,基本格式为:

```
<A href = "链接地址"target = "超链接窗口打开方式">超链接名称</A>
```

如:<a href = "http://www.bcpl.cn/zf.htm" target = "_blank" >软件技术服务</a>

(2) 超级链接属性

标签<a>的属性 href 指定了链接到的目标地址,该地址可以是文件所在位置,也可以是一个 URL,只有正确指定目标地址,才能正确访问需要的资源。

属性 target 用于指定打开链接的目标窗口,其默认方式是原窗口。

3. 表单

在网页中,表单是最常用的网页元素,主要用来收集客户端信息,特别是在制作动态网页的时候,使得网页具有交互的功能。通常将表单设计在一个 html 文档中,当用户填写完信息后提交,表单的内容就可以从客户端发送到服务器端,经过服务器端处理,将用户所需信息传送给客户端的浏览器上,完成一次交互。

表单通常由窗体和控件组成,一般包括文本框、单选按钮、复选、单复选框按钮等。

(1) 表单标签语法格式

表单是由<form>和</form>标签配对创建,在这两个标签之间的一切定义都属于表单的内容,可以包含所有的表单控件和伴随的数据。

创建表单的语法为:

```
<form action = "url" method = get|post name = "myform" target = "_blank">…</form>
```

(2) 表单标签属性

在表单<form>标签中主要属性有 action、method 和 target 等。

action 属性的值是表单提交的处理程序的程序名,这个值是程序或脚本的一个完整 URL,这个地址可以是绝对地址,也可以是相对地址,表示接收该表单信息的 URL,如果这个属性是空值,则当前文档的 URL 将被默认使用。当表单被提交时,服务器将执行该地址里的程序,完成提交数据的处理。如果该地址是一个邮件地址,则程序运行后会把提交的数据以邮件形式发送给指定的地址。

method 属性是定义处理程序从表单中获得信息的方式,可取值为 GET 或 POST,表示收集到的表单数据以何种形式发送。当 method 值为 GET 时,表示表单数据会被 CGI 或 ASP 程序从 HTML 文档中获得,并将数据附加在 URL 后,由用户端直接发送至服务器,所以速度较快,但是这种方式传送的数据量是有所限制的。如果不指定 method 的值,默认就是 GET;如果 method 的值取为 POST,表单数据是和 URL 分开发送的,用户端会通知服务器来读取数据,这样传送的数据量要比 GET 方式大得多,缺点是速度相对较慢。

target 属性是用来指定目标窗口的打开方式。表单的目标窗口是用来显示表单的返回信息,如是不是提交成功或出错等。目标窗口的打开方式有 4 个取值:_blank 表示返回信息在新打开的窗口显示;_self 表示在当前浏览器窗口显示;_top 表示在顶级浏览器窗口中显示;_parent 表示在父级窗口显示。

4. 表单控件

在表单控件中,可以按照填写方式不同,分为输入类和菜单列表类两类。

输入类的控件一般就以 input 开始,说明这个控件需要输入。

该标签的语法为:

```
<form>
    <input name="控件名称"type="控件类型">
</form>
```

控件名称是来标识当前选择的控件的,而类型的值及控件属性主要有文本框(text)、密码域(password)和按钮(button、submit、reset)。

2.2 JavaScript 技术

2.2.1 JavaScript 简介

1. JavaScript 简介

JavaScript 是一种解释性的,基于对象的脚本语言(An Interpreted, Object-based Scripting Language)。其前身是 LiveScript,JavaScript 的正式名称是“ECMAScript”,是由 Netscape (Navigator 2.0)公司的 Brendan Eich 发明了这门语言,从 1996 年开始,它就出现在所有的 Netscape 和 Microsoft 浏览器中。现在几乎所有浏览器都支持 JavaScript,如 Internet Explorer(IE)、Firefox、Netscape、Mozilla 和 Opera 等。

JavaScript 正式的标准是 ECMA-262,这个标准基于 JavaScript (Netscape) 和 JScript (Microsoft)。标准由 ECMA 组织发展和维护。

JavaScript 主要是基于客户端运行的,用户点击带有 JavaScript 的网页,网页里的 JavaScript 就传到浏览器,由浏览器对此作处理。前面提到的下拉菜单、验证表单有效性等大量互动性功能,都是在客户端完成的,不需要和 Web Server 发生任何数据交换, 因此,不会增加 Web Server 的负担。JavaScript 可以创建直接运行于 Internet 上的应用。使用 JavaScript,可以创建你所需要的动态 HTML 页面,用于处理用户输入及使用特殊的对象、文件和关系数据库维护稳固的数据。从内部的协作信息管理和内联网发布到大型超市的电子交易和商务,都可得到应用。通过 JavaScript 的 LiveConnect 功能,程序还可以访问 Java 和 CORBA 发布的应用程序。

2. 在 HTML 页面中使用 JavaScript

JavaScript 是一种解释性编程语言,其源代码在被网络传送到客户端执行之前不需经过编译,而是将文本格式的字符代码发送给客户,由浏览器解释执行。JavaScript 的代码是一种文本字符格式,可以直接嵌入 HTML 文档中,并且可动态装载。

在 JavaScript 中,可以添加注释来对 JavaScript 进行解释,或者提高其可读性。

JavaScript 的注释分为两种：一种是单行注释，以//开始；一种是多行注释，以/*开始，以*/结束，/*和*/配对使用。

在 HTML 页面中使用 JavaScript 的方法有两种：一种是直接加入到 HTML 文件中；另外一种是引用方式。简单的 JavaScript 应用通常都采用第一种方式，即直接加入到 HTML 文档中。复杂的 JavaScript 会采用引用的方式，采用引用会使网页代码结构更为清晰、易用。实践中，大型网页的应用通常都采用这种方式。

2.2.2 JavaScript 表单应用

1. JavaScript 对象类型

JavaScript 对象的类型分为 4 种：JavaScript 本地对象和内置对象、Browser 对象(BOM)、HTML DOM 对象和自定义对象。

在 JavaScript 中，常用的内置对象有数组对象(Array)、字符串对象(String)、数学对象(Math)和日期对象(Date)等。

Browser 对象(BOM)包括 Window 对象、Navigator 对象、Screen 对象、History 对象和 Location 对象。

HTML DOM 对象包括 Document 对象、Event 对象、Anchor 对象、Form 对象、Frame 对象、Link 对象和 Table 对象等。

2. 对象的使用

在 JavaScript 中提供了几个用于操作对象的语句、关键词和运算符。主要通过 for…in 语句、with 语句、this 关键词和 new 运算符来使用。

3. Form 对象

表单对象可以使设计人员用表单中不同的元素与客户机用户相交互，就可以实现动态改变 Web 文档的行为，而用不着在之前先进行数据输入。

(1) Form 表单

表单(Form)：它构成了 Web 页面的基本元素。通常一个 Web 页面有一个或几个表单，使用 Forms[]数组来实现不同表单的访问。

```
<form Name = Form1>
<INPUT type = text…>
<Input type = text…>
<Inpup byne = text…>
</form>

<form Name = Form2>
<INPUT type = text…>
<Input type = text…>
</form>
```

在 Forms[0]中共有 3 个基本元素，而 Forms[1]中只有两个元素。

表单对象最主要的功能就是能够直接访问 HTML 文档中的表单，它封装了相关的 HTML 代码：

```
<Form
name = "表的名称"
target = "指定信息的提交窗口"
action = "接收表单程序对应的 URL"
method = "信息数据传送方式(get/post)"
enctype = "表单编码方式"
[onsubmit = "JavaScript 代码"]>
</Form>
```

（2）表单对象的方法

表单对象的方法有 submit()、reset()两种方法，submit()方法主要用于实现表单信息的提交，reset()方法主要是实现信息的重置，如提交 Mytest 表单，则使用下列格式：

```
document.mytest.submit()
```

（3）表单属性

表单对象中的属性主要包括 elements、name、action、target、encoding 和 method。

除 elements 外，其他几个均反映了表单标识中相应属性的状态，这通常是单个表单标识；而 elements 常常是多个表单元素值组成的数组，如：

```
elements[0].Mytable.elements[1]
```

（4）访问表单对象

在 JavaScript 中访问表单对象可由两种方法实现：

① 通过表单名访问表单

在表单对象的属性中首先必须指定其表单名，然后就可以通过下列标识访问表单。

```
document.Mytable()
```

② 通过数组来访问表单

除了使用表单名来访问表单外，还可以使用表单对象数组来访问表单对象。但需要注意一点，因表单对象是由浏览器环境提供的，而浏览器环境所提供的数组下标是由 0 到 n，所以可通过下列格式实现表单对象的访问：

```
document.forms[0]
document.forms[1]
document.forms[2]
...
```

4. 表单验证

用户在 Form 中输入数据后，触发 Sumbit 事件，在该事件处理中常常设定数据校验的操作，如核对是否有些文本框未输入数据、电话号码是否为 8 个数字和电邮地址是否有@符号等。如果校验成功，返回 true，JavaScript 向服务器提交 Form；否则，提示用户出错，让用户重新输入数据。

检查文本框输入的合法性应用，代码如下：

```
<html>
<head>
<title></title>
<script language="javascript">
     function checkIt( ){
     var re=document.fm.tx.value;
     for(i=0;i<re.length;i++){
          if(re.charAt(i)>="a"&&re.charAt(i)<="z") {
               return true;
          }
          else {
               alert("您输入的字符串超出范围!")
               return false;
          }
     }
}
</script>
</head>
<body>
<form name="fm" method="post" action="#" enctype="text/plain" onSubmit="return checkIt( )">
姓名:
<input type="text" name="tx"s size="20">
<input type="submit" value="提交">
</form>
</body>
</html>
```

2.2.3 JavaScript 正则表达式

1. 正则表达式简介

正则表达式(Regular Expression),又称正规表示法或常规表示法(在代码中常简写为 regex、regexp 或 RE),正则表达式使用单个字符串来描述、匹配一系列符合某个句法规则的字符串。在很多文本编辑器里,正则表达式通常被用来检索、替换那些符合某个模式的文本。正则表达式是一种可以用于模式匹配和替换的强有力的工具。

其作用主要表现在以下两个方面。

(1) 测试字符串的某个模式:例如,可以对一个输入字符串进行测试,看在该字符串里是否存在一个电话号码模式或一个信用卡号码模式,这种测试称为数据有效性验证。

(2) 替换文本:可以在文档中使用一个正则表达式来标识特定文字,然后可以全部将其删除,或者替换为别的文字。

根据模式匹配从字符串中提取一个子字符串,可以用来在文本或输入字段中查找特定文字。

2. 正则表达式语法格式

正则表达式的形式一般如下:

```
/wjj/
```

其中,位于"/"定界符之间的部分就是将要在目标对象中进行匹配的模式,用户只要把希望查找匹配对象的模式内容放入"/"定界符之间即可。

(1) 元字符:为了能够使用户更加灵活地定制模式内容,正则表达式提供了专门的"元字符"。所谓元字符就是指那些在正则表达式中具有特殊意义的专用字符,可以用来规定其前导字符(即位于元字符前面的字符)在目标对象中的出现模式。

常用的元字符包括"+"、"*"和"?"。

- "+"元字符规定其前导字符必须在目标对象中连续出现一次或多次。
- "*"元字符规定其前导字符必须在目标对象中出现零次或连续多次。
- "?"元字符规定其前导对象必须在目标对象中连续出现零次或一次。

除了元字符之外,用户还可以精确指定模式在匹配对象中出现的频率,如/jim {2,6}/,上述正则表达式规定字符 m 可以在匹配对象中连续出现 2~6 次,因此,上述正则表达式可以同 jimmy 或 jimmmmmy 等字符串相匹配。

- "\s"用于匹配单个空格符,包括 tab 键和换行符。
- "\S"用于匹配除单个空格符之外的所有字符。
- "\d"用于匹配从 0 到 9 的数字。
- "\w"用于匹配字母、数字或下划线字符。
- "\W"用于匹配所有与\w 不匹配的字符。
- ". "用于匹配除换行符之外的所有字符。

注意:可以把\s 和\S 以及\w 和\W 看作互为逆的运算。

(2) 限定符:有时候不知道要匹配多少字符,为了能适应这种不确定性,正则表达式支持限定符的概念。这些限定符可以指定正则表达式的一个给定组件,必须要出现多少次才能满足匹配。

- {*n*}:*n* 是一个非负整数,匹配确定的 *n* 次,例如,"o{2}"不能匹配"Bob"中的"o",但是能匹配"food"中的两个"o"。
- {*n*,}:*n* 是一个非负整数,至少匹配 *n* 次,例如,"o{2,}"不能匹配"Bob"中的"o",但能匹配"foooood"中的所有"o"。"o{1,}"等价于"o+","o{0,}"则等价于"o*"。
- {*n*,*m*}:*m* 和 *n* 均为非负整数,其中 $n \leqslant m$。最少匹配 *n* 次且最多匹配 *m* 次,例如,"o{1,3}"将匹配"fooooood"中的前 3 个"o","o{0,1}"等价于"o?"。请注意在逗号和两个数之间不能有空格。

(3) 定位符:正则表达式中还具有另外一种较为独特的专用字符,即定位符。定位符用于规定匹配模式在目标对象中的出现位置。

常用的定位符包括"^"、"$"、"\b" 和"\B"。

- "^"定位符规定匹配模式必须出现在目标字符串的开头。
- "$"定位符规定匹配模式必须出现在目标对象的结尾。
- "\b"定位符规定匹配模式必须出现在目标字符串的开头或结尾的两个边界之一。
- "\B"定位符则规定匹配对象必须位于目标字符串的开头和结尾两个边界之内,即匹配对象既不能作为开头,也不能作为结尾,匹配非单词边界。

(4) 范围匹配:为了更加灵活地设定匹配模式,正则表达式允许使用者在匹配模式中指定某一个范围而不局限于具体的字符。

- /[A~Z]/:正则表达式将会与从 A 到 Z 范围内任何一个大写字母相匹配。
- /[a~z]/:正则表达式将会与从 a 到 z 范围内任何一个小写字母相匹配。
- /[0~9]/:正则表达式将会与从 0 到 9 范围内任何一个数字相匹配。
- /([a~z][A~Z][0~9])+/:正则表达式将会与任何由字母和数字组成的字符串,如

“aB0” 等相匹配。

注意:在正则表达式中可以使用“()”把字符串组合在一起,“()”符号包含的内容必须同时出现在目标对象中。但上述范围正则表达式将无法与像“abc”等一样的字符串匹配,因“abc”中的最后一个字符为字母而非数字。

(5) 管道符:管道符可以在正则表达式中实现类似编程逻辑中的“或”运算,在多个不同的模式中任选一个进行匹配。

如/to|too|2/,此正则表达式将会与目标对象中的 “to”、“too”或“2”相匹配。

(6) 否定符:正则表达式中还有一个较为常用的运算符,即否定符 “[^]”,与前述的定位符 “^” 不同,否定符 “[^]”规定目标对象中不能存在模式中所规定的字符串。

如/[^A～C]/,此字符串将会与目标对象中除 A,B 和 C 之外的任何字符相匹配。一般来说,当“^”出现在 “[]”内时就被视作否定运算符;而当“^”位于“[]”之外,或没有“[]”时,则应当被视作定位符。

(7) 转义符:当在正则表达式的模式中加入元字符,并查找其匹配对象时,可以使用转义符“\”,如/Th*/,正则表达式将会与目标对象中的“Th*”而非“The”等相匹配。

(8) 正则表达式对象:包含正则表达式模式以及表明如何应用模式的标志。

```
语法格式 1:bc = /pattern/[flags]
语法格式 2:bc = new RegExp("pattern",["flags"])
```

参数

• Pattern(必选项):表示要使用的正则表达式模式。

```
语法格式 1:用 "/" 字符分隔模式。
语法格式 2:用引号将模式引起来。
```

• Flags(可选项):使用语法 2 要用引号将 flags 引起来。标志可以组合使用,可选用的参数如下所示。

```
g(全文查找出现的所有 pattern)
i(忽略大小写)
m(多行查找)
```

正则表达式优先级:构造正则表达式之后,可以如数学表达式求值顺序一样来求值,即可以从左至右并按照一个优先级顺序来求值。优先级顺序如下:

- \
- (), (?:), (?=), []
- *, +, ?, {n}, {n,}, {n,m}
- ^, $, \,任何元字符,任何字符
- |

3. 正则表达式应用

(1) 元字符应用

“/fo+/”:因为上述正则表达式中包含“+”元字符,表示可以与目标对象中的 “fool”、“fo”或“football”等在字母 f 后面连续出现一个或多个字母 o 的字符串相匹配。

“/eg*/”:因为上述正则表达式中包含“*”元字符,表示可以与目标对象中的 “easy”、“ego”或“egg”等在字母 e 后面连续出现零个或多个字母 g 的字符串相匹配。

“/Wil?/”:因为上述正则表达式中包含“?”元字符,表示可以与目标对象中的 “Win”或

“Wilson”等在字母 i 后面连续出现零个或一个字母 l 的字符串相匹配。

“/\s+/”:上述正则表达式可以用于匹配目标对象中的一个或多个空格字符。

“/\d000/”:如果我们手中有一份复杂的财务报表,那么我们可以通过上述正则表达式轻而易举地查找到所有总额达千元的款项。

(2) 定位符应用

如上所述,“^”和“$”以及“\b”和“\B”看作是互为逆运算的两组定位符,如下所示。

“/^hell/”:正则表达式中包含“^”定位符,可以与目标对象中以“hell”、“hello”或“hellhound”开头的字符串相匹配。

“/ar$/”:正则表达式中包含“$”定位符,可以与目标对象中以“car”、“bar”或“ar”结尾的字符串相匹配。

“/\bbom/”:正则表达式模式以“\b”定位符开头,可以与目标对象中以“bomb”或“bom”开头的字符串相匹配。

“/man\b/”:正则表达式模式以“\b”定位符结尾,可以与目标对象中以“human”、“woman”或“man”结尾的字符串相匹配。

(3) 验证用户输入的邮件地址的有效性

验证用户输入邮件的有效性代码如下:

```
<html>
    <head>
        <script language = "Javascript1.2">
            function verifyAddress(obj){
                var email = obj.email.value;
                  var pattern = /^([a-zA-Z0-9_-]) + @([a-zA-Z0-9_-]) + (¥.[a-zA-Z0-9
_-]) + /;
                flag = pattern.test(email);
                if(flag){
                    alert("Your email address is correct!");
                    return true;
                }else {
                    alert("Please try again!");
                    return false;
                }
            }
        </script>
      </head>
      <body>
        <form onSubmit = "return verifyAddress(this);">
            <input name = "email" type = "text">
            <input type = "submit">
        </form>
      </body>
</html>
</form>
</body>
</html>
```

2.3　Servlet 技术

1. Servlet 简介

Servlet 是在服务器上运行的 Java 小应用程序，实际上为一个在服务器端运行 Java 类程序，通过请求、响应模型来访问的应用程序，即一个 Servlet 就是 Java 编程语言中的一个类，它被用来扩展服务器的性能，服务器上驻留着可以通过“请求-响应”编程模型来访问的应用程序。虽然 Servlet 可以对任何类型的请求产生响应，但通常只用来扩展 Web 服务器的应用程序。Servlet 是一个运行在程序模块内部的，并且增强了面向请求和响应服务应用的程序模块，用来替代公共网关接口（Common Gateway Interface，CGI）应用程序完成的功能。

Servlet 是一个运行在 Servlet 容器内部的，并且增强了面向请求和响应服务应用的对象，通过它可以访问远程对象，可以跟踪大量的信息，允许多用户之间的协作，可以生成动态的 HTML 页面内容。

Servlet 容器（以前称 Servlet 引擎），实际上是执行 Servlet 的软件。所有支持 Servlet 的服务器都包括一个 Servlet 容器（集成的或通过插件）。通常而言支持 Java 的服务器常为增强的 ServletHTTP 服务器（即它包括一个用于运行 Servlet 的 Servlet 容器）。

2. Servlet 与 CGI 区别

在通信量大的服务器上，Servlet 的优点在于它们的执行速度更快于 CGI 程序。各个用户请求被激活成单个程序中的一个线程，而无须创建单独的进程，即当 Servlet 处于服务器进程中，它通过多线程方式运行其 service 方法，一个实例可以服务于多个请求，并且其实例一般不会销毁，而 CGI 对每个请求都产生新的进程，服务完成后就销毁，所以效率上低于 Servlet，使用 Servlet 意味着服务器端处理请求的系统开销将明显降低。

通用网关接口（Common Gateway Interface，CGI）程序实现数据在 Web 上的传输，使用的是如 Perl 这样的语言编写的，它对于客户端的每个请求，必须创建 CGI 程序的一个新实例，这样占用了大量的内存资源，而 Servlet 技术不用。

Servlet 是一个用 Java 编写的应用程序，在服务器上运行、处理请求信息并将其发送到客户端。对于客户端的请求，只需要创建 Servlet 的实例一次，因此节省了大量的内存资源。Servlet 在初始化后就保留在内存中，因此每次作出请求时无须加载。

3. Servlet 生命周期

Servlet 运行在 Servlet 容器中，其生命周期由容器来管理。Servlet 的生命周期通过 javax. servlet. Servlet 接口中的 init()、service()和 destroy()方法来表示。

Servlet 的生命周期分为 4 个阶段过程，分别如下所示。

（1）加载和实例化

Servlet 容器负责加载和实例化 Servlet。当 Servlet 容器启动时，或者在容器检测到需要这个 Servlet 来响应第一个请求时，创建 Servlet 实例。当 Servlet 容器启动后，它必须要知道所需的 Servlet 类在什么位置，Servlet 容器可以从本地文件系统、远程文件系统或其他的网络服务中通过类加载器加载 Servlet 类，成功加载后，容器创建 Servlet 的实例。因为容器是通过 Java 的反射 API 来创建 Servlet 实例，调用的是 Servlet 的默认构造方法（即不带参数的构造方法），所以在编写 Servlet 类的时候，不应该提供带参数的构造方法。

(2) 初始化

在 Servlet 实例化之后,容器将调用 Servlet 的 init()方法初始化这个对象。初始化的目的是为了让 Servlet 对象在处理客户端请求前完成一些初始化的工作,如建立数据库的连接、获取配置信息等。对于每一个 Servlet 实例,init()方法只被调用一次。在初始化期间,Servlet 实例可以使用容器为它准备的 ServletConfig 对象从 Web 应用程序的配置信息(在 web. xml 中配置)中获取初始化的参数信息。在初始化期间,如果发生错误,Servlet 实例可以抛出 ServletException 异常或 UnavailableException 异常来通知容器。ServletException 异常用于指明一般的初始化失败,如没有找到初始化参数;而 UnavailableException 异常用于通知容器该 Servlet 实例不可用,如数据库服务器没有启动、数据库连接无法建立,Servlet 就可以抛出 UnavailableException 异常向容器指出它暂时或永久不可用。

(3) 请求处理

Servlet 容器调用 Servlet 的 service()方法对请求进行处理,注意在 service()方法调用之前,init()方法必须成功执行。在 service()方法中,Servlet 实例通过 ServletRequest 对象得到客户端的相关信息和请求信息,在对请求进行处理后,调用 ServletResponse 对象的方法设置响应信息。在 service()方法执行期间,如果发生错误,Servlet 实例可以抛出 ServletException 异常或者 UnavailableException 异常。如果 UnavailableException 异常指示了该实例永久不可用,Servlet 容器将调用实例的 destroy()方法,释放该实例。此后对该实例的任何请求,都将收到容器发送的 HTTP 404(请求的资源不可用)响应。如果 UnavailableException 异常指示了该实例暂时不可用,那么在暂时不可用的时间段内,对该实例的任何请求,都将收到容器发送的 HTTP 503(服务器暂时忙,不能处理请求)响应。

(4) 服务终止

当容器检测到一个 Servlet 实例应该从服务中被移除的时候,容器就会调用实例的 destroy()方法,以便让该实例可以释放它所使用的资源,保存数据到持久存储设备中。当需要释放内存或容器关闭时,容器就会调用 Servlet 实例的 destroy()方法。在 destroy()方法调用之后,容器会释放这个 Servlet 实例,该实例随后会被 Java 的垃圾收集器所回收。如果再次需要这个 Servlet 处理请求,Servlet 容器会创建一个新的 Servlet 实例。

在整个 Servlet 的生命周期过程中,创建 Servlet 实例、调用实例的 init()和 destroy()方法都只进行一次,当初始化完成后,Servlet 容器会将该实例保存在内存中,通过调用它的 service()方法,为接收到的请求服务。

4. 服务器调用 Servlet 的过程

Servlet 请求调用过程如图 2-2 所示。

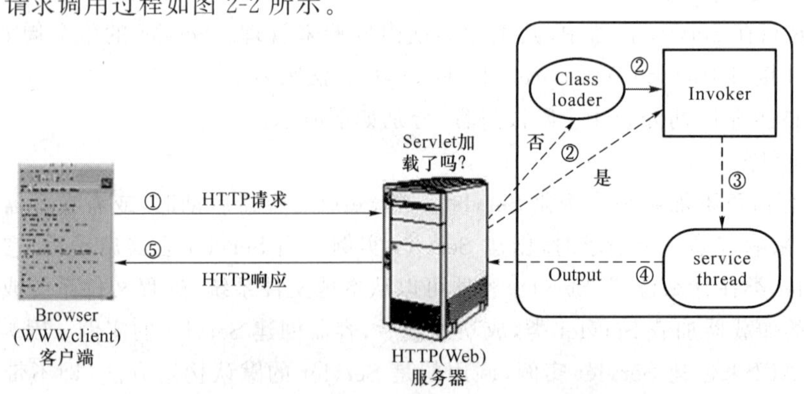

图 2-2　Servlet 请求调用过程

详细调用步骤如下所示。

(1) 在服务器启动时,当 Servlet 被客户首次请求或被配置好,这时由服务器加载 Servlet。Servlet 的加载可以通过合用一个自定义的 Java 类加载工具(可允许自动 Servlet 重载)从本地或远程地址来实现。这一步等同于:

```
Class c = Class.forName("com.sourceStream.MyServlet");
```

这里的加载指同时加载和初始化 Servlet 的过程。

(2) 服务器创建一个 Servlet 类实例来为所有请求服务,利用多线程,可以由单个 Servlet 实例来服务于并行的请求。唯一例外的是,实现 SingleThreadModel 接口的 Servlet,服务器会创建一个实例池并从中选择一个来服务于每一个新的请求。此步相当于以下 Java 代码:

```
Servlet s = (Servlet)c.newInstance();
```

(3) 服务器调用 Servlet 的 init()方法,它用来保证完成在首次请求 Servlet 处理以前的执行过程。如果 Servlet 创建了多个 Servlet 实例,则为每个实例调用一次 init()方法。

(4) 服务器从包括在客户请求中的数据里构造一个 ServletRequest 或 HttpServletRequest 对象,还构造一个 ServletResponse 或 HttpServletResponse 对象来为返回响应提供方法。其参数类型依赖于 Servlet 是否分别扩展 GenericServlet 或 HttpServlet(注意:如果服务器没有收到对这一 Servlet 的请求,本步骤与以下 5~7 步将不会执行)。

(5) 服务器调用 Servlet 的 service()方法(对于 Httpservlet,service()将调用更为具体的方法,如 doGet()、doPost()),在此步骤中作为参数传递结构化对象。当并行的请求到来时,多个 service()方法能够同时运行在独立的线程中(除非 Servlet 实现了 SingleThreadModel 接口)。

(6) 通过分析 ServletReuest 或 HttpServletRequest 对象,service()方法处理客户的请求,并调用 ServletResponse 或 HttpServletResponse 对象来响应。

(7) 如果服务器收到另一个对该 Servlet 的请求,这个处理过程从步骤 5 重复。

(8) 一旦 Servlet 容器检测到一个 Servlet 要被卸载,这可能是因为要回收资源或因为它正在被关闭,服务器会在所有 Servlet 的 service()线程完成之后(或在服务器规定时间后)调用 Servlet 的 destroy()方法,然后 Servlet 就可以进行无用存储单元收集清理。Servlet 容器不需要为保留一个 Servlet 而指定时间。

5. Servlet 应用实例

(1) 创建 Servlet 类

```
packagewww.bcpl.cn;
  import java.io.*;
  import javax.servlet.*;
  import javax.servlet.http.*;

public class HelloWorldServlet extends HttpServlet{
  public void doGet (HttpServletRequest req, HttpServletResponse res) {
       res.setContentType("text/html");
       try{
          PrintWriter out = res.getWriter();
          out.println("<HTML>");
          out.println("<HEAD><TITLE>Hello World! </TITLE></HEAD>");
```

```
            out.println("<BODY>");
            out.println("<h1>Hello World</h1>");
            out.println("</BODY></HTML>");
            out.close();
        } catch(IOException ioe) {
            getServletContext().log (ioe,"Error in HelloWorld");
        }
    }

    public void doPost(HttpServletRequest req, HttpServletResponse res){
        doGet(req, res);
    }
```

（2）访问 Servlet 程序

在浏览器地址栏中输入 http://hostname:port/project_name/servlet_url，即可实现访问。

hostname：主机名称或者 IP 地址，本机为 localhost。

port：通讯端口号，如服务器为 Tomcat，默认端口为 8080。

project_name：应用程序根目录名称。

servlet_url：查找 Servlet 程序所使用的 URL 地址。

6. Servlet 程序部署

Servlet 的部署可以采用两种方式：一种是人工手动部署的方式，另外一种是采用集成开发工具 MyEclipse 来进行开发部署。

（1）人工部署方式，选择 Web 服务器 Tomcat、WebLogic 或 Websphere 等，本书简化采用 Tomcat，作为 Web 服务器。

在 Tomcat 安装目录的 webapps 中，新建一个文件夹 helloworld，然后在 helloworld 中新建一个 WEB-INF 文件夹，并在 WEB-INF 中新建一个 classes 文件夹。

将编译好的 Servlet 连同编译产生的包(package)路径(文件夹)，一同复制到 classes 目录下。

在 WEB-INF 中，新建一个 XML 文件，命名为 web.xml，在 web.xml 中输入以下内容：

```
<? xml version = "1.0" encoding = "gbk"? >
<! DOCTYPE web-app
    PUBLIC "-//Sun Microsystems, Inc.//DTD Web Application 2.3//EN"
    "http://java.sun.com/dtd/web-app_2_3.dtd">
  <web-app>  <! --根元素 表示该文件是 Web 应用的配置文件-->
    <servlet> <! -- Servlet 元素 ，表示为该 Web 应用配置一个 Servlet 程序 -->
        <servlet-name>myServletName</servlet-name> <! --为 Servlet 类，在该配置文件中
        起一个名称，便于引用 -->
            <servlet-class>package.class_name</servlet-class> <! -- Servlet 类在
            classes 目录中的路径，如果有包名，则用“.”分割 -->
    </servlet> <! -- 结束声明 Servlet -->
    <servlet-mapping> <! --配置 Servlet 映射信息，为该 Servlet 分配一个虚拟的 URL 地址 -->
        <servlet-name>myServletName</servlet-name> <! --表示要为哪一个 Servlet 程序
        分配 URL-->
        <url-pattern>/servlet/URL</url-pattern> <! --分配一个虚拟的 URL 给该 Servlet
        程序 -->
    </servlet-mapping>  <! --结束映射 -->
  </web-app> <! --结束 webapp 声明 -->
```

在 web. xml 文件中，服务器执行从 web. xml 查找的顺序为：url-pattern→servlet-name→servlet-class。

在 Tomcat 安装目录中的 conf 的 catalina 中的 localhost 目录中，新建一个名称为 helloworld. xml文件，并输入以下内容：

```
<Context docBase = "${catalina.home}/webapps/helloworld" privileged = "true"
        antiResourceLocking = "false" antiJARLocking = "false">
  </Context>
```

然后，启动 Tomcat 服务器，在 IE 地址栏中输入 http://localhost:8080/helloworld/servlet/URL，即可访问上述 Servlet 程序。

(2) MyEclipse 来进行开发部署 Servlet 程序并访问，与前述的创建部署 Web 程序一样，只是需要在 src 下创建包，然后创建 Servlet 类程序，其他步骤可参考前述步骤进行部署。

2.4　JSP 技术

2.4.1　JSP 技术简介

1. JSP 语言简介

JSP(Java Server Pages)是一种动态页面技术，在传统的网页 HTML 文件(*. htm,*. html)中加入 Java 程序片段(Scriptlet)和 JSP 标签，就构成了 JSP 网页。JSP 页面由 HTML 代码和嵌入其中的 Java 代码所组成。JSP 网页中的 Java 程序片段可以操纵数据库、重新定向网页以及发送 E-mail 等，实现建立动态网站所需要的功能。所有程序操作都在服务器端执行，网络上传送给客户端的仅是得到的结果，大大降低了对客户浏览器的要求，即使客户浏览器端不支持 Java，也可以访问 JSP 网页。它实现了 HTML 语法中的 Java 扩张(以 <%, %>形式)，这种扩张是在服务器端执行的，通常返回给客户端的就是一个 HTML 文本，因此客户端只要有浏览器就能浏览。Web 服务器在遇到访问 JSP 网页的请求时，首先执行其中的程序段，然后将执行结果连同 JSP 文件中的 HTML 代码一起返回给客户端。

前述的 Servlet 是 JSP 的技术基础，而且大型的 Web 应用程序的开发需要 Servlet 和 JSP 配合才能完成。JSP 具备了 Java 技术的简单易用，完全的面向对象，具有平台无关性且安全可靠。自 JSP 推出后，众多大公司都支持 JSP 技术的服务器，如 IBM、Oracle 和 Bea 公司等，所以 JSP 迅速成为商业应用的服务器端语言。

JSP 2.0 以上版本的一个主要特点是它支持表达语言(Expression Language)。JSTL 表达式语言可以使用标记格式方便地访问 JSP 的隐含对象和 JavaBeans 组件，JSTL 的核心标记提供了流程和循环控制功能。自制标记也有自定义函数的功能，因此基本上所有 Scriptlet 能实现的功能都可以由 JSTL 替代。在 JSP 2.0 以上的版本中，建议尽量使用 EL 而使 JSP 的格式更一致。

2. JSP 与 CGI、ASP、PHP 区别

(1) CGI 与 JSP 区别

CGI 是一个进程处理一个请求，假如有 10 个人在线，那么就得开 10 个进程；而 PHP、

ASP 和 JSP 都是一个进程处理多个请求，无论多少人在线，都只有一个进程。

JSP 与传统的 CGI 方式相比具有以下几方面的优势。

- 后台实现逻辑是基于 Java Component 的，具有跨平台的特点。
- 将应用逻辑与页面表现分离，使得应用逻辑能够最大程度得到复用，从而提高开发效率。
- 运行比 CGI 方式高，尤其对于数据库访问时，提供了连接池缓冲机制，使运行所需资源最小。
- 安全，由于后台是完全基于 Java 技术的，安全性由 Java 的安全机制予以保障。
- 由于与 ASP 很近似，不需要太多的编程知识就可以动手编写 JSP。
- 内置支持 XML，使用 XML 从而使页面具有更强的表现力并减少了编程工作量。

(2) JSP 与 ASP 和 PHP 的区别

目前，最常用的 3 种动态网页语言有 ASP(Active Server Pages)、JSP(Java Server Pages)和 PHP(Hypertext Preprocessor)。

ASP 是一个 Web 服务器端的开发环境，利用它可以产生和运行动态的、交互的和高性能的 Web 服务应用程序。ASP 采用脚本语言 VB Script(Java Script)作为自己的开发语言。

PHP 是一种跨平台的服务器端的嵌入式脚本语言，它大量地借用 C、Java 和 Perl 语言的语法，并耦合 PHP 自己的特性，使 Web 开发者能够快速地写出动态生成页面，它支持目前绝大多数数据库。还有一点，PHP 是完全免费的，不用花钱，可以从 PHP 官方网站(http://www.php.net)自由下载，而且可以不受限制地获得源码，甚至可以从中加进自己需要的特色。

JSP 是 Sun 公司推出的新一代站点开发语言，它解决了目前 ASP、PHP 的缺陷——脚本级执行(目前 PHP4 也已经在 Zend 的支持下，实现编译运行)，JSP 可以在 Serverlet 和 JavaBean的支持下，完成功能强大的服务器端站点程序。

三者都提供在 HTML 代码中混合某种程序代码和由语言引擎解释执行程序代码的能力。但 JSP 代码被编译成 Servlet 并由 Java 虚拟机解释执行，这种编译操作仅在对 JSP 页面的第一次请求时发生。在 ASP、PHP 和 JSP 环境下，HTML 代码主要负责描述信息的显示样式，而程序代码则用来描述处理逻辑。普通的 HTML 页面只依赖于 Web 服务器，而 ASP、PHP 和 JSP 页面需要附加语言引擎分析和执行程序代码。程序代码的执行结果被重新嵌入到 HTML 代码中，然后一起发送给浏览器。ASP、PHP 和 JSP 三者都是面向 Web 服务器的技术，客户端浏览器不需要任何附加的软件支持。

2.4.2 JSP 页面元素

JSP 页面中常见的页面元素有注释元素、模板元素、脚本元素、指令元素和动作元素，其分别详述如下。

1. 注释元素

JSP 页面中注释分为隐藏性注释、输出性注释和 Scriptlet 注释。

输出性注释是指会在客户端显示的注释，与 HTML 页面的注释一样，其表示形式为：

```
<! -- comments  <% expression here %> -->
```

隐藏性注释是指在 JSP 页面中写好的注释，但经过编译后不会发送到客户端，不在客户端显示的注释，其作用主要是为了方便开发人员使用，其表示形式为：

```
<%-- comments here --%>
```

隐藏性注释，不生成页面内容。

Scriptlet 注释是指注释 Java 程序代码的注释，其表示形式为：

```
<% // comment %>单行注释
<%/* * comment */%> 块注释
```

2. 模板元素

模板元素指在 JSP 页面文件中出现的静态 HTML 标签、XML 内容、XSL、XSLT 和 JavaScript 等。

3. 脚本元素

脚本元素包含声明、表达式和 Scriplets 片段，即 JSP 页面中的 Java 代码。

(1) 声明是指在页面中声明合法的成员变量和成员方法，可以为 Servlet 声明成员变量或者方法，也可以重写 JSP 引擎父类的方法。

声明变量：

```
<%! String name = ""; %>
<%! public String getName(){return name;} %>
```

在 JSP 标签元素中<jsp:declaration>相当于<%!%>，它们的作用相同，可以采用任意一种方式进行使用。

```
<jsp:declaration> String greetingStr = "Hello, World!"; </jsp:declaration>
```

声明方法：

```
<%!
    public void print(){
        greetingStr = "welcome";
        int i = 0;
                    }
  %>
```

(2) 表达式就是位于 <%= 和 %>之间的代码，通常是变量或者是有返回类型的方法，输出字符串内容到页面。如：

```
<% = greetingStr %>
```

注意：表达式后面没有“;”。

在 JSP 标签元素中<jsp:expression> 相当于 <%=%>中的“=”号，表示输出表达式，它们的作用相同，可以采用任意一种方式进行使用。

```
<jsp:expression> greetingStr</jsp:expression>
```

(3) Scriptlets 就是位于<% 和 %>之间的、合法的 Java 代码，如业务逻辑代码等。

```
<% greetingStr + = " Best wishes to you! ";   %>
```

在 JSP 标签元素中<jsp:scriptlet> 相当于<% %>，它们的作用相同，可以采用任意一种方式进行使用。

```
<jsp:scriptlet> greetingStr + = " Best wishes to you!"; </jsp:scriptlet>
```

4. 指令元素

指令元素是指出现在<%@ 和 %> 之间，包含 page 指令、include 指令和 taglib 指令。

(1) page 指令:用于定义 JSP 文件的全局属性。其语法格式如下:

```
<%@ page [language = "java"] //声明脚本语言采用 Java,目前只能是 Java
        [import = "java.util.ArrayList"] //导入其他包中的 Java 类文件

        [contentType = "text/html;charset = GBK"] //页面的格式和采用的编码,格式见 MIME 类型
        [session = "{true|false}"] //这个页面是否支持 Session,即是否可以在这个页面中使用 Session
        [buffer = "none|8kb|size kb]"] //指定到客户端的输出流采用的缓冲大小
        [autoFlush = "true|false"] //如果为 true,表示当缓冲区满了,到客户端的输出会自动刷新,如果为 false,则抛出异常
        [isThreadSafe = "true|false"] //如果为 true,表示一个 JSP 页面可以同时处理多个用户的请求,否则只能一次处理一个
        [pageEncoding = "encodingStr"] //页面的字符编码
        [isELIgnored = "true|false"] //是否支持 EL 表达式语言 "${}"
        [isErrorPage = "true|false"] //该页面是否为错误信息页面,如果是则可以直接使用 exception 对象
        [errorPage = "page_url"] //页面出现错误后,跳转的页面与[isErrorPage]不能同时出现"]"
        [info = "description"] //有关页面的描述信息
        [extends = "package.class"] //继承了什么样的类
        [method = "service"]//生成一个 service 方法来执行 JSP 中的代码
%>
```

Page 指令常见的语法格式为:

```
<%@page contentType = "text/html;charset = GBK" language = "java" %>
<%@page import = "java.util.ArrayList" %>
```

(2) include 指令:将指定位置上的资源代码在编译的过程中包含到当前页面中,称为静态包含,在编译时就要包含进来,随同当前的页面代码一同进行编译。其语法格式如下:

```
<%@include file = "file_url" %>
```

(3) taglib 指令:JSP 页面中使用自定义或者其他人已经定义好的标签文件。其语法格式如下:

```
<%@ taglib uri = "tld_url" prefix = "prefix_name" %>
```

在 JSP 标签元素中<jsp:directive. page />相当于 <%@page%>,它们的作用相同,可以采用任意一种方式进行使用。

5. 动作元素

JSP 动作元素包含<jsp:include>、<jsp:forward>、<jsp:useBean >、<jsp:setProperty>、<jsp:getProperty>、<jsp:param>、<jsp:plugin>等。

(1) <jsp:include>:即引入页面,在运行代码时,将指定的外部资源文件导入到当前 JSP 中,外部资源不能设置头信息和 Cookie。如:

```
<jsp:include page = "page_URL" flush = "true" />
<jsp:include page = "./public/header.jsp" flush = "true" />
```

<%@ include >与 jsp:include 的区别如下所示。

两者都可以用于包含静态内容或者动态内容,差别在于,前者是静态包含,也就是在生成

Servlet 之前就被包含进来了，生成的是单个文件，不会为被包含者生成单独的 Servlet(假如被包含的是动态的)；后者是动态包含，也就是说生成 Servlet 的时候只是添加一个引用，并不真正将内容包含进来，内容是在运行时才被包含进来的，容器会为被包含的文件生成独立的 Servlet(假如包含的是动态的)。所以，前者常用于包含固定不变的、多个页面共用的内容片段，后者则用于经常变化的内容片段，无论是否被多个页面共用。

(2) <jsp:forward>：即跳转，将客户的请求重定向到某个资源，该资源文件必须与该 JSP 文件在同一个 Context 环境中。如：

```
<jsp:forward page = "uri"/>
```

(3) <jsp:useBean>：在 JSP 页面中创建一个 JavaBean 的实例，并可以存放在相应的 Context 范围之中。如：

```
<jsp:useBean id = "beanName" class = "ClassPath"|BeanName = "ClassName" scope = "{page|request
|session|application}"typeName = "typeName"/>
```

常用的属性是 id 和 class。

(4) <jsp:setProperty>：嵌套在<jsp:useBean>标签体中，用来设置 JavaBean 的实例中的属性的值。如：

```
<jsp:setProperty name = "beanId" [ property = "bean_propertyname" value = "property_value"|
property = " * " ]
```

(5) <jsp:getProperty>：获得用<jsp:useBean >设置的 JavaBean 的对象的某个属性，并将其输出到相应的输出流中。如：

```
<jsp:getProperty name = "beanId" property = "bean_propertyname" />
```

(6) <jsp:param>：一般和<jsp:forward>、<jsp:include>以及<jsp:plugin> 配合使用，用来向引入的 URL 资源传递参数。如：

```
<jsp:param name = "param_name" value = "param_value" />
```

(7) <jsp:plugin>：产生客户端浏览器的特殊标签(<object> 或者 <embed>)。如：

```
<jsp:plugin type = "applet" code = "sample.AppletTest" codebase = "." >

            <jsp:param name = "username" value = "guo" />

            <jsp:param name = "password" value = "1234" />

            <jsp:fallback> Cann't Load Applet from Certain URL </jsp:fallback>
</jsp:plugin>
```

(8) <jsp:fallback>：只能嵌套在 <jsp:plugin>中使用，表示如果找不到资源，则显示其他的信息。

2.4.3 JSP 内置对象

JSP 内置对象是在编写 JSP 时，可以直接使用某些已经创建好的对象，不需要人工创建的对象。JSP 的内在对象共 9 个，分别为 out 对象、request 对象、response 对象、session 对象、

pageContext 对象、application 对象、config 对象、page 对象和 exception 对象。

通常将.jsp 文件编译为 Servlet 后，在 Servlet 中的 service 方法中定义，其对应的 Servlet 类中方法如表 2-1 所示。

表 2-1 内置对象与 Servlet 方法对应

序 号	内置对象	Servlet 方法	说 明
1	request	HttpServletRequest	封装了来自客户端的请求信息
2	response	HttpServletResponse	封装了服务程序向客户端做出响应的信息
3	session	HttpSession	用来维持客户端状态的会话对象
4	application	ServletContext	表示一个 JSP 的 Web 环境 ServletContext
5	out	PrintWriter	用来向客户端输出字符流的内置对象
6	exception	Exception	保存 JSP 运行时的异常信息，只能在 isErrorPage="true"页面中使用
7	config	ServletConfig，Servlet 配置	读取一个 web.xml 中 JSP 的配置<jsp-config />
8	pageContext	PageContext（JSP 运行环境）	获取其他的内置对象，保存数据，在页面的其他位置使用，表示 JSP 的上下文，可以获取上述 8 个对象
9	page		表示的是当前的 JSP 页面对象

在 JSP 被编译成 Servlet 之后，在 Servlet 源代码中预先生成了某些对象，即可在页面里使用这些对象的引用。

1. out 对象

out 对象是 JspWriter 类的实例，是向客户端输出内容常用的对象，其常用的方法如表 2-2 所示。

表 2-2 out 对象常用方法

序 号	方法名	作 用
1	void clear()	清除缓冲区的内容
2	void clearBuffer()	清除缓冲区的当前内容
3	void flush()	清空流
4	int getBufferSize()	返回缓冲区以字节数的大小，如不设缓冲区则为 0
5	int getRemaining()	返回缓冲区还剩余多少可用
6	boolean isAutoFlush()	返回缓冲区满时，是自动清空还是抛出异常
7	void close()	关闭输出流

JSP 的输出流等同于表达式(Java 代码)out. println("test")。

2. request 对象

request 对象封装客户端的请求信息，通过它获取到客户端的数据，然后作出响应，它是 HttpServletRequest 类的实例。

(1) request 对象获取表单提交参数，在请求完成后，作用域结束，变量失效。

```
String userName = request.getParameter("userName");
```

(2) 获取表单多个参数，返回数组，如 checkbox 多选的参数。

```
String[] chooseOs = request.getParameterValues("os");
```

（3）设置请求的编码：

```
request.setCharacterEncoding("UTF-8");
```

（3）设置和访问共享属性：

```
request.setAttribute("error", "用户名不能为空.");//设置 request 属性
request.getAttribute("error");//得到 request 属性
request.removeAttribute("error");//移除 request 属性
```

（4）得到访问用户的 IP 地址：

```
request.getRemoteAddr();
```

获取当前目录或 JSP 文件等的绝对路径(返回页面部署的路径，参数同样可以查找文件的路径)。

```
request.getRealPath("");//当前目录
request.getRealPath("a.jsp");//JSP 文件
```

3. response 对象

response 对象包含了响应客户请求的有关信息，它是 HttpServletResponse 类的实例。

（1）增加 header 属性，如通知浏览器不缓存响应：

```
response.addHeader("pragma", "no-cache");
response.addHeader("cache-control", "no-cache");
response.addHeader("expires", "0");
```

（2）页面重定向(发送两次请求，返回两次响应)：

```
response.sendRedirect("url 地址");
```

4. session 对象

session 对象指的是客户端与服务器的一次会话，从客户连到服务器的一个 WebApplication 开始，直到客户端与服务器断开连接为止，它是 HttpSession 类的实例。在会话中，session 对象属于每个访问的用户专用。

（1）设置和访问共享属性：

```
session.getAttribute("bean");//得到 session 属性
session.setAttribute("bean",bean);//设置 session 属性
session.removeAttribute("bean");//移除 session 属性
```

（2）销毁 session 会话(键、值)：

```
session.invalidate();
```

5. pageContext 对象

pageContext 对象提供了对 JSP 页面内所有的对象及名字空间的访问，也就是说它可以访问到本页所在的 session，也可以取本页面所在的 application 的某一属性值，它相当于页面中所有功能的集成，它的本类名也叫 pageContext，其他常用的方法如表 2-3 所示。

表 2-3 pageContext 对象常用方法

序 号	方法名称	作 用
1	JspWriter getOut()	返回当前客户端响应被使用的 JspWriter 流(out)
2	HttpSession getSession()	返回当前页中的 HttpSession 对象(session)
3	Object getPage()	返回当前页的 Object 对象(page)
4	ServletRequest getRequest()	返回当前页的 ServletRequest 对象(request)
5	ServletResponse getResponse()	返回当前页的 ServletResponse 对象(response)
6	Exception getException()	返回当前页的 Exception 对象(exception)
7	ServletConfig getServletConfig()	返回当前页的 ServletConfig 对象(config)
8	ServletContext getServletContext()	返回当前页的 ServletContext 对象(application)

6. application 对象

application 对象实现了用户间数据的共享,可被用来存放全局变量。它的作用域开始于服务器的启动,结束于服务器的关闭,在此期间,此对象将一直存在。因而在用户的前后连接或不同用户之间的连接中,可以对此对象的同一属性进行操作,在任何地方对此对象属性的操作,都将影响到其他用户对此的访问。服务器的启动和关闭决定了 application 对象的生命周期,它是 ServletContext 类的实例。

通常该对象用来保存一个应用系统中一些公用的数据,与 session 对象相比,application 对象是所有客户共享的。

(1) 获取 Servlet 的版本号:

```
application.getMajorVersion() + "." + application.getMinorVersion();
```

(2) 可配置的全局信息(通过 web.xml):

```
application.getInitParameter("company.name");
```

在 web.xml 中加入:

```
<context-param>
        <param-name>company.name</param-name>
        <param-value>北京首都政法 xxx 信息技术有限公司</param-value>
    </context-param>
```

其他方法:

```
this.getMajorVersion();//得到 Servlet 大版本号
this.getMinorVersion();//得到 Servlet 小版本号
this.getServerInfo();//得到服务器信息
this.setAttribute("","");//设置共享全局属性信息
this.getAttribute("");//得到共享全局属性信息
```

7. config 对象

config 对象是在一个 Servlet 初始化时,JSP 引擎向它传递信息所用的,此信息包括 Servlet 初始化时所要用到的参数(通过属性名和属性值构成)以及服务器的有关信息(通过传递一个 ServletContext 对象)。

设置 JSP 页面的常量(通过 web.xml):

```
config.getInitParameter("page.title");
```

在 web.xml 中加入：

```
<servlet>
        <servlet-name>configServlet</servlet-name>
        <jsp-file>/config.demo.jsp</jsp-file>
        <init-param>
            <param-name>page.title</param-name>
            <param-value>config 对象使用</param-value>
        </init-param>
    </servlet>
    <servlet-mapping>
        <servlet-name>configServlet</servlet-name>
        <url-pattern>/config.demo.jsp</url-pattern>
    </servlet-mapping>
```

8. page 对象

page 对象就是指向当前 JSP 页面本身，JSP 页面的 this 指针，它是 java.lang.Object 类的实例。

9. exception 对象

exception 对象用来表示 JSP 页面中的异常。exception 对象是一个例外对象，当一个页面在运行过程中发生了异常，就产生这个对象。如果一个 JSP 页面要应用此对象，就必须把 isErrorPage 设为 true，否则无法编译。它实际上是 java.lang.Throwable 的对象，其页面用法如下：

```
<%@ page errorPage="relativeURL" isErrorPage="true|false" %>
```

2.4.4　JSP 异常处理

通常，HTTP 常见页面服务访问错误，其在浏览器客户端显示的错误代码有以下几种。

401：验证出错。

404：路径错误或资源不存在错误。

405：访问资源不允许，一般是 Servlet 中，如请求方式 post，而只写了 doGet()方法，出现 405 错误。

500：服务器端运行错误(语法异常等)。

通常 JSP 页面在执行时，在两个阶段下会发生错误。一个是 JSP 网页 → Servlet 类，另一个是 Servlet 类处理每一个请求时。在第一阶段时，产生的错误称为 Translation Time Processing Errors；在第二阶段产生的错误，称为 Client Request Time Processing Errors。

1. Translation Time Processing Errors

Translation Time Processing Errors 产生的主要原因：在编写 JSP 页面时发生语法错误，导致 JSP Container 无法将 JSP 网页编译成 Servlet 类文件(.class)，如 500 Internal Server Error，500 是指 HTTP 的错误状态码，因此是 Server Error。通常产生这种错误时，可能是 JSP 的语法有错误，或是 JSP Container 在开始安装、设定时，有不适当的情形发生。解决的方法就是再一次检查程序是否有写错的语法，如无，有可能是 JSP Container 的 bug。

2. Client Request Time Processing Errors

Client Request Time Processing Errors 错误的发生，通常不是语法错误，而可能是逻辑上的错误，如写一个计算除法的程序，当用户输入的分母为零时，程序会发生错误并抛出异常

(Exception),交由异常处理(Exception Handling)机制做适当的处理。对于这种错误的处理,通常会交给 errorPage 去处理。

3. errorPage 应用实例

(1) 编写的 errorPage. jsp 代码如下:

```
<%@ page contentType = "text/html;charset = GB2312" errorPage = "myError.jsp" %>
//设置 myError.jsp 页为本 JSP 页面的错误处理页面
<html>
      <head>
          <title>错误处理应用</title>
      </head>
      <body>
          <h2>errorPage 的应用</h2>
          <%!
                private double MyDouble(String value)
                {
                      return(Double.valueOf(value).doubleValue());
                }
          %>
          <%
                double num1 = MyDouble(request.getParameter("num1"));
                double num2 = MyDouble(request.getParameter("num2"));
          %>
          两个数字分别为:<% = num1 %> 和 <% = num2 %><br>
          两数相除结果为:<% = (num1/num2) %>
      </body>
</html>
```

在 errorPage. jsp 中,page 指令的属性 errorPage 设为 MyError. jsp 页面,当 errorPage. jsp 发生错误时,MyError. jsp 页面为产生的错误所跳转的错误处理页面。因而,若 errorPage. jsp 有错误发生时,会自动转到 MyError. jsp 来显示。

(2) MyError. jsp 页面的代码如下:

```
<%@ page contentType = "text/html;charset = GB2312" isErrorPage = "true" %>   //该页为错
误处理页,设置 isErrorPage 为 true
<%@ page import = "java.io.PrintWriter" %>
<html>
     <head>
          <title>错误显示页面</title>
     </head>
     <body>
          <h2>MyError 的错误</h2>
          <p>页面错误产生的原因为:<I><% = exception %></I>
          </p><br>
          <pre>
               问题是:<% exception.printStackTrace(new PrintWriter(out)); %>
          </pre>
     </body>
</html>
```

其他相关技术

(1) JDBC

JDBC(Java Data Base Connectivity)是 Java 应用程序访问关系数据库的接口,可以跨平台的语言,在其上面的应用(Pure Java)可以在任何操作系统、应用服务器上运行,如随着用户和访问量的增加,可能会考虑改用商业 Unix 服务器和商业的应用服务器,基于 Java 的解决方案的应用可以不加任何改动平滑移植到新系统中,可以最大限度地保护现有投资。

借助于 EJB(Enterprise JavaBeans)的支持,可以实现基于组件和负载平衡的分布式计算环境。

各厂商提供了很多高效的开发工具,如 IBM WebSphere Studio 和一些 Java IDE 等。考虑到性能上的影响,建议 Oracle Server 和 Web Server 采用分布式结构。

(2) JavaBeans

JavaBeans 是一种可重用的 Java 组件,它可以被 Applet、Servlet 和 JSP 等 Java 应用程序调用,也可以可视化地被 Java 开发工具使用。它包含属性(Properties)、方法(Methods)和事件(Events)等特性。

(3) Servlet

Servlet 是一种在服务器端运行的 Java Application,它可以作为一种插件(Plug-ins)嵌入到 Web Server 中去,提供诸如 HTTP、FTP 等以及用户定制的协议服务。如果已经用过 Java Servlet,那么就会知道 Servlet 可以建立动态生成的网页,而网页中包含有从服务器方的 Java 对象中所获得的数据。但是也得知道 Servlet 生成网页的方法就是在 Java 类中嵌入 HTML 标签和表述代码,这就意味着改变表述代码需要修改和重新编译 Servlet 源文件。因为设计 HTML 页面的设计人员可能与编写 Servlet 代码的开发人员不是同一个人,更新基于 Servlet 的 Web 应用程序就成了一件非常棘手的事情。

Enter JavaServer Page 是 Servlet API 的一个扩展。事实上,JSP 网页在编译成 Servlet 之前也可以使用,所以它们也具有 Servlet 的所有优势,包括访问 Java API。由于 JSP 是嵌入到 Servlet 中关于应用程序的一般表述代码,所以它们能够被看成一种"彻底"的 Servlet。

JSP 网页主要提供了一种建立 Servlet 的高水平方法,它还带来了其他的优点。即使已经为 Web 应用程序编写了 Servlet,使用 JSP 仍然有很多优势,如下所示。

① JSP 网页可以非常容易地与静态模板结合,包括 HTML 和 XML 片段,以及生成动态内容的代码。

② JSP 网页可以在被请求的时候动态地编译成 Servlet,所以网页的设计人员可以非常容易地对表述代码进行更新。如果需要的话,JSP 网页还可以进行预编译。

③ 为了调用 JavaBean 组件,JSP 标签可以完全管理这些组件,避免网页设计人员复杂化应用程序。

④ 开发人员可以提供定制化的 JSP 标签库。

⑤ 网页设计人员能够改变和编辑网页的固定模板部分而不影响应用程序。同样,开发人员也无须一个个编辑页面而只需对组件进行合理的改变。

⑥ 通常,JSP 允许开发人员向许多网页设计人员分发功能性应用程序,这些设计人员也不必知道 Java 编程语言或任何 Servlet 代码,所以他们能够集中精力去编写 HTML 代码,而编程人员就可以集中精力去建立对象和应用程序

2.5 小　　结

在 Web 的基础应用技术中，HTML 语言是最为基础，经过技术的不断变迁和信息展示的需求推动，HTML 语言已经从 HTML 4 过渡发展到 HTML 5，由于 HTML 5 的应用需要浏览器内核技术支持等原因，虽然得到了应用，但并没有广泛替代 HTML 4，且大多基础的信息展示，并不需要 HTML 5 特殊的功能进行展示，本书由于篇幅的原因，没有对 HTML 5 进行介绍，只是对常用的 Web 前端技术 JavaScript、JSP 及中间技术 Servlet 进行了详细阐述，并对重要的技术点进行了应用例举。

第3章　JDBC技术

3.1　JDBC技术简介

3.1.1　JDBC简介

1. JDBC简介

JDBC是Java Data Base Connectivity的简称，又称Java数据库连接，是一种用于执行SQL语句的Java API，可以为多种关系数据库提供统一访问，它由一组用Java语言编写的类和接口组成。JDBC对Java程序员而言是API，对实现与数据库连接的服务提供商而言是接口模型。作为API，JDBC为程序开发提供了标准的接口，并为数据库厂商及第三方中间件厂商实现与数据库的连接提供了标准方法。JDBC使用已有的SQL标准并支持与其他数据库连接标准，如ODBC之间的桥接。JDBC为数据库开发人员提供了一个标准的API，使数据库开发人员能够用纯Java API编写数据库应用程序，并在此基础上构建更为高级的工具和接口。

JDBC API既支持数据库访问的两层模型(C/S)，同时也支持三层模型(B/S)。在两层模型中，Java Applet或应用程序将直接与数据库进行对话，需要一个JDBC驱动程序来与所访问的特定数据库管理系统进行通讯。用户的SQL语句被送往数据库中，而其结果将被送回给用户。数据库可以位于另一台计算机上，用户通过网络连接到上面。通过客户机/服务器配置进行实现，其中，用户的计算机为客户机，提供数据库的计算机为服务器。

在三层模型中，命令先是被发送到服务的"中间层"，然后由它将SQL语句发送给数据库。数据库对SQL语句进行处理并将结果送回到中间层，中间层再将结果送回给用户。使用中间层的优势，用户可以利用它创建高级API，而中间层将把它转换为相应的低级调用。

2. JDBC与ODBC

开放数据库互连(Open Data Base Connectivity，ODBC)是微软公司开放服务结构(Windows Open Services Architecture，WOSA)中有关数据库的一个组成部分，它建立了一组规范，并提供了一组对数据库访问的标准API(应用程序编程接口)，这些API利用SQL来完成其大部分任务。ODBC本身也提供了对SQL语言的支持，用户可以直接将SQL语句发送给ODBC。

基于ODBC的应用程序对数据库的操作不依赖任何DBMS，也不直接与DBMS打交道，所有的数据库操作由对应的DBMS的ODBC驱动程序完成。即不论是FoxPro、Access还是Oracle数据库，均可用ODBC API进行访问，其最大优点是能以统一的方式处理所有的数据库。

JDBC与ODBC的相同之处：它们都是基于X/Open的SQL调用级接口；JDBC的总体结构类似于ODBC，也有4个组件(应用程序、驱动程序管理器、驱动程序和数据源)；JDBC工作原理类同于ODBC。

① JDBC 保持了 ODBC 的基本特性,也独立不依赖于特定数据库。

② 使用相同源代码的应用程序,通过动态加载不同的 JDBC 驱动程序,可以访问不同的 DBMS。连接不同的 DBMS 时,各个 DBMS 之间仅通过不同的 URL 进行标识。

③ JDBC 的 DatabaseMetaData 接口提供了一系列方法,可以检查 DBMS 对特定特性的支持,并相应确定有什么特性,从而能对特定数据库的特性进行支持。

④ JDBC 也支持在应用程序中同时建立多个数据库连接,采用 JDBC 可以很容易地用 SQL 语句同时访问多个异构的数据库,实现异构数据库之间的相互操作。

JDBC 与 ODBC 的不同之处:JDBC 与 ODBC 相比较,更为容易理解;JDBC 数据库驱动程序是面向对象的,了解 Java 的用户,一般都可以在较短的时间内掌握 JDBC 驱动程序的架构,而 ODBC 内部功能较为复杂,且需要 C 语言驱动,对编程者有一定的难度;JDBC 的移植性与 ODBC 相比较好,ODBC 驱动程序安装完后,通常需要进行一定的配置,不同的配置在不同的数据库服务器之间并不能够通用,而 JDBC 只需加载驱动,并不需要进行额外的配置,在安装过程中,JDBC 数据库驱动程序会完成自己相关的配置;ODBC 的优势是可以为不同的数据库提供相应的驱动程序。

3.1.2 JDBC API

JDBC API 是由一组 Java 语言编写的类和接口组成,使用内嵌式的 SQL,主要实现三方面的功能:建立与数据库的连接或访问数据源、执行 SQL 声明以及处理 SQL 执行结果。JDBC API包括 java. sql 和 javax. sql 两个包。

JDBC API 包含两个主要的接口集合:面向程序开发人员的 JDBC API 和面向底层驱动人员的 JDBC Driver API。

Java 应用程序通过 JDBC API 界面访问 JDBC 驱动管理器,JDBC 管理器通过 JDBC 驱动程序 API 访问不同的 JDBC 驱动程序,从而实现对不同数据库的访问。

直接连接数据库完全 Java 驱动的 JDBC 的体系结构,如图 3-1 所示。

JDBC-ODBC 桥连接数据库或利用数据库 Client Libraries 连接数据库的 JDBC 体系结构,如图 3-2 所示。

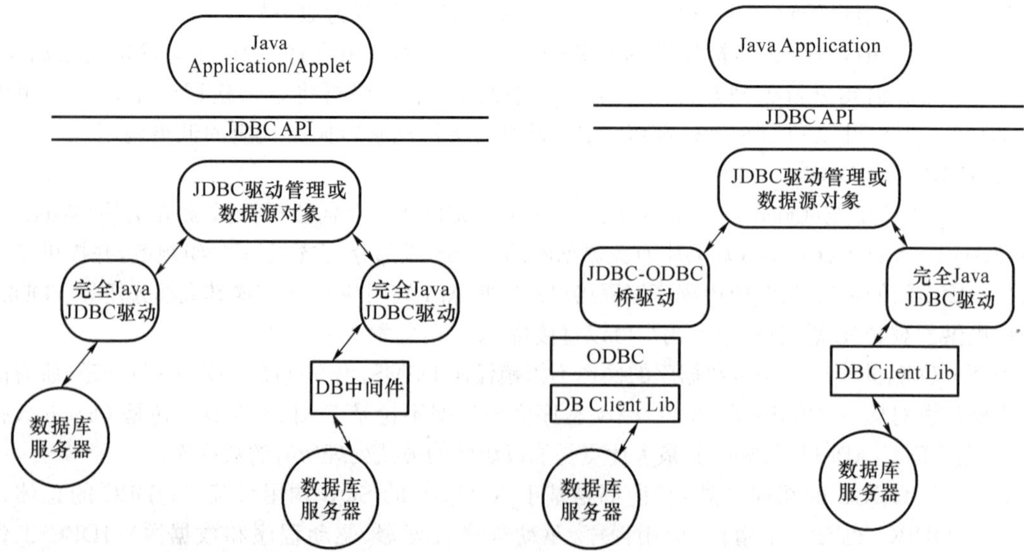

图 3-1 纯 Java 驱动的 JDBC 的体系结构

图 3-2 JDBC-ODBC 桥连接及 Client Lib 连接数据库

从上可以看出，一个 JDBC DriverManager 可以管理多个 Driver，不同数据库的驱动程序是不同的。

3.2　JDBC 驱动和数据库访问

3.2.1　JDBC 驱动

1. JDBC 驱动原理

JDBC 驱动是实现 JDBC 类和接口方法的类集合，数据库驱动程序中必须实现在 JDBC API 中定义的抽象类，尤其是对 java. sgl. Connection、java. sgl. Prepared-Statement、java. sgl. CallableStatement 和 java. sgl. ResultSet。java. sgl. Driver 用于通用的 java. sgl. DriverManager 类，使其在对一个指定的数据库 URL 访问时，可以查找相应的驱动程序。

JDBC 的驱动管理器负责管理针对各种类型 DBMS 的 JDBC 驱动程序，也负责和用户的应用程序交互，为 Java 应用程序建立数据库连接。Java 应用程序通过 JDBC API 向 JDBC 驱动管理器发出请求，指定要装载的 JDBC 驱动程序类型和数据源。驱动管理器会根据这些要求装载合适的 JDBC 驱动程序并使该驱动连接相应的数据源，一旦连接成功，该 JDBC 驱动程序就会负责 Java 应用与该数据源的一切交互。

2. JDBC 驱动类型

JDBC 驱动有 4 种类型：JDBC-ODBC 桥、本地 API 部分 Java 驱动、网络协议完全 Java 驱动和本地协议完全 Java 驱动。

① JDBC-ODBC 桥：利用了现有的 ODBC，将 JDBC 调用转换为 ODBC 的调用，此类型的驱动使 Java 应用可以访问所有支持 ODBC 的 DBMS。

② 本地 API 部分 Java 驱动：将 JDBC 调用转换成对特定 DBMS 客户端 API 的调用。

③ 网络协议完全 Java 驱动（Net-protocol Java Driver）：此类型的驱动将 JDBC 的调用转换为独立于任何 DBMS 的网络协议命令，并发送给一个网络服务器中的数据库中间件，该中间件进一步将网络协议命令转换成某种 DBMS 所能理解的操作命令。

④ 本地协议完全 Java 驱动（Native-protocol Fully Driver）：此类型的驱动直接将 JDBC 的调用转换为特定的 DBMS 所使用的网络协议命令，并且完全由 Java 语言实现，平台独立，但其缺陷是不同的数据库需要下载不同的驱动程序。其允许一个客户端程序直接调用 DBMS 服务器，在网络环境下，此方法经常被使用，此驱动是通过数据库厂商提供的一个 jar 包来完成的。

3.2.2　JDBC 访问数据库

1. JDBC 常用的类、接口

JDBC 进行数据库开发的接口主要在如下两个包中：

① java. sql——JDBC 主要功能在 Java 2 平台标准版(J2SE)；

② javax. sql——拓展功能在 Java 2 平台企业版 (J2EE)。

JDBC 常用的类和接口方法如下。

(1) java. sql(J2SE)包常用的类、接口方法如表 3-1 所示。

表 3-1 java. sql 包中常用的类、接口方法

序 号	类、接口方法名称	说 明
1	Driver	驱动，用来连接数据库
2	DriverManager	驱动管理，从驱动列表中找到合适的驱动去连接数据库
3	Connection	数据库的连接是其他数据操作对象的基础
4	Statement，PreparedStatement	向数据库发送 SQL 语句
5	CallableStatement	调用数据库中的存储过程
6	ResultSet	获取 SQL 查询语句的结果集

上述表中的 DriverManager 类是数据库驱动管理，它的主要功能是获取数据库的连接，其常用的连接数据库的方法如表 3-2 所示。

表 3-2 连接数据库的方法

返回类型	方法名	作 用
static Connection	getConnection(String url)	与给定的服务器数据库 URL 建立连接
static Connection	getConnection(String url，Properties info)	与给定的服务器数据库 URL 建立连接，数据库用户名、密码可以通过类 Properties 的属性设置
static Connection	getConnection(String url，String username，String password)	与给定的服务器数据库 URL 建立连接，给定数据库用户名、密码

(2) javax. sql(J2EE) 包常用的类、接口方法如表 3-3 所示。

表 3-3 javax. sql(J2EE) 包中常用的类、接口方法

序 号	类、接口方法名称	说 明
1	DatabaseMetaData	数据库的元数据，包含数据库的版本号、名称、含有的表、用户等
2	ResultSetMetaData	查看查询结果集的一些信息，包含列数、每个列的类型等

此外，还有 TYPE，实现 Java 语言的数据类型与数据库的数据类型之间的映射。

2. JDBC 连接创建

(1) 使用 JDBC-ODBC 进行桥连步骤如下。

① 配置数据源：控制面板→管理工具→ODBC 数据源→系统 DSN。

② 编程，通过桥连方式与数据库建立连接。

```
Class.forName("sun.jdbc.odbc.JdbcOdbcDriver");//JDBC-ODBC 桥驱动类的完全限定类名
Connection con = DriverManager.getConnection("jdbc:odbc:tw","tt","t2");//数据源名称
```

(2) 本地协议完全 Java 驱动连接建立。

本地协议完全 Java 驱动把 JDBC 调用转换为符合数据库系统规范的请求，即直接转为 DBMS 所使用的网络协议，完全由 Java 实现，其连接步骤如下。

首先，下载数据库厂商提供的驱动程序包(本例以 MySQL 数据库为例，MySQL 数据库驱

动目前为 mysql-connector-java-5. x-bin. jar)，然后将驱动程序包加入到工程目录。

通过完全 Java 驱动方式与数据库建立连接的方式如下所示。

```
Class.forName("com.mysql.jdbc.Driver");  //或 Class.forName("org.gjt.mm.mysql.Driver");
String url = jdbc:mysql://IP:3306/test;  //IP 为数据库 IP 地址,如本机是 localhost,test 为数据库名
//jdbc:mysql://[<IP/Host>][<PORT>]/<DB>;
Connection connection = DriverManager.getConnection("jdbc:mysql://IP:3306/test","root","wjj");
```

不同类型数据连接 URL 不同，常用的数据库连接 URL 如下所示。

① SQLServer 数据库连接：

```
Connection connection = DriverManager.getConnection("jdbc:microsoft:sqlserver://IP:1433;DatabaseName = test","wjj","123");
```

② Oracle 数据库连接：

```
Connection connection = DriverManager.getConnection("jdbc:oracle:thin:@IP:1521:test","root","123");
```

或

```
Connection connection = DriverManager.getConnection("jdbc:oracle:oci:@IP:1521:test","root","123");
```

③ Weblogic MS-SQL 数据库连接：

```
Connection connection = DriverManager.getConnection("jdbc:weblogic:mssqlserver4:test@//IP:port","wjj","123");
```

3. JDBC 访问数据库

在进行 JDBC 访问数据库操作前，如果采用上述的完全 Java 方式访问数据库，需要提前加载数据库厂商的驱动 jar 包，JDBC 访问操作数据库共 6 个步骤，本书以 MySQL 数据库为例，其步骤如下所示。

(1) 注册驱动

```
Class.forName("com.mysql.jdbc.Driver");  //注意有可能抛出异常,需异常处理
```

(2) 利用 DriverManager 获取一个数据库连接

```
String url = "jdbc:mysql://IP:3306/test";
String user = "wjj";
String password = "123";
Connection con = DriverManager.getConnection(url,user,password);
```

(3) 获取各种 Statement

```
Statement sta = con.createStatement();
PreparedStatement ps = con.prepareStatement(sql);
CallableStatement cs = con.prepareCall(sql_pl); //sql_pl: {call PL_NAME(?,?)}
```

(4) 可执行相应的 SQL

① 查询操作

```
sta.executeQuery(sql); ps.executeQuery();
cs.registerOutParameter(index,Types);
cs.getXXTypes(index);
```

② 非查询的操作

```
sta.execute(sql);
ps.execute();
sta.executeUpdate(sql);
ps.executeUpdate();
cs.execute();
```

(5) 处理 ResultSet

```
  ResultSet rs = ps.executeQuery();
  while(rs.next()){
  Types var = rs.getTypes(columnIndex||columnName);
}
```

(6) 关闭打开的资源

```
sta.close();
connection.close();
```

4. JDBC 应用实例

下述代码包含了访问数据库的 6 步,也包含了创建 MyTable 数据库表,具体如下所示。

```
packagewww.bcpl.cn;

import java.sql.*;

public classMyDataTest {

  public static void main(String[ ] args) {
    Connection con;
    Statement st;
    ResultSet rset;

    try {
      //step 1 Register a driver
      Class.forName("org.gjt.mm.mysql.Driver");

      //step 2 Establish a connection to the database
      con = DriverManager.getConnection("jdbc:mysql://localhost:3306/test","root","wjj");
      /*第二种连接方式示例
        * Properties properties = new Properties();
        * properties.setProperty("username","root");
        * properties.setProperty("password","wjj");
        * Connection connection =
        * DriverManager.getConnection("jdbc:mysql://IP:3306/test",properties);
        */
      //step 3 create a statement
      st = con.createStatement();
```

```
//step 4 & 5 Execute SQL statement & Process the Resultset

 try{
     st.execute("create table MyTable(id int not null auto_increment,name varchar(20)
     not null,age integer,primary key(id));");
     System.out.println("create table successfully");

   }catch(SQLException e){
     System.out.println("The table has been created. You can't create it again, please use
     the old table");
     st.execute("delete from MyTable;");
     System.out.println("The all old data have been deleted successfully");
     System.out.println();
   }

/*   st.execute( "create table  if not exists MyTable (name varchar(20) not null primary
     key, age integer);");
   System.out.println("create table successfully"); */


st.execute("insert into MyTable values('lisi', 22);");
st.execute(str);

System.out.println("insert successfully");


st.execute("select * from MyTable;");
rset = st.getResultSet();
String name;
int age;

while(rset.next()){  //Process the Resultset
    name = rset.getString("name");
    age = rset.getInt("age");
    System.out.println("name is:  " + name + " age is:" + age);
}
System.out.println();

st.execute("update MyTable set age = 12 where name = 'linan';");
System.out.println("upDate successfully");

st.execute("delete from MyTable where name = 'Li';");
System.out.println("Delete some data successfully");
System.out.println();

st.execute("select * from MyTable");
```

```
        rset = st.getResultSet();
        while(rset.next()){
            name = rset.getString("name");
            age = rset.getInt("age");
            System.out.println("name is:  " + name + " age is:" + age);
        }

        //step 6 close down JDBC objects
        rset.close();
        st.close();
        con.close();
    }


    catch(Exception e){
        System.out.println("failed");
        e.printStackTrace();
    }

  }

}
```

3.3 JDBC数据库高级应用

3.3.1 JDBC SQL异常处理

JDBC SQL异常分为两种:一种是SQL异常(SQL Exception),一种是SQL警告(SQL Warning)。SQL异常通常在以下情况下会发生:

① 与服务器失去连接时产生;

② SQL命令格式不正确时产生;

③ 使用了底层数据库不支持的功能时产生;

④ 引用了不存在的列时产生。

其错误返回的主要方法有:

```
getErrorCode(); //返回错误代码
getNextException() ; //返回下一个异常对象
getMessage(); //得到异常信息
```

SQL警告(SQL Warning)是在产生非致命错误的SQL状态而产生的SQL警告,通常SQL Warning都是相互关联的,与SQL Exception的方法相似,用getNextWarning来代替getNextException。

JDBC SQL异常处理通常分为异常的捕获和异常的抛出,下面分别进行介绍。

1. 异常的捕获

通常使用 try{} catch{}进行捕获,如下:

```
try{
    …

}catch(Exception e){
    System.out.print(e.getMessage());
    e.printStackTrace();
}
```

采用方法 getErrorCode()来获取异常信息。下面采用异常捕获的方式,捕获 SQL 异常。首先定义异常类和处理异常的方法,然后在 main 方法中进行创建对象和调用,代码如下:

```
package exception;

import java.sql.Connection;
import java.sql.DriverManager;
import java.sql.ResultSet;
import java.sql.SQLException;
import java.sql.Statement;

import util.DBUtil;
public class SQLExceptionTest {
    private Connection conn;
        public SQLExceptionTest(){
        conn = DBUtil.getConnection();
            }
    public void lostConnection(){
                try {
            Class.forName("com.mysql.jdbc.Driver");

            conn = DriverManager.getConnection(
                    "jdbc:mysql://localhost:3306/test",
                    "root","root");
        } catch (ClassNotFoundException e) {
            // TODO Auto-generated catch block
            e.printStackTrace();
        } catch (SQLException e) {
            // TODO Auto-generated catch block
            //e.printStackTrace();
            System.out.println(e.getErrorCode());
            System.out.println(e.getMessage());

        }


    }

  public void sendErrorCommand(){

     Statement statement = null;
```

```
        try {
            statement = conn.createStatement();

            ResultSet rs = statement.executeQuery(
                    "select max(personid) as '最大值' from person ");
            rs.next();
            Integer id = rs.getInt(1);
            System.out.println("id = " + id);
            rs.close();
        } catch (SQLException e) {

            System.out.println(e.getErrorCode());
            System.out.println(e.getMessage());
            e.printStackTrace();
        }finally{

            if(statement! = null)
                try {
                    statement.close();
                } catch (SQLException e) {
                    // TODO Auto-generated catch block
                    e.printStackTrace();
                }

        }

    }

    public static void main(String[ ] args) {

        SQLExceptionTest st = new SQLExceptionTest();

        //st.lostConnection();
         st.sendErrorCommand();

    }

}
```

下面采用连接 Oracle 数据库的方式,进行 SQL 警告的测试。

```
import java.sql.Connection;
import java.sql.DriverManager;
import java.sql.ResultSet;
import java.sql.SQLException;
import java.sql.SQLWarning;
import java.sql.Statement;

public class SQLWarningTest {
```

```
    public static void main(String[ ] args) {
      try {
        Class.forName("oracle.jdbc.driver.OracleDriver").newInstance();

        String jdbcUrl = "jdbc:oracle:thin:@localhost:1521:ORCL";
        Connection conn = DriverManager.getConnection(jdbcUrl, "yourName", "mypwd");
        Statement stmt = conn.createStatement(ResultSet.TYPE_SCROLL_SENSITIVE, ResultSet.CON-
CUR_UPDATABLE);
        SQLWarning sw = null;
        ResultSet rs = stmt.executeQuery("Select * from employees");
        sw = stmt.getWarnings();
        System.out.println(sw.getMessage());
        while (rs.next()) {
          System.out.println("Employee name: " + rs.getString(2));
        }
        rs.previous();
        rs.updateString("name", "Jon");
      } catch (SQLException se) {
        System.out.println("SQLException occurred: " + se.getMessage());
      } catch (Exception e) {
        e.printStackTrace();

      }
    }
  }
```

2. 异常的抛出

SQL 异常属于受检查异常，如果不进行捕获，必须向上声明抛出（使用 throws 关键字），否则无法通过编译，通常使用 throws 关键词抛出 SQLException、ClassNotFoundException 等，由调用的程序进行异常处理或捕获。如下方法所示：

```
public boolean insertstatis_topoly_bvk(int bv,int category)throws ClassNotFoundException;
```

3.3.2　事务处理

事务处理（Transactions）：处理一组互相依赖的操作行为，数据库事务是指由一个或多个 SQL 语句组成的工作单元。事务（Transaction）是并发控制的单位，一个事务是一个连续的一组数据库操作，就好像它是一个单一的工作单元进行。换言之，永远不会是完整的事务，除非该组内的每个单独的操作是成功的。如果在事务的任何操作失败，则整个事务将失败。事务是用户定义的一个操作序列，操作序列要么全部都做，要么全部都不做，是一个不可分割的工作单位。通过事务，数据库 SQL 操作能将逻辑相关的一组操作绑定在一起，以便服务器保持数据的完整性。

MySQL 的事务支持不是绑定在 MySQL 服务器本身，而是与存储引擎相关，如下所示。

- MyISAM：不支持事务，用于只读程序提高性能。
- InnoDB：支持 ACID 事务、行级锁和并发。
- Berkeley DB：支持事务。

1. 事务的特性

数据库事务具有 ACID 特性，由关系数据库管理系统(RDBMS)实现，保证数据的正确性，事务通常通过以下 4 个标准属性(缩写为 ACID)来进行数据的保证。

(1) 原子性(Atomic)：确保工作单元内的所有操作都成功完成，否则事务将被中止在故障点，以前的操作失效，事务将回滚到以前的状态。

(2) 一致性(Consistence)：确保数据库正确地改变状态后，成功提交事务，事务处理过程是一致的保持不变的，指数据库事务不能破坏关系数据的完整性以及业务逻辑上的一致性。

(3) 隔离性(Isolation)：指在并发环境中，当不同的事务同时操纵相同的数据时，每个事务都有各自的完整数据空间，使事务操作彼此独立和透明，事务操作数据的中间状态对其他事务是不可见的。

(4) 持久性(Duration)：指只要事务成功结束，它对数据库所作的更新就必须永久保存下来。即使发生系统崩溃，重新启动数据库系统后，数据库还能恢复到事务成功结束时的状态。确保提交的事务的结果或效果的系统出现故障的情况下仍然存在，完成事务的结果是持久的。

事务终止的两种方式：

① 提交，一个事务使其结果永久不变；

② 回滚，撤销所有更改回到原来状态。

2. 数据同时读取(同步)

(1) 数据同时读取存在的问题

脏读取：一个事务读取了另外一个并行事务未提交的更新数据。

不可重复读取：一个事务再次读取之前的数据时得到的数据不一致，被另外一个事务修改。

虚读：一个事务重新执行一个查询，返回的记录包含了其他事务提交的新记录，即一个事务读到另一个事务已提交的新插入的数据，同一个事务里读两次。

第一类丢失更新：撤销一个事物时，把其他事务已提交的更新数据进行覆盖。

第二类丢失更新：这是不可重复读中的特例，一个事务覆盖另一个事务已提交的更新数据。

(2) 处理方法

可以通过设定事务的隔离级别(con. setTransactionIsolation (Connection. isolationLevel))来防止上述数据同时读取中存在的问题。

事务的隔离级别共有 5 种，分别如下所示。

① 没有事务隔离，上述(1)中，数据同时读取存在的 4 种情况都有可能发生。

```
con.setTransactionIsolation(Connection.TRANSACTION_NONE)
```

② 最底级别，只保证不会读到非法数据，上述(1)中 3 个问题有可能发生。

```
con.setTransactionIsolation(Connection.TRANSACTION_READ_UNCOMMITTED)
```

③ 默认级别，可以防止脏数据读取。

```
con.setTransactionIsolation(Connection.TRANSACTION_READ_COMMITTED)
```

④ 可以防止脏数据读取和不可重复读取。

```
con.setTransactionIsolation(Connection.TRANSACTION_REPEATABLE_READ)
```

⑤ 最高级别，防止上述(1)中 3 种情况，事务串行执行(一个接一个排队)。

```
con.setTransactionIsolation(Connection.TRANSACTION_SERIALIZABLE)
```

各种隔离级别所能避免的并发问题，如表 3-4 所示。

表3-4　各种隔离级别所能避免的并发问题

级别号	隔离级别	数据丢失	脏　读	虚　读	不可重复读
0	TRANSACTION_NONE	是	是	是	是
1	TRANSACTION_READ_UNCOMMITTED	否	是	是	是
2	TRANSACTION_READ_COMMITTED	否	否	是	是
4	TRANSACTION_REPEATABLE_READ	否	否	是	是
8	TRANSACTION_SERIALIZABLE	否	否	否	否

3. 事务处理步骤

(1) 常规事务处理

```
try{     //1.设置自动提交为 false
            con.setAutoCommit(false);
        //2.创建 SQL 语句
            PreparedStatement st = con.prepareStatement("update book set name = ? where
            id = ?");
            st.setString(1,"hibernate4");
            st.setInt(2,2);
            st.executeUpdate();
        //3.提交
            con.commit();
            }catch(Exception e){
        //4.如果有异常发生回滚到原来的状态
                con.rollback();
            }finally{
          //5.设置成为默认状态
                con.setAutoCommit(true);
            }
```

(2) 设置隔离级别的事务处理

```
try{
//step 1 设置连接事务的隔离级别
conn.setTransactionIsolation(Connection.TRANSACTION_SERIALIZABLE);
// step 2 同(1)
    …
}
```

3.3.3 元数据

JDBC中有两种元数据，一种是数据库元数据，另一种是ResultSet元数据。元数据是描述存储用户数据容器的数据结构。

数据库元数据用来获取具体的表的相关信息，例如，数据库的版本、名称等，以及数据库中有哪些表，表中有哪些字段和字段的属性等。

JDBC中的两种元数据如下所示。

(1) DatabaseMetaData：用来获得数据库的相关信息。

```
DatabaseMetaData  dbmd = con.getMetaData();//通过 connection 对象获得
System.out.println(dbmd.getDatabaseProductName()); //获得数据库的产品名称
System.out.println(dbmd.getDriverName());  //获得数据库的驱动名称
System.out.println(dbmd.getSchemas()); //获得数据的 Schema (MySql 为 database 名称,Oracle 为用户名称)
public String getDriverName() throws SQLException
public String getDriverVersion() throws SQLException
```

(2) ResultSetMetaData:用来获得表的信息。

```
ResultSet rs = ps.executeQuery(sql);
ResultSetMetaData rsmd = rs.getMetaData();//通过 ResultSet 对象获得
    int column = rsmd.getColumnCount();//获得总列数(有多少列)
    System.out.println(rsmd.getColumnName(1));//获得列名(参数为列的索引号,从 1 开始)
    System.out.println(rsmd.getColumnTypeName(1));//获得列的数据类型名
    System.out.println(rsmd.getTableName(1));//获得表名
//打印结果集

public static void printRS(ResultSet rs)throws SQLException{
    ResultSetMetaData rsmd = rs.getMetaData();
    while(rs.next()){
    for(int i = 1 ; i <= rsmd.getColumnCount() ; i++){
    String colName = rsmd.getColumnName(i);
    String colValue = rs.getString(i);
    if(i>1){
    System.out.print(",");
    }
    System.out.print(name + " = " + value);
    }
    System.out.println();
    }
}
```

3.3.4 数据源应用

1. 创建数据源应用

```
package sample;

import java.sql.Connection;
import java.sql.SQLException;
import javax.sql.DataSource;
import oracle.jdbc.pool.OracleDataSource;
import com.mysql.jdbc.jdbc2.optional.MysqlDataSource;
public class DataSourceTest {

    public static Connection getConnMysql(){
        MysqlDataSource ds = new MysqlDataSource();
```

```
        ds.setServerName("localhost");
        ds.setPortNumber(3306);
        ds.setDatabaseName("test");

        ds.setUser("root");
        ds.setPassword("123");
        Connection con = null ;
        try {
            con = ds.getConnection();
            System.out.println(con);
        } catch (SQLException e) {
            e.printStackTrace();
        } finally{
            try {
                con.close();
            } catch (SQLException e) {
                e.printStackTrace();
            }
        }
        return con ;
    }
    public static Connection getConnOracle(){
        OracleDataSource ds = null ;
        Connection con = null ;
        try {
            ds = new OracleDataSource();
            //ds.setURL("jdbc:oracle:thin:@localhost:1521:orcl");

            ds.setDriverType("thin");
            ds.setServerName("localhost");
            ds.setPortNumber(1521);
            ds.setDatabaseName("orcl");

            ds.setUser("iie");
            ds.setPassword("123");
            con = ds.getConnection();
            System.out.println(con);
        } catch (SQLException e) {
            e.printStackTrace();
        } finally{
            try {
                con.close();
            } catch (SQLException e) {
                e.printStackTrace();
            }
        }
        return con;
```

```
    }

    /**
     * @param args
     */
    public static void main(String[] args) {
        DataSourceTest.getConnMysql();
    }

}
```

2. JNDI 应用

Java 命名和目录接口(Java Naming and Directory Interface,JNDI)是一组在 Java 应用中访问命名和目录服务的 API。命名服务将名称和对象联系起来,使我们可以用名称访问对象。它是一个应用程序设计的 API,为开发人员提供了查找和访问各种命名和目录服务的通用、统一的接口,类似 JDBC 都是构建在抽象层上。

常用的 JNDI 程序包如下所示。

(1) Javax. naming:包含了访问命名服务的类和接口。例如,它定义了 Context 接口,是命名服务执行查询的入口。

(2) Javax. naming. directory:对命名包的扩充,提供了访问目录服务的类和接口。例如,它为属性增加了新的类,提供了表示目录上下文的 DirContext 接口,定义了检查和更新目录对象的属性的方法。

(3) Javax. naming. event:提供了对访问命名和目录服务时的事件通知的支持。例如,定义了 NamingEvent 类,这个类用来表示命名/目录服务产生的事件,定义了侦听 NamingEvents 的 NamingListener 接口。

(4) Javax. naming. ldap:提供了对 LDAP 版本 3 扩充的操作和控制的支持,通用包javax. naming. directory 没有包含这些操作和控制。

(5) Javax. naming. spi:通过 javax. naming 和有关包动态增加了对访问命名和目录服务的支持,是为有兴趣创建服务提供者的开发者提供的。

3. JNDI 创建步骤

(1) 创建 JNDI Properties,为初始化上下文(InitialContext)做准备。

```
java.util.Properties ps = new java.util.Properties();
ps.put(Context.INITIAL_CONTEXT_FACTORY, "com.sun.jndi.fscontext.RefFSContextFactory");
ps.put(Context.PROVIDER_URL, "file:\\e:\\temp");
```

(2) 使用 JNDI Properties 创建上下文。

```
Context cx = new InitialContext(p);
```

(3) 创建 DataSource(ConnectionPoolDataSource) 对象,有两种方式,如下:

```
XxxDataSource ds = new XDataSource();
```

```
XxxxConnectionPoolDataSource ds = new XxxxConnectionPoolDataSource();
ds.setURL(databaseURL);
ds.setUser(userName);
ds.setPassword(password);
```

(4) 绑定数据源对象与名称。

```
cx.bind(name, ds);
```

4. 查询上下文步骤

(1) 创建 JNDI Properties 为初始化上下文(InitialContext)做准备。

```
java.util.Properties p = new java.util.Properties();
p.put(Context.INITIAL_CONTEXT_FACTORY, "com.sun.jndi.fscontext.RefFSContextFactory");
    p.put(Context.PROVIDER_URL, "file:\\e:\\temp");
```

(2) 使用 JNDI Properties 创建上下文。

```
Context cx = new InitialContext(p);
```

(3) 通过名称查询上下文,有两种方式,如下:

```
DataSource ds = (DataSource) cx.lookup(name);
```

```
ConnectionPoolDataSource ds = (ConnectionPoolDataSource) cx.lookup(name);
```

(4) 获取连接,有两种方式,如下:

```
Connection con = ds.getConnection(); // connection-> statement-> Resultset
```

```
PooledConnection pc = ds.getPooledConnection();
Connection con = pc.getConnection(); // connection-> statement-> Resultset
```

5. JNDI 创建具体应用案例

```
package sample;

import java.sql.*;
import javax.naming.Context;
import javax.naming.InitialContext;
import javax.naming.NamingException;
import com.mysql.jdbc.jdbc2.optional.MysqlDataSource;

public class MySqlJNDILBindTest {
    public static void main(String[] args) {
        // 1. create JNDI properties for initial context
        Context ctx = null;

        java.util.Properties jndiEnv = new java.util.Properties();
jndiEnv.setProperty(Context.INITIAL_CONTEXT_FACTORY,"com.sun.jndi.fscontext.RefFSContext-
Factory"); // 1. parameter
        jndiEnv.setProperty(Context.PROVIDER_URL, "file:\\C:"); // 2. parameter

        try {
            // 3. create initial context with JNDI properties
            ctx = new InitialContext(jndiEnv);

            //4. call bindDS method to bind
            MysqlDataSource ocpds = new MysqlDataSource();
```

```
            ocpds.setURL("jdbc:mysql://localhost:3306/wjj");
            ocpds.setUser("root");
            ocpds.setPassword("123");
            // 5. bind datasource object with a name
            ctx.rebind("jdbc/mySql", ocpds);

            System.out.println("bind successfully");
        } catch (NamingException ne) {
            ne.printStackTrace();
        }
    }

}
```

6. 查询上下文具体应用案例

```
package sample;
import java.sql.*;
import javax.sql.*;

import javax.naming.Context;
import javax.naming.InitialContext;
import javax.naming.NamingException;

public class MySqlJNDILookupTest {

    static final String DSNAME = "jdbc/mySql";

    public static void main(String[] args) {
        Context ctx = null;

        // 1. create JNDI properties for initial context
        java.util.Properties jndiEnv = new java.util.Properties();
        jndiEnv.setProperty(Context.INITIAL_CONTEXT_FACTORY,"com.sun.jndi.fscontext.RefFS-
        ContextFactory");
        jndiEnv.setProperty(Context.PROVIDER_URL, "file:\\C:");

        try {
            // 2. create initial context with JNDI properties
            ctx = new InitialContext(jndiEnv);

            Connection con = null;
            Statement stm = null;
            ResultSet rs = null;

            // 3. lookup the dataSource through the name
            DataSource ds = (DataSource) ctx.lookup(DSNAME);
```

```
            System.out.println("lookup successfully");

            // 4. get connection using datasource
            try {
                con = ds.getConnection();

                stm = con.createStatement();
                rs = stm.executeQuery("select * from book");
                while (rs.next()) {
                    System.out.println("price is: " + rs.getDouble("price"));
                    System.out.println("title is: " + rs.getString("title"));
                }
            } catch (SQLException sqle) {
                sqle.printStackTrace();
            } finally {
                if(rs != null)
                    try {
                        rs.close();
                    } catch (SQLException sqle) {
                    }
                if (stm != null)
                    try {
                        stm.close();
                    } catch (SQLException sqle) {
                    }
                if (con != null)
                    try {
                        con.close();
                    } catch (SQLException sqle) {
                    }
            }

            // System.out.println("connection successfully");
        } catch (NamingException ne) {
            ne.printStackTrace();
        } catch (Exception sqle) {
            sqle.printStackTrace();
        }
    }

}
```

3.4 小　结

JDBC 作为 Java 与数据库连接、交互的关键技术，在不同的应用场景下，其应用的方式并不相同，简单应用的情况下，通常的加载 JDBC 驱动和连接数据库步骤基本相同，但在数据量较大、数据库交互比较频繁的情况下，尤其在 Java Web 工程应用系统开发中，数据库连接采用 JDBC 的高级应用数据源连接配置 XML 文件的方式，进行数据库连接的共享，此外，也可以采用配置属性文件的方式，进行 JDBC 数据库连接的配置和连接共享。在此过程中，需要注意的是不同的数据库，其数据库连接的方式、JDBC 驱动都不相同，目前并没有超数据库的 JDBC 驱动，但它们共同的 SQL 语句处理、事务处理以及数据源的应用方式本质上基本相同。

JDBC 技术的实现与所使用的编程语言、数据库具有强依赖关系，所以，在创建 JDBC 应用连接、加载驱动时，以及配置连接方式时，都需要开发者详细考虑数据库的类型、数据源的大小和交互是否频繁等因素。

第 4 章　XML 技术

4.1　XML 技术简介

4.1.1　XML 简介

1. XML 简介

可扩展标记语言(EXtensible Markup Language,XML)类似于 HTML,它被设计的初衷是用来传输和存储数据,而不是像超文本标记语言一样用来显示数据。其标签没有被预定义,需要自行定义标签,具有自我描述性,在 1998 年 2 月成为 W3C 的推荐标准。

XML 是 SGML 的另一个扩展协议,它的主要作用是采用统一的格式定义和保存数据。XML 可以对文档和数据进行结构化处理,从而能够在部门、客户和供应商之间进行交换,实现动态内容生成、企业集成和应用开发。可以使我们能够更准确地搜索、更方便地传送软件组件和更好地描述一些事物,如电子商务交易等。目前 XML 技术在各个领域应用非常广泛,XML 的应用主要有:

① 存储数据,可以用来存储一些定义的系统配置或组参数,如前述的 Servlet 应用的 web. xml 配置文件;

② 共享数据,通过 XML 文件,可以实现在同构系统之间或异构系统之间共享某些定义数据;

③ 分离数据,利用 XML 可以实现将数据的存储和显示展现进行分离,XML 取消了所有标识,包括 font、color、p 等风格样式定义标识,XML 全部是采用类似 DHTML 中 CSS 的方法来定义文档风格样式,而其只是专注于数据的存储和组织;

④ 传输数据,通过 XML 可以实现在异构系统之间保存并传递数据,降低了数据传输交换的复杂性和转换时间,方便不同的应用方便地传输、交换数据,如 Ajax 等。

XML 不能直接用来显示网页,即使包含了数据的 XML 文件,也要转换成 HTML 格式才能在浏览器上显示,它只是用来结构辨识和意义说明。下面为 XML 示例文档,用来表示本文的信息。

```
<? xml version = "1.0" encoding = "utf-8"? >
 <Book>
     <type>数据库系统</type>
        <authors>
            <author>Peter Pan</author>
            <author>John</author>
        </authors>
     <title>Database systems</title>
     <price>30</price>

 </Book>
```

第一行是一个 XML 声明，表示文档遵循的是 XML 的 1.0 版的规范。

第二行定义了文档里面的第一个元素(Element)，也称为根元素(Root Element)，即<Book>，它类似于 HTML 的<HTML>开头标记，这里<Book>是用户自定义。<authors>称为元素，也可称为元素的数据块。需要注意的是，元素必须在使用完后封闭</element_name>。

从上可以看出，通过 XML 存储数据有以下优势：

- XML 是标注型的数据格式，能够让业务人员非常容易理解；
- XML 层次型的数据格式，更能实际地反映出对象和业务的层次关系；
- XML 灵活的数据存储方式，更能反映业务的变化，能够存储相对更广泛的数据。

2. XML 标记格式

在 XML 文档中，所有的标记都必须要有一个相应的结束标记；所有的 XML 标记都必须合理嵌套；所有的 XML 标记都区分大小写；所有标记的属性必须用双引号("") 或单引号 ('')括起来。

XML 标记必须遵循下面的命名规则：

- 名字中可以包含字母、数字以及其他字母；
- 名字不能以数字或"_"（下划线）开头；
- 名字不能以字母 xml（或 XML 或 Xml 或…）开头；
- 名字中不能包含空格。

3. XML 与 HTML 的差异

XML 与 HTML 都继承于通用标识语言标准(Standard Generalized Markup Language，SGML)规范，该规范指导数据的定义者采用统一的格式来描述和定义数据，不依赖于具体的技术平台。XML 不是超文本标记语言的替代，它是对超文本标记语言的补充。XML 同 HTML 一样，都属于标签语言。

XML 和 HTML(超文本标记语言)为不同的目的而设计。XML 被设计用来传输和存储数据，其焦点是数据的内容，它是一种可以按照该语言的规范与语法，能够随意地设计、定义和创造出其他"新的标记语言"的标记语言，或称为元语言。它仅仅规定了严谨的语法规范和弹性的标记语义。XML 定义的标记语言制作的文件，特性是将来不易受到软件版本和文件格式变动的影响，可以长期的保存纯文本数据。HTML 应用的领域主要是为浏览器描述如何显示标签内的数据，并没有定义数据的含义，它是被设计用来显示数据，其焦点是数据的外观。超文本标记语言旨在显示信息，而它旨在传输信息。XML 是独立于软件和硬件的信息传输工具。

4.1.2 XML 特性

1. 扩展性

XML 的扩展性体现在，使用 XML 可以为文档建立自定义的标记(tags)，而这正是 XML 强大的功能和广泛应用的关键。

在 HTML 里有许多固定的标记，必须熟记并使用定义好的标记，不能使用 HTML 规范里没有的标记。而在 XML 中，可以建立任务需要的标记。可以根据任务的类型，给文档定义一些便于记忆的标记名称。例如，文档里包含一些书籍的目录，可以建立一个名为<Books>的标记，然后在<Books>下再根据书籍类别建立<book>、<type>等标记。此外，标记的数量没有限制，只要清晰，易于理解即可。

使用 XML 建立文档之前，需要理解掌握所定义的文档对象相互之间的关系，以及如何识别它们，然后再进行定义 XML 标记。

标识是描述数据的类型或特性的，如高度<height>、年龄<age>和姓名<username>等，而不是数据的内容。标识可以理解为字段的名称，如<15dip>、<20>和<李四>，这些都是无用的标记，标识就是一种字段名。

2. 标识性

XML 的标识性体现在其目的是标识文档中的元素，不论是 HTML，还是 XML，标识的实质在于便于理解，通过标识，文档才便于阅读和理解，文档可以划分段落、列明标题。XML 中，更可以利用其扩展性来为文档建立更合适的标识。但是标识仅仅是用来识别信息，它本身并不传达信息，如 HTML 代码：<b>xml sart</b>。

这里<b>表示粗体，只用来说明是用粗体来显示"xml start"字符，<b>本身并不包含任何实际的信息，在页面上看不到<b>，真正传达信息的是"xml start "。

3. 结构化规则性

作为语言的 XML 遵循一定的规则，其有明确的语法，特定的结构和定义，XML 文档结构化，所有的信息按某种关系排列。XML 虽然灵活，但也需要一些规则、规范类约束 XML 文档的结构，这样才能最大程度地增加灵活性。所有这些规则被 W3C 组织负责维护，XML 结构化过程中，应遵循的原则：

① 定义的每一个元素都和其他元素有逻辑上关联，关联的层级就自发地形成了结构；

② 标识本身的含义与它描述的信息相分离，并不进行绑定。

4. XML 显示

XML 是将数据和格式分离的，XML 取消了所有标识，包括 font、size 和 br 等风格样式定义标识，因此 XML 文档本身并没有规定内容如何显示，其显示必须借助于辅助文件来实现。在 XML 中用来设定显示风格样式的文件类型有 CSS、XSL，通过 CSS 或 XSL，规定 XML 文档显示的格式，才能显示 XML 标记创建的文档。XML 文档全部是采用类似 DHTML 中 CSS 的方法来定义文档风格样式，如图 4-1 所示。

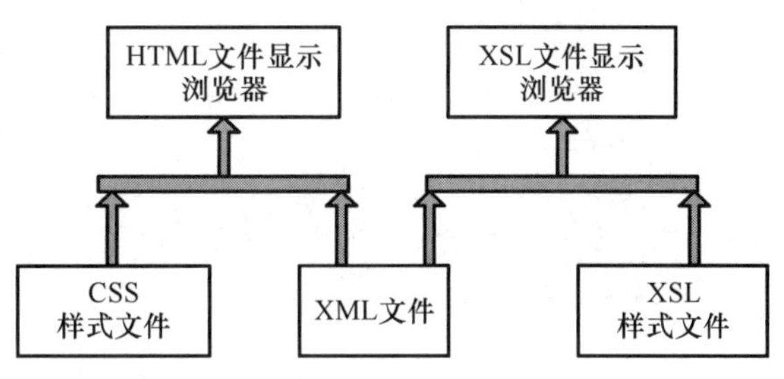

图 4-1　XML 数据显示

4.2　XML 组成、规范

XML 作为可扩展的标记语言，虽然允许用户进行自定义标记，但其本身具有一定的规范，自定义的标记也需遵循一定的规则，其有明确的语法，特定的结构和定义，下面就其基本的规范和语法要求进行介绍。

4.2.1　XML 文档结构

一个文档被标识为 XML 文档，除了文档的后缀名必须命名为. xml 外，其文档内容也有固定的格式要求，如图 4-2 所示。

XML 文件头声明了本文档为 XML 文件，<?　　?>是处理命令的表示；version 属性定义了该文档所遵循的 XML 标准的版本为 1.0；encoding 属性声明了该文档所使用的字符编码为 utf-8;standalone 定义了外部定义的 DTD 文件的存在性。standalone 元素(Element),取值范围是 yes 和 no。本文档值 no 表示这个 XML 文档不是独立的而是依赖于外部所定义的 myfile. dtd;值 yes 表示这个 XML 文档是自包含的(Self-contained)。

```
<?xml version="1.0" encoding="utf-8" standalone="no"?>
<!DOCTYPE myfile system "myfile.dtd">
<employees>
<employee id="0">
<name>wjj</name>
<age>12</age>
<salary>12343</salary>
<department value="">信息技术</department>
<company>BCPL</company>
</employee>

</employees>
```

XML文档头，文档声明

文档体

图 4-2　XML 文件基本结构

XML 文件只能有一个根元素,如上图根元素为<employees></employees>,根元素下可以有多个子元素<employee>,代码如下所示:

```
<? xml version = "1.0" encoding = "utf-8" standalone = "no"? >
<! DOCTYPE myfile system "myfile.dtd">
<employees>
    <employee id = "0">
        <name>wjj</name>
        <age>12</age>
        <salary>12343</salary>
        <department value = "">信息技术</department>
        <company>BCPL</company>
    </employee>
    <employee id = "1">
        ...
    </employee>

</employees>
```

4.2.2　XML 基本语法

如上所述,XML 的文档和 HTML 的文档类似,也是用标识(标签)来标识内容。因此,创建 XML 文档需要遵循一些基本的语法规则,其主要有以下 4 个语法规则。

1. XML 文档必须有声明语句

```
<? xml version = "1.0" encoding = "UTF-8" standalone = "yes/no"? >
```

XML 文档头声明的作用是通知浏览器或者其他处理程序,本文档为 XML 文档。声明语句中的 version 表示该文档遵守的 XML 规范的版本;standalone 表示文档是否附带 DTD 文件,如果有 DTD 文件,该参数为 no,否则为 yes;encoding 表示文档所用的语言编码,默认是 UTF-8。

2. 标记大小写区分

在 XML 文档中,标记大小写分别表示不同的意义,是有区别的,如<p>和<P>表示不同的标识。此外,标识的前后要保持一致。

如将<Title>wjj</Title>错写为<Title>wjj</title>,将导致错误发生。

一般情况下，标记全部大写，或者全部小写，或者大写第一个字母，以免发生错误。

3. DTD 文件声明

DTD 是 Document Type Definition 的简写，其主要作用是约束 DTD 文件的使用者，在编写 XML 时应该按照 DTD 中定义的标签和顺序来编写 XML 文档，不可随意编写标签。如果文档为一个标准的 XML 文档（有效的 XML 文档），那么该文档应具有相应 DTD 文件，并且严格遵守 DTD 文件制定的规范。DTD 文件的声明语句放置在 XML 声明语句之后，DTD 是用来定义 XML 文档中元素、属性以及元素之间关系的。通过 DTD 文件可以检测 XML 文档的结构是否正确，但建立 XML 文档并不一定需要 DTD 文件。DTD 声明格式如下：

```
<! DOCTYPE type-of-doc SYSTEM/PUBLIC "dtd-name">
```

其中：

“! DOCTYPE”是表明要定义一个 DOCTYPE；

“type-of-doc”是文档类型的名称，由用户自定义，通常与 DTD 文件名相同；

“SYSTEM/PUBLIC”这两个参数选用其一，SYSTEM 是指文档使用的是私有 DTD 文件的网址，而 PUBLIC 则指文档调用一个公用的 DTD 文件的网址；

“dtd-name”就是 DTD 文件的网址和名称，所有 DTD 文件的后缀名为“. dtd”。

DTD 的写法如下：

```
<? xml version = "1.0" encoding = "utf-8" standalone = "no"? >
<! DOCTYPE myfile system "myfile.dtd">
```

4. 属性值必须加引号标识

在 XML 中则规定，所有属性值必须加引号，可以是单引号，也可以为双引号，如短缺引号，与 HTML 不一样的地方是，在 XML 文档中将被视为错误。

4.2.3 XML 标记

1. 标记的类型

标记用于区分字符数据和描述该数据的一些字符数据，通常标记有下几种类型，如表 4-1 所示。

表 4-1　标记的类型

标记类型	说　明
Processing Command（处理指令）	在解析阶段，向应用程序提供信息
Comments（注释）	提供文档注释
Entity References（实体引用）	引用相似字符数据或 XML 的实体，通常插入标记类的字符数据
CDATA sections（字符数据）	不想当作标记或者不被解析的字符数据
DOCTYPE declarations（文档类型声明）	使用 DTD 文档，其包含了对该 XML 的约束
Start and End tags（起始和结束标记）< > </>	
Empty tags（空元素标记） </>	

2. 标记成对

一个 XML 文档中元素开始标记和结束标记，必须成对出现，进行匹配。

```
<element>
</element>
```

3. 空元素标记

XML 文档中标记之间如果没有元素，就采用空元素的标记<element/>。

4. 注释

XML 文档中的注释类似 HTML，即注释开始使用“<! --”，“-- >”标识注释的结束 。

```
<!-- 注释内容 -->
```

5. 字符数据段

当 XML 文档中包含的内容需要作为普通的文本来进行，以免引起 XML 解析器的解析错误，XML 解析器会解析“<”为一个新的元素的开始，可以使用 CDATA 来进行处理，CDATA 以“<! [CDATA”开始，以“]>”结束，其中间包含的所有文本都会当作普通文本进行处理，所有的特殊符号都会按普通文本的含义进行处理。CDATA 不能嵌套，]]>之间不能有空格，否则会报错，CDATA 示例如下所示：

```
<![CDATA[
    Int i = 0;
    While(i ++ < = 100){System.out.println(i);}
]]>
```

4.2.4 XML 元素和属性

1. XML 元素

在 XML 中，元素是指标记(标签)之间的内容，一个元素由一个标记来定义，包括开始和结束标记以及其中的内容，如：

```
<title>story</title>
```

XML 中，标记需要用户自定义创建，而在 HTML 中，标记(标签)是固定的，已经定义好(用户无法自定义)，使用即可。

通过元素之间具有逻辑关系，它们之间或者是并列关系(兄弟元素)，或者是上下关系(父子元素)，用户根据它们之间的关系，即可遍历 XML 文档。例如，<value>标签是并列关系，<result>与<value>是父子关系，<value>与<no>、<addr>都是父子关系。如下例子：

```
<? xml version = "1.0" encoding = "GB2312"? >
<result>
    <value>
        <no>11024</no>
        <addr>北京市 xx 区 xx 镇 xx 路 x 段 xx 号</addr>
    </value>
    <value>
        <no>11032</no>
        <addr>北京市 xx 区 xx 乡 xx 村 xx 组</addr>
    </value>
</result>
```

2. XML 属性

属性是对标记进一步的描述和说明，一个标记可以有多个属性，如 font 的属性有 size、color。

XML 中的属性与 HTML 中的属性是一样的，每个属性都有它自己的名字和数值，属性是标记的一部分。如下代码所示：

```
<? xml version = "1.0" encoding = "UTF-8"? >
<! DOCTYPE dbconfig SYSTEM "dbconfig.dtd" >
<dbconfig>
   <database name = "mysql">
      <driver>com.mysql.jdbc.Driver</driver>
      <url>jdbc:mysql://localhost:3306/test</url>
      <user>root</user>
      <password>root</password>
   </database>
</dbconfig>
```

上述代码中标记<database>有一个属性 name，属性是个“名值对”。

4.2.5　XML DTD 格式

1. DTD 类型

通常按引用方式的不同，把 DTD 分为内部 DTD 和外部 DTD。

内部 DTD：是指 DTD 写在 XML 文档内部，其作用范围仅限于当前的 XML 文件。外部 DTD：是指 DTD 写在 XML 文件外部，可以被其他的 XML 文件重用。上述 DTD 声明部分使用的是外部的 DTD，内部的 DTD 写法如下：

```
<? xml version = " 1.0" encoding = " utf-8" ? >
    <! DOCTYPEemployees[
<! ENTITY createDate "2014-05-31">
<! ELEMENTemployees(employee + )>
<! ELEMENTemployee(name)>
<! ELEMENT name( # PCDATA)>
]>
```

外部 DTD 可以用文本编辑器创建单独的后缀为“.dtd”的文件，其写法如下：

```
<! --定义元素  -->
<! ELEMENT database (driver,url,user,password)>
<! ELEMENT driver ( # PCDATA)>
<! ATTLIST database name CDATA  # FIXED oracle >
```

2. DTD 元素声明

在内部 DTD 中，其元素声明的格式如下：

```
<!ELEMENT element_name(content)>
```

具体含义如表 4-2 所示。

表 4-2　DTD 元素说明

组成部分	说　明
<! ELEMENT>	定义声明的元素
element_name	定义元素的名称
(content)	元素的内容，取值范围是 Empty，ANY，Mixed，children

```
<! ELEMENT employees (employee)>
```

表示声明一个元素 employees，该元素下必须有一个子元素 employee。

```
<! ELEMENT name ( #PCDATA)>
```

表示声明一个元素 name，该元素必须包含一个字符数据的子元素。

Empty：表示元素体可以为空，如< employee>　</ employee>或< employee/>。

ANY：表示元素可以包含子元素，也可以包含任何字符数据的任何形式，此参数一般在根元素使用。

Mixed：混合组织元素中的内容，其并不是真正的值。

```
<! ELEMENT employee(name, #PCDATA)>
```

Children：表示该元素包含子元素，并为每个子元素起个名称，并规定了每一个子元素出现的顺序，也并不是表示真正的值。

```
<! ELEMENT employee(name,age,id)>
```

在上面描述的元素内容中子元素出现的次数和顺序，可以用特殊的字符进行标识，具体含义如表 4-3 所示。

表 4-3　子元素出现的次数和顺序符号说明

字　符	说　明	字　符	说　明
无	子元素必须出现，并只能出现一次	*	子元素可以不出现，或者出现多次
?	子元素可以不出现，或者出现一次	,	子元素必须按出现的先后顺序出现
+	子元素至少出现一次，也可以出现多次	|	子元素可以按任意顺序出现，但至少必须出现一个

3. DTD 元素属性的声明

DTD 元素属性声明的语法格式如下：

```
<! ATTLIST element_name attribute_name TYPE default_value >
```

属性声明中每一个组成部分的说明如表 4-4 所示。

表 4-4　属性声明组成部分说明

属性列表中的组成部分	说　明
element_name(元素名称)	属性所属的元素的名称
attribute_name(属性名称)	属性的名称
TYPE(类型)	属性的类型，该成分可以有多种值，常用的值是 CDATA，表示不是标记的字符数据
default value	属性类型为 CDATA 时的默认值，或者是 XML 定义的关键字 #REQUIRED、#IMPLIED或 #FIXED

default value 中 XML 定义的关键词具体含义如下所示。

- #REQUIRED：表示该属性是必须出现的，并且只有一个值。
- #FIXED：表示该属性值是必须的，并且必须有固定值(指定的值)。

- #IMPLIED:表示属性不是必须的,并且没有值。

```
<!ATTLISTemployee name CDATA #REQUIRED
                  id CDATA #REQUIRED
                  Company CDATA #FIXED"lenovo">
```

注意:符合 XML 语法规则及上述 XML 规范的文档称为具有“良好格式”的 XML 文档。

- 它必须以 XML 声明开头。
- 它必须拥有唯一的根元素。
- 开始标签必须与结束标签相匹配。
- 元素对大小写敏感。
- 所有的元素都必须关闭。
- 所有的元素都必须正确地嵌套。
- 必须对特殊字符使用实体。

符合 XML 语法规则,并遵守相应 DTD 文件规范的 XML 文档称为有效的 XML 文档。良好格式的 XML 文档和有效的 XML 文档,其差别在于前者完全遵守 XML 规范,后者则有自己的“文件类型定义(DTD)”。

4.2.6 XML Schema 格式

1. XML Schema 简介

XML Schema,简称 XSD,是基于 XML 的 DTD 替代者,用来描述 XML 文档的结构。定义可出现在文档中的元素、属性、元素的子元素、子元素的次序和数目、元素和属性的数据类型以及元素和属性的默认值、固定值等。XML Schema 是可扩展的,因为它们由 XML 编写,XML Schema 语言也可作为 XSD(XML Schema Definition)来引用,支持数据类型、命名空间。它主要是被用来定义 XML 文档的合法构建模块,类似 DTD,但其功能比 DTD 更强大。

XML Schema 可定义 XML 文件的元素,分为简易元素和复合元素。

2. Schema 简易元素

简易元素指那些仅包含文本的元素,它并不包含任何其他的元素或属性,简易元素常用的类型有 xs:string、xs:decimal、xs:integer、xs:boolean、xs:date 和 xs:time。

定义一个简单元素格式,如下所示。

```
<xs:element         //表示要定义一个元素
name="color"        //表示要定义元素的名称为 color
type="xs:string"    //表示要定义元素的数据类型,类型为 xs:string
default="red"       //表示要定义元素的默认值,默认值为 red
fixed="red"/>       //表示要定义元素的固定值,此元素只可以取“red”值
```

定义一个属性格式,如下所示。

```
<xs:attribute
        name="birthday"          //表示要定义属性的名字
        type="xs:date"           //表示要定义属性的数据类型
        default="2001-01-11"     //表示要定义属性的默认值
        fixed="2001-01-11"       //表示要定义属性的固定值
        use="required"/>         // 表示此属性是否是必须指定的,即如果不指定就不符合
                                 Schema,默认没有 use="required"属性表示属性可有可无。
```

如一些 XML 文档中元素定义,如下所示。

```
<lastname>Smith</lastname>
<age>28</age>
<dateborn>1980-03-27</dateborn>
```

而相应的 Schema 简易元素代码,如下所示。

```
<xs:element name = "lastname" type = "xs:string"/>
<xs:element name = "age" type = "xs:integer"/>
<xs:element name = "dateborn" type = "xs:date"/>
```

3. Schema 复合元素

复合元素指包含其他元素及/或属性的 XML 元素。复合元素有 4 种类型,分别为空元素、包含其他元素的元素、仅包含文本的元素、包含元素和文本的元素。

如复合 XML 元素"employee",仅包含其他元素。

```
<employee>
<firstname>John</firstname>
<lastname>Smith</lastname>
</employee>
```

在 XML Schema 中,有两种方式来定义复合元素。

(1) 通过命名此元素,可直接对"employee"元素进行声明,代码如下。

```
<xs:element name = "employee">
    <xs:complexType>
        <xs:sequence>
            <xs:element name = "firstname" type = "xs:string"/>
            <xs:element name = "lastname" type = "xs:string"/>
        </xs:sequence>
    </xs:complexType>
</xs:element>
```

用上面所描述的方法,那么仅有"employee"可使用所规定的复合类型。注意其子元素,"firstname"以及"lastname",被包围在指示器 <sequence>中,意味着子元素必须以它们被声明的次序出现。

(2) "employee" 元素可以使用 type 属性,这个属性的作用是引用要使用的复合类型的名称,代码如下。

```
<xs:element name = "employee" type = "personinfo"/>
<xs:complexType name = "personinfo">
    <xs:sequence>
        <xs:element name = "firstname" type = "xs:string"/>
        <xs:element name = "lastname" type = "xs:string"/>
    </xs:sequence>
</xs:complexType>
```

使用上面所描述的方法,其若干元素均可以使用相同的复合类型,代码如下。

```
<xs:element name="employee" type="personinfo"/>
<xs:element name="student" type="personinfo"/>
<xs:element name="member" type="personinfo"/>

<xs:complexType name="personinfo">
  <xs:sequence>
    <xs:element name="firstname" type="xs:string"/>
    <xs:element name="lastname" type="xs:string"/>
  </xs:sequence>
</xs:complexType>
```

4. Schema 根元素

在 XML 文档中，Schema 根元素的定义如下。

```
<xs:schema xmlns:xs="http://www.w3.org/2001/XMLSchema"  //表示数据类型等定义来自 W3C
targetNamespace=http://www.w3schools.com  //表示文档中要定义的元素来自什么命名空间，此命
                                            名空间为 http://www.w3schools.com
xmlns=http://www.w3schools.com  //表示此文档的默认命名空间是什么，此默认命名空间为 ht-
                                 tp://www.w3schools.com
elementFormDefault="qualified">  //表示要求 XML 文档的每一个元素都要有命名空间指定
```

4.3　XML 技术应用

4.3.1　XML DTD 应用

按照 XML 语法规则，定义简单的 XML 文件。

```
<? xml version="1.0" encoding="UTF-8"? >
  <employees>
<employee id="0">
    <name >John</name>
    <age>20</age>
    <salary>1000</salary>
    <department>trainingCenter</department>
    <company>bcpl </company>
  </employee>
  <employee id="1">
    <name>Gron</name>
    <age>34</age>
    <salary>1000</salary>
    <department>Elect</department>
    <company>bcpl</company>
  </employee>
</employees>
```

按照 DTD 语法规则，命名为“employees.dtd”的 DTD 文件，它对上面那个 XML 文档的

元素进行了定义。

```
<! ELEMENT employees (employee + ) >
<! ELEMENT employee (name,age,salary,department,company) >
<! ELEMENT name ( # PCDATA)>
<! ELEMENT age ( # PCDATA)>
<! ELEMENT salary ( # PCDATA)>
<! ELEMENT department ( # PCDATA)>
<! ELEMENT company ( # PCDATA)>
<! ATTLIST employee id CDATA # REQUIRED >
<! ATTLIST department value CDATA # REQUIRED>
<! ENTITY company "bcpl" >
```

第 2 行定义 employees 元素有 5 个子元素 name、age、salary、department 和 company。

第 3~7 行定义了 name、age、salary、department 和 company 元素的类型是"#PCDATA "。

第 8 行定义了属性 id 必须出现。

第 10 行定义了实体为 bcpl。

在上述 XML 文件中对 DTD 的引用,代码如下。

```
<? xml version = "1.0" encoding = "UTF-8"? >
  <! DOCTYPE employees SYSTEM " employees.dtd" >
  <employees>
  <employee id = "0">
     <name >John</name>
     <age>20</age>
     <salary>1000</salary>
     <department>trainingCenter</department>
     <company>bcpl </company>
  </employee>
  <employee id = "1">
     <name>Gron</name>
     <age>34</age>
     <salary>1000</salary>
     <department>Elect</department>
     <company>bcpl</company>
  </employee>
</employees>
```

4.3.2 XML Schema 应用

同样使用上述的 XML 文件内容,按照 XML Schema 语法规则,创建上述 XML 文件的 Schema 文件,代码如下。

```
<? xml version = "1.0" encoding = "UTF-8"? >
<xsd:schema xmlns:xsd = "http://www.w3.org/2001/XMLSchema"
    targetNamespace = "http://www.example.org/employees"
    xmlns:tns = "http://www.example.org/employees"
    elementFormDefault = "unqualified">

<xsd:element name = "employees">
```

```
    <xsd:complexType>
        <xsd:all>
            <xsd:element name = "employee"  minOccurs = "1" maxOccurs = "unbounded">
                <xsd:complexType>
                    <xsd:all>
                        <xsd:element name = "name" type = "xsd:string" minOccurs = "1"
                        maxOccurs = "1"></xsd:element>
                        <xsd:element name = "age" type = "xsd:int"></xsd:element>
                          <xsd:element name = "salary" type = "xsd:double"></xsd:
                          element>
                        <xsd:element name = "department"
                            type = "xsd:string">
                        </xsd:element>
                        <xsd:element name = "company" type = "xsd:string"
                            fixed = "bcpl">
                        </xsd:element>
                    </xsd:all>
                    <xsd:attribute name = "id" type = "xsd:int"></xsd:attribute>
                </xsd:complexType>
            </xsd:element>
        </xsd:all>
        </xsd:complexType>
    </xsd:element>
</xsd:schema>
```

在上述XML文件中对Schema文件的引用,代码如下。

```
<?xml version = "1.0" encoding = "UTF-8"? >
<employees
xmlns = "http://www.example.org/employees"
xmlns:xsi = "http://www.w3.org/2001/XMLSchema-instance"
xsi:schemaLocation = "http://www.example.org/employees employees.xsd "
>
<employee id = "0">
    <name >John</name>
    <age>20</age>
    <salary>1000</salary>
    <department>trainingCenter</department>
    <company>bcpl </company>
  </employee>
  <employee id = "1">
     <name>Gron</name>
     <age>34</age>
     <salary>1000</salary>
     <department>Elect</department>
     <company>bcpl</company>
  </employee>
</employees>
```

4.4 XML 解析

XML 作为数据存储、交换和传输的格式文件，如何读取或者修改其中的数据信息，成为 XML 应用的关键，XML 文档解析在不同的语言里其解析方式基本相同，但实现不同。XML 文档解析方式有 4 种，它们分别是 DOM 解析、SAX 解析、JDOM 解析和 DOM4J 解析，下面就它们对 XML 文档解析的基本的方法和过程进行介绍。

4.4.1 DOM 解析

DOM 是 Document Ojbect Model 的简写，又称文档对象模型，在 DOM 解析中 XML 解析开发包使用 JAXP，它是基于树或对象的 API，它把 XML 文档的内容加载到内存，以层次结构组织并生成一个与 XML 文档对应的树结构，以层次结构组织的节点或信息片断的对象集合。这个层次结构允许开发人员在树中寻找特定信息。由于树在内存中是持久的，因此可以修改它以便应用程序能对数据和结构作出更改。

下面的 DOM 解析数据程序，用到 xerces. parsers. DOMParser 类来解析 XML 文件路径，需要加载 xerces. jar 包。由于是数据库连接的 XML 文件解析，需要加载 mysql-connector-java-5. * -bin. jar 包。

(1) 创建 XML 文件 database. config. xml，代码如下。

```
<?xml version="1.0" encoding="UTF-8"?>
<tns:database
xmlns:tns="http://www.example.org/database.schema"
xmlns:xsi="http://www.w3.org/2001/XMLSchema-instance"
xsi:schemaLocation="http://www.example.org/database.schema ../../schema/database.schema.xsd ">
 <tns:db driver="com.mysql.jdbc.Driver" name="MySQL">
    <tns:driver>com.mysql.jdbc.Driver</tns:driver>
    <tns:url>jdbc:mysql://localhost:3306/twitter_20140516</tns:url>
    <tns:user>root</tns:user>
    <tns:password>123456</tns:password>
    <tns:description >Datasource Info</tns:description>
  </tns:db>
</tns:database>
```

(2) 创建 DatabaseBean. java 类，代码如下。

```
package com.bcpl.cn;

import java.sql.Connection;
import java.sql.DriverManager;
import java.sql.SQLException;

public class DatabaseBean {

   private String driver;
```

```
private String url;
private String user;
private String password;
private Connection con;

public DatabaseBean() {
    super();
    // TODO Auto-generated constructor stub
}
public DatabaseBean(String driver, String password, String url, String user) {
    super();
    this.driver = driver;
    this.password = password;
    this.url = url;
    this.user = user;
}
public String getDriver() {
    return driver;
}
public void setDriver(String driver) {
    this.driver = driver;
}
public String getUrl() {
    return url;
}
public void setUrl(String url) {
    this.url = url;
}
public String getUser() {
    return user;
}
public void setUser(String user) {
    this.user = user;
}
public String getPassword() {
    return password;
}
public void setPassword(String password) {
    this.password = password;
}

//业务方法得到连接
public Connection getCon(){

    try {
        Class.forName(driver);
        con = DriverManager.getConnection(url,user,password);
    } catch (ClassNotFoundException e) {
        // TODO Auto-generated catch block
```

```
            e.printStackTrace();
        } catch (SQLException e) {
            // TODO Auto-generated catch block
            e.printStackTrace();
        }

        return con;

    }


    public void close() throws SQLException {
        con.close();
    }
    @Override
    public int hashCode() {
        final int prime = 31;
        int result = 1;
        result = prime * result + ((driver == null) ? 0 : driver.hashCode());
        result = prime * result + ((password == null) ? 0 : password.hashCode());
        result = prime * result + ((url == null) ? 0 : url.hashCode());
        result = prime * result + ((user == null) ? 0 : user.hashCode());
        return result;
    }
    @Override
    public boolean equals(Object obj) {
        if (this == obj)
            return true;
        if (obj == null)
            return false;
        if (getClass() != obj.getClass())
            return false;
        DatabaseBean other = (DatabaseBean) obj;
        if (driver == null) {
            if (other.driver != null)
                return false;
        } else if (! driver.equals(other.driver))
            return false;
        if (password == null) {
            if (other.password != null)
                return false;
        } else if (!password.equals(other.password))
            return false;
        if (url == null) {
            if (other.url != null)
                return false;
        } else if (! url.equals(other.url))
            return false;
        if (user == null) {
```

```
            if (other.user != null)
                    return false;
        } else if (! user.equals(other.user))
            return false;
        return true;
    }
}
```

(3) 创建 DOM 解析类 DatabaseConfigDOM,代码如下。

```
package com.bcpl.cn;

import java.io.IOException;
import java.sql.Connection;
import java.sql.SQLException;

import org.apache.xerces.parsers.DOMParser;
import org.w3c.dom.Document;
import org.w3c.dom.Element;
import org.w3c.dom.Node;
import org.w3c.dom.NodeList;
import org.xml.sax.InputSource;
import org.xml.sax.SAXException;

public class DatabaseConfigDOM {
    //封装数据的实体 Bean
    private DatabaseBean dbBean;
    //已经封装了数据的 Document 对象
    private Document document;

    public DatabaseConfigDOM(String xmlFilePath) {
        // 1.用 DOMParser 解析 xmlFilePath

        DOMParser domParser = new DOMParser();

        try {
            // 2.解析 classpath 路径下 config 字目录中的 XML 文件
            domParser.parse(new InputSource(DatabaseConfigDOM.class
                    .getResourceAsStream(xmlFilePath)));
            // 3.获取包含了数据的 Document 树
            this.document = domParser.getDocument();

        } catch (SAXException e) {
            // TODO Auto-generated catch block
            e.printStackTrace();
        } catch (IOException e) {
            // TODO Auto-generated catch block
            e.printStackTrace();
        }
        // 初始化 Document

    }
```

```
public void getDatabaseBean() {

    this.dbBean = new DatabaseBean();

    // 提取 Document 中的数据并存放到 DatabaseBean
    // Document 常用的 API
    // 1.获取根节点<database>
    Element root = document.getDocumentElement();
    // 2.获取某个元素下所有的字节点
    // 3.根据节点名称获取节点对象集合
    NodeList dbNodes = root.getElementsByTagName("tns:db");

    for (int j = 0; j < dbNodes.getLength(); j++) {

        Node dbnode = dbNodes.item(j);

        if (dbnode instanceof Element) {

            NodeList nodeList = dbnode.getChildNodes();

            for (int i = 0; i < nodeList.getLength(); i++) {

                Node node = nodeList.item(i);

                if (node instanceof Element) {
                    System.out.println(node.getNodeName() + ">>>");
                    if (node.getNodeName().equals("tns:driver")) {
                        // 将该节点值保存到 DatabaseBean 的 driver 属性

                        Node nodeValue = node.getFirstChild();
                        String value = nodeValue.getNodeValue();
                        dbBean.setDriver(value);
                    }
                    if (node.getNodeName().equals("tns:url")) {
                        // 将该节点值保存到 DatabaseBean 的 driver 属性

                        Node nodeValue = node.getFirstChild();
                        String value = nodeValue.getNodeValue();
                        dbBean.setUrl(value);
                    }
                    if (node.getNodeName().equals("tns:user")) {
                        // 将该节点值保存到 DatabaseBean 的 driver 属性

                        Node nodeValue = node.getFirstChild();
                        String value = nodeValue.getNodeValue();
                        dbBean.setUser(value);
                    }
                    if (node.getNodeName().equals("tns:password")) {
                        // 将该节点值保存到 DatabaseBean 的 driver 属性

                        Node nodeValue = node.getFirstChild();
                        String value = nodeValue.getNodeValue();
```

```
                                    dbBean.setPassword(value);

                              }

                        }

                  }

            }
      }

      walkNode(root);

}

//遍历一棵树
  private void walkNode(Element root) {

  // 判断如果 root 下面没有子元素
  if (!root.hasChildNodes()) {
        // System.out.println(root.getNodeName());
        return;
  }
  // 获取某个节点下所有的字节点集合
  NodeList childs = root.getChildNodes();

  for (int i = 0; i < childs.getLength(); i++) {
        // 3. 返回节点集合中某个索引位置上的节点信息
        Node node = childs.item(i);

        // 4. 判断如果节点类型是元素,只有元素才能包含其他子节点
        if (node instanceof Element) {

              // 5. 转换 Element
              Element el = (Element) node;
              System.out.print(node.getNodeName() + "\t");

              // 获取某个元素下的第一个子节点
              Node textNode = (Node) el.getFirstChild();
              // 获取该字节点的字符串值
              System.out.println(textNode.getNodeValue());

              if (el.hasChildNodes()) {

                    this.walkNode(el);

              }
        }

        // 4.Node 常用的方法
         System.out.print(node.getNodeName() + "\t");
         System.out.println(node.getNodeValue());
```

```
        }

    }

    public DatabaseBean getDbBean() {
        return dbBean;
    }

    public void setDbBean(DatabaseBean dbBean) {
        this.dbBean = dbBean;
    }

    public static void main(String[] args) throws SQLException {

        DatabaseConfigDOM databaseConfigDOM = new DatabaseConfigDOM("./database.config.xml");
        databaseConfigDOM.getDatabaseBean();
        DatabaseBean dbBean = databaseConfigDOM.getDbBean();
        Connection con = dbBean.getCon();
        System.out.println(con == null);
        dbBean.close();
    }

}
```

（4）运行结果，如图 4-3 所示。

```
Problems  Tasks  Web Browser  Console  Servers
<terminated> DatabaseConfigDOM [Java Application] C:\java\jdk1.7.0_25\bin\javaw.exe (2014-10-7 下午7:51:51)
tns:driver>>>
tns:url>>>
tns:user>>>
tns:password>>>
tns:description>>>
tns:db

tns:driver      com.mysql.jdbc.Driver
tns:url jdbc:mysql://localhost:3306/twitter_20140516
tns:user        root
tns:password    123456
tns:description Datasource Info
false
```

图 4-3　DOM 解析结果输出

4.4.2　SAX 解析

SAX 解析器采用了基于事件的模型，它在解析 XML 文档的时候，根据定义好的事件解析器，可以触发一系列的事件，来决定当前所解析的部分，当发现给定的 tag 的时候，它可以激活一个回调方法，告诉该方法制定的标签已经找到。SAX 与 DOM 解析不同，它对内存的要求通常会比较低。

（1）创建解析的 XML 文档 employees. xml，代码如下。

```
<? xml version = "1.0" encoding = "UTF-8"? >
<employees
xmlns = "http://www.example.org/employees"
xmlns:xsi = "http://www.w3.org/2001/XMLSchema-instance"
xsi:schemaLocation = "http://www.example.org/employees employees.xsd "
>
  <employee id = "0">
    <name >John</name>
    <age>23</age>
  </employee>
  <employee id = "1">
    <name>张三</name>
    <age>25</age>
  </employee>
</employees>
```

（2）创建 employees. xml 对应的 Javabean 文件 Employee. java，代码如下。

```
package com.bcpl.cn;

public class Employee {
   private String name;
   private int age;
   private int id;

public int getId() {
   return id;
}
public void setId(int id) {
   this.id = id;
}
public String getName() {
   return name;
}
public void setName(String name) {
   this.name = name;
}
public int getAge() {
   return age;
}
public void setAge(int age) {
   this.age = age;
}
@Override
public String toString() {
   //TODO Auto-generated method stub
   String s = "id:" + this.id + " 姓名:" + this.name + " " + "年龄: " + this.age;
   return s;
}

}
```

(3) 创建 SAX 解析文件 EmployeeSAXReader.java,代码如下。

```
package com.bcpl.cn;

import java.io.File;
import java.io.IOException;
import java.util.List;
import java.util.Stack;
import java.util.Vector;

import javax.xml.parsers.ParserConfigurationException;
import javax.xml.parsers.SAXParser;
import javax.xml.parsers.SAXParserFactory;

import org.xml.sax.Attributes;
import org.xml.sax.SAXException;
import org.xml.sax.helpers.DefaultHandler;

class EmployeeHandler extends DefaultHandler {

    private List<Employee> list = new Vector<Employee>();

    private Employee employee = null;

    private Stack stack = new Stack();

    @Override
    public void startDocument() throws SAXException {
          System.out.println("开始读取文档…");
    }

    @Override
    public void startElement(String uri, String localName, String qName,
                Attributes attributes) throws SAXException {
          stack.push(qName);
          System.out.println("开始读取起始标签" + qName);
          if (qName.equalsIgnoreCase("employee")) {
                employee = new Employee();
                String id = attributes.getValue("id");
                System.out.println(id);
                employee.setId(Integer.parseInt(id));

          }

    }

    @Override
    public void characters(char[] ch, int start, int length)
                throws SAXException {

        String qName = (String) stack.peek();
```

```
        if (qName.equals("name")) {
                String content = new String(ch, start, length);
                employee.setName(content);

        }else if(qName.equals("age")){
                String content = new String(ch, start, length);
                employee.setAge(Integer.parseInt(content));

        }else {

        }

    }

    @Override
    public void endElement(String uri, String localName, String qName)
                throws SAXException {
        String befName = (String) stack.pop();

        if (! befName.equals(qName)) {
                System.out.println("标签没有闭合");
                System.exit(1);
        }

        if (qName.equalsIgnoreCase("employee")) {

                list.add(employee);

        }

        System.out.println("读取闭合标签" + qName);
    }

    @Override
    public void endDocument() throws SAXException {
        System.out.println("文档读取结束");

    }

    public List<Employee> getEmployees(){

        return this.list;

    }

}

public class EmployeeSAXReader {

    /**
     * @param args
```

```
     * @throws SAXException
     * @throws ParserConfigurationException
     * @throws IOException
     */
    public static void main(String[] args) throws ParserConfigurationException,
                SAXException, IOException {
        // GARY Auto-generated method stub

        SAXParserFactory factory = SAXParserFactory.newInstance();

        SAXParser sp = factory.newSAXParser();

        EmployeeHandler handler = new EmployeeHandler();

        sp.parse(new File("./employees.xml"), handler);

        List<Employee> list = handler.getEmployees();

        for(Employee e:list){

            System.out.println("" + e);

        }

    }

}
```

(4) SAX 解析运行结果,如图 4-4 所示。

```
Problems  Tasks  Web Browser  Console  Servers
<terminated> EmployeeSAXReader [Java Application] C:\java\jdk1.7.0_25\bin\javaw.exe (2014-10-7 下午8:11:44)
开始读取起始标签employees
开始读取起始标签employee
0
开始读取起始标签name
读取闭合标签name
开始读取起始标签age
读取闭合标签age
读取闭合标签employee
开始读取起始标签employee
1
开始读取起始标签name
读取闭合标签name
开始读取起始标签age
读取闭合标签age
读取闭合标签employee
读取闭合标签employees
文档读取结束
id:0 姓名: John 年龄: 23
id:1 姓名: 张三 年龄: 25
```

图 4-4　SAX 解析运行结果

4.4.3 DOM4J 解析

DOM4J 是一个良好的 Java XML API,具有性能优异、功能强大和易用的特点,包括集成的 XPath 支持、XML Schema 支持以及用于大文档或流化文档的基于事件的处理。下面就 DOM4J 的解析和 XML 文档生成进行介绍。

(1) 创建 XmlDom4J.java 文件,代码如下。

```
package com.dom4j;

import org.dom4j.Document;
import org.dom4j.DocumentHelper;
import org.dom4j.Element;
importorg.dom4j.io.XMLWriter;
import java.io.*;

public class XmlDom4J{
    public void generateDocument(){
        //使用 DocumentHelper 类创建一个文档实例,DocumentHelper 是生成 XML 文档节点的 DOM4J API
          工厂类
        Document document = DocumentHelper.createDocument();

        //使用 addElement() 方法创建根元素 catalog,addElement() 用于向 XML 文档中增加元素
        Element catalogElement = document.addElement("catalog");

        //在 catalog 元素中使用 addComment() 方法添加注释“An XML catalog”
        catalogElement.addComment("An XML Catalog");

        //在 catalog 元素中使用 addProcessingInstruction() 方法增加一个处理指令
        catalogElement.addProcessingInstruction("target","text");

        //在 catalog 元素中使用 addElement() 方法增加 journal 元素
        Element journalElement =   catalogElement.addElement("journal");

        //使用 addAttribute() 方法向 journal 元素添加 title 和 publisher 属性
        journalElement.addAttribute("title", "XML Zone");
        journalElement.addAttribute("publisher", "IBM developerWorks");

        //向 journal 元素中添加 article 元素
        Element articleElement = journalElement.addElement("article");

        //为 article 元素增加 level 和 date 属性
        articleElement.addAttribute("level", "Intermediate");
        articleElement.addAttribute("date", "December-2014");

        //向 article 元素中增加 title 元素
        Element   titleElement = articleElement.addElement("title");

        //使用 setText() 方法设置 article 元素的文本
        titleElement.setText("Java configuration with XML Schema");
```

```
        //在 article 元素中增加 author 元素
        Element authorElement = articleElement.addElement("author");

        //在 author 元素中增加 firstname 元素并设置该元素的文本
        Element  firstNameElement = authorElement.addElement("firstname");
        firstNameElement.setText("Ajax");

        //在 author 元素中增加 lastname 元素并设置该元素的文本
        Element lastNameElement = authorElement.addElement("lastname");
        lastNameElement.setText("dom4j");

        try{
            java.io.Writer output = new java.io.OutputStreamWriter(new java.io.FileOutputStream(new
            File("catalog.xml")),"UTF-8");
          document.write(output);
           output.close();
        }catch(IOException e){
             System.out.println(e.getMessage());
        }
    }
    public static void main(String[ ] args) {
        XmlDom4J dom4j = new XmlDom4J();
        dom4j.generateDocument();

    }

}
```

(2) 程序运行后，在工程的根目录下自动生成 catalog.xml 文件，文件内容代码如下。

```
<?xml version="1.0" encoding="UTF-8"?>
<catalog><!--An XML Catalog-->
<?target text?>
<journal title="XML Zone" publisher="IBM developerWorks">
   <article level="Intermediate" date="December-2014">
      <title>Java configuration with XML Schema</title>
      <author>
       <firstname>Ajax</firstname>
       <lastname>dom4j</lastname>
      </author>
    </article>
</journal>
</catalog>
```

(3) 创建解析 XML 文件的程序并进行 XML 文件修改,程序代码如下。

```
package com.dom4j;

import org.dom4j.Document;
import org.dom4j.Element;
import org.dom4j.Attribute;
import java.util.Iterator;
import org.dom4j.io.XMLWriter;
import java.io.*;
import org.dom4j.DocumentException;
import org.dom4j.io.SAXReader;

public class Dom4JParser {

    public void modifyDocument(File inputXml) {

        try {
            // 使用 SAXReader 解析 XML 文档, SAXReader 包含在 org.dom4j.io 包中
            SAXReader saxReader = new SAXReader();
            Document document = saxReader.read(inputXml);

            // inputXml 是从 c:/catalog/catalog.xml 创建的 java.io.File
            // 使用 XPath 表达式从 article 元素中获得 level 属性节点列表
            // 如果 level 属性值是"Intermediate",则改为"Introductory"
            java.util.List list = document.selectNodes("//article/@level");
            Iterator iter = list.iterator();
            while (iter.hasNext()) {
                Attribute attribute = (Attribute) iter.next();
                if (attribute.getValue().equals("Intermediate"))
                    attribute.setValue("Introductory");
            }

            // 使用 XPath 表达式从 article 元素中获得 date 属性节点列表
            list = document.selectNodes("//article/@date");
            iter = list.iterator();
            while (iter.hasNext()) {
                Attribute attribute = (Attribute) iter.next();
                if (attribute.getValue().equals("December-2014"))
                    attribute.setValue("October-2015");
            }

            // 获取 article 元素列表,从 article 元素中的 title 元素得到一个迭代器,并修改
              title 元素的文本
            list = document.selectNodes("//article");
            iter = list.iterator();
            while (iter.hasNext()) {
                Element element = (Element) iter.next();
                Iterator iterator = element.elementIterator("title");
                while (iterator.hasNext()) {
                    Element titleElement = (Element) iterator.next();
```

```
                    titleElement.setText("Create flexible and extensible XML schema");
                }
            }

            // 从 article 元素中的 title 元素节点列表
            list = document.selectNodes("//article/author/firstname");
            iter = list.iterator();
            while (iter.hasNext()) {
                Element element = (Element) iter.next();
                element.setText("wjj");

            }


            XMLWriter output = new XMLWriter(new FileWriter(new File("catalog-modified.xml")));
            output.write(document);
            output.close();
        } catch (DocumentException e) {
            System.out.println(e.getMessage());
        } catch (IOException e) {
            System.out.println(e.getMessage());
        }// try
    }// modifyDocument

    public static void main(String[] argv) {
        Dom4JParser dom4jParser = new Dom4JParser();
        dom4jParser.modifyDocument(new File("catalog.xml"));
    }
}
```

(4) 程序运行后，在工程的根目录下生成 catalog-modified.xml 文件，文件内容代码如下。

```
<?xml version="1.0" encoding="UTF-8"?>
<catalog><!--An XML Catalog-->
<?target text?>
<journal title="XML Zone" publisher="IBM developerWorks">
    <article level="Introductory" date="October-2015">
    <title>Create flexible and extensible XML schema</title>
    <author>
        <firstname>wjj</firstname>
        <lastname>dom4j</lastname>
    </author>
    </article>
</journal>
</catalog>
```

4.5 小　　结

XML 技术的出现实现了互联网数据的统一格式定义和存储，极大地便利了网络数据的存储、共享、分离和传输，使数据和格式实现了真正意义上的分离，但 XML 本身具有自己的结构、语法以及标记，同时，也可自定义标记，并就 XML 的元素和属性、DTD 格式、Schema 格式，以及它们的技术应用进行了介绍，关于 XML 应用中的解析方式 DOM、SAX、DOM4J、JDOM 的解析原理和方法进行了实现和详述，XML 作为数据存储、交换和传输的格式文件，如何读取或者修改其中的数据信息，是 XML 应用的关键，也是项目中应用的重点。

第5章 Struts技术

5.1 Struts基础

5.1.1 Struts技术简介

Struts技术起源于Apache的Jakarta项目,2001年由Craig McClanahan发布,2004年成为Apache软件基金会的顶级项目,它通过采用JavaServlet/JSP技术,实现了基于J2EE Web应用的MVC设计模式的应用框架,经过多年的发展,Struts 1已经成为了一个高度成熟的框架。程序开发人员使用Struts为业务应用的每一层提供支持,其目的是为了减少程序开发人员运用MVC设计模型来开发Web应用的时间,改进和提高了JavaServer Pages 、Servlet、标签库以及面向对象的技术应用水平。但是随着时间的流逝和技术的进步,Struts 1的局限性也越来越多地暴露出来,并且制约了Struts 1的继续发展。Struts 1的主要缺陷表现在以下几个方面:

① Struts 1支持的表现层技术比较单一,Struts 1推出的时候,并没有FreeMarker、Velocity等技术,因此,Struts 1并不支持后续的新技术,并不能与这些视图层的模板技术进行整合;

② Struts 1与Servlet API的耦合严重,使其应用难于进行测试;

③ Struts 1代码严重依赖于Struts 1 API,属于侵入性框架。

此外,后续出现了许多与Struts 1具有竞争性的视图层框架,如JSF、Tapestry和Spring MVC等。它们都应用了最新的设计理念,同时也从Struts 1中吸取了经验,克服了很多不足,同时也促进了Struts本身的发展和升级。

Struts 2是在Struts 1的基础上发展而来,但其与Struts 1体系结构有本质的不同,它是以WebWork为核心,整合了Struts 1的技术框架而发展起来的技术框架。采用拦截器的机制来处理用户的请求,使得业务逻辑控制器能够与Servlet API完全分离,因而Struts 2也可以说是WebWork框架的升级。

Struts是基于经典的MVC(模型-视图-控制器)模型的Web应用变体,其主要是由于网络应用特性——HTTP协议——的无状态性如引起的,可以用来提高系统灵活性、复用性和可维护性。

5.1.2 Struts模型映射

Struts的体系结构实现了模型-视图-控制器设计模式,它将设计模式映射并实现了Web

应用程序的组件和应用。在MVC设计模式中，Model(模型)是执行某些任务的代码，而这部分代码并没有任何逻辑决定它对用户端的表示方法。Model只有纯粹的功能性接口，也就是一系列的公开方法，通过这些公开方法，便可以取得模型端的所有功能。View(视图)指界面显示，一个Model可以有几个View端，使用MVC模式可以允许多于一个的View端存在，并可以在需要的时候动态地登记上所需要的View。Controller(控制器)，用于控制用户与视图的交互，当用户端与相应的视图发生交互时，用户可以通过视图更新模型的状态，而这种更新是通过控制器端进行的。控制器端通过调用模型端的方法更改其状态值，同时，控制器端会通知所有的登记了的视图刷新并显示给用户。而在Struts中，按MVC设计模式，可以把Struts框架中的组件分为3个部分，分别为模型、视窗和控制器。

1. 控制器层(Controller)

在Struts中基本的控制器组件是ActionServlet类中的实例Servlet，Servlet的调用，在配置文件中由一组映射(由ActionMapping类进行描述)进行定义。与Struts 1使用ActionServlet作为控制器不同，Struts 2使用了filter技术，FilterDispatcher是Struts框架的核心控制器，该控制器负责拦截和过滤所有的用户请求。如果用户请求以action结尾，该请求将被转入Struts框架来进行处理。Struts框架获得了*.action请求后，将根据*.action请求的前面名称部分决定调用哪个业务控制Action类，例如，对于wjj.action请求，调用名为wjj的Action来处理该请求。

Struts应用中的Action都被定义在struts.xml文件中，在该文件中配置Action时，主要定义了该Action的name属性和class属性，其中，name属性决定了该Action处理哪个用户请求，而class属性决定了Action的实现类，如：

```
<action name = "registAction" class = "com.bcpl.action.RegistAction">
```

用于处理用户请求的Action实例，并没有与Servlet API耦合，所以无法直接处理用户请求。为此，Struts框架提供了系列拦截器，该系列拦截器负责将HttpServletRequest请求中的请求参数解析出来，传入到Action中，并回调Action的execute方法来处理用户请求。

2. 显示层(View)

Struts的显示层由JSP建立，Struts包含扩展自定义标签库，可以简化创建完全国际化用户界面的过程，Struts 2框架改变了Struts 1只能使用JSP作为视图技术的现状，它允许使用其他的视图技术，如FreeMarker、Velocity等作为显示层。当Struts 2的控制器调用业务逻辑组件处理完用户请求后，会返回一个字符串，该字符串代表逻辑视图，它并未与任何的视图技术关联。当我们在struts.xml文件中配置Action时，还要为Action元素指定系列result子元素，每个result子元素定义上述逻辑视图和物理视图之间的映射。一般情况下我们使用JSP技术作为视图，故配置result子元素时没有type属性，默认使用JSP作为视图资源，如<result name="error">/app/register.jsp</result>。如果需要在Struts 2中使用其他视图技术，则可以在配置result子元素时，指定相应的type属性即可。type属性指定值可以是velocity或者其他。此外，Struts显示层包含上述讲到的一个便于创建用户界面的自定义标签库，对国际化和表达式语言进行支持。

3. 模型层(Model)

在Struts框架中，模型分为两个部分：系统的内部状态和改变状态的操作(事务逻辑)。内部状态通常由一组ActionForm JavaBean表示，根据设计或应用程序复杂度的不同，这些

Bean 可以是自包含的并具有持续的状态，或只在需要时才获得数据(从某个数据库)。大型应用程序通常在方法内部封装事务逻辑(操作)，这些方法可以被拥有状态信息的 Bean 调用，例如，购物车 Bean，它拥有用户购买商品的信息，可能还有 checkOut()方法用来检查用户的信用卡等信息。小型程序中，操作可能会被内嵌在 Action 类，它是 Struts 框架中控制器角色的一部分，适用于逻辑简单的情况。用户将事务逻辑(要做什么)与 Action 类所扮演的角色(决定做什么)分开。模型层会被 Action 调用来处理用户请求。当控制器需要获得业务逻辑组件实例时，通常并不会直接获取业务逻辑组件实例，而是通过工厂模式来获得业务逻辑组件的实例，或者利用其他 IoC 容器(如 Spring 容器)来管理业务逻辑组件的实例。

5.2　Struts 2 框架及工作流程

5.2.1　Struts 2 框架

Struts 2 体系结构中主要的组件及其框架结构如图 5-1 所示。

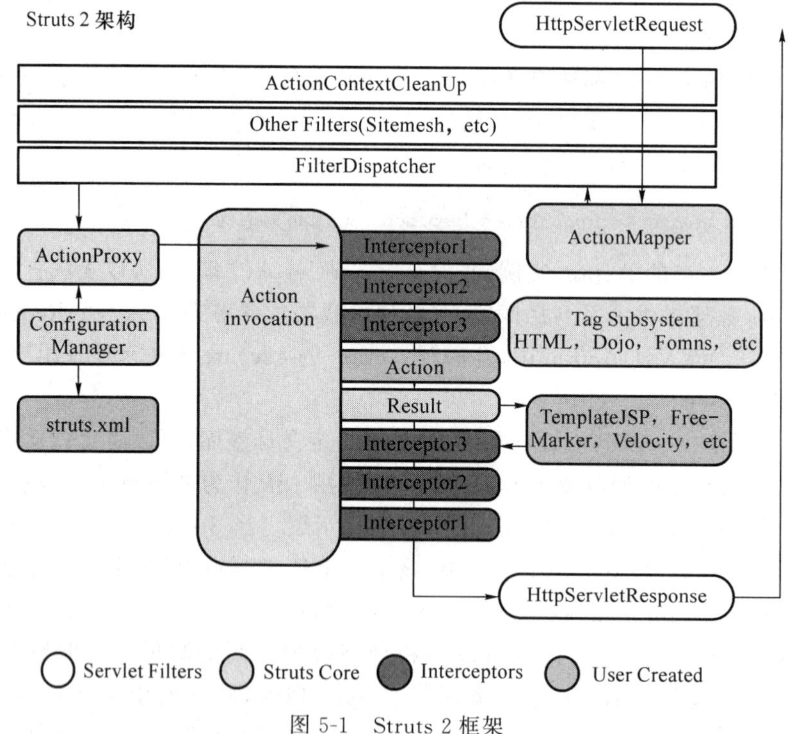

图 5-1　Struts 2 框架

5.2.2　Struts 2 的工作流程

Struts 2 的工作流程是 WebWork 的升级，而不是 Struts 1 的升级。Struts 2 的工作流程步骤如下：

① 客户端浏览器发送请求，如请求/regist. action、/reports/myreport. pdf 等；

② 核心控制器 FilterDispatcher 接收到请求，根据请求决定调用合适的映射 Action；

③ WebWork 的拦截器链自动对请求应用通用功能，如验证、工作流或文件上传等功能；

④ 回调 Action 的 execute 方法，该 execute 方法先获取用户请求参数，然后执行某种业务操作，既可以是将数据保存到数据库，也可以从数据库中检索信息，实际上，因为 Action 只是一个控制器，它会调用业务逻辑组件(Model)来处理用户的请求；

⑤ Action 的 execute 方法处理结果信息将被输出到浏览器中，可以是 HTML 页面、图像，也可以是 PDF 文档或者其他文档，Struts 2 支持的视图技术非常多，既支持 JSP，也支持 Velocity、FreeMarker 等模板技术。

5.2.3　Struts 2 基本配置及简单应用

在进行 Struts 2 工程应用之前，首先要进行基本环境配置，从 http://struts.apache.org/ 下载 Struts 2 需要的所有 jar 包，目前的最新版本为 struts-2.3.16.3-all 包，如图 5-2 所示。

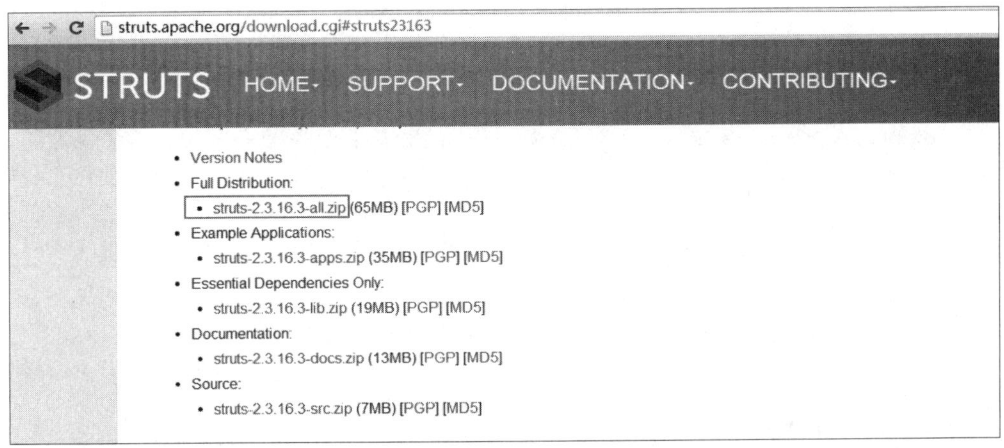

图 5-2　Struts 2 最新的包

创建 Struts 2 应用的基本步骤如下所示。

1. 建立 Web 项目

工程名为 Struts 2Hello，在 lib 目录里放入下载的 Struts 2 需要的 jar 包。

给项目添加外部引用包(project-properties-Java Build Path-Add External Jars)。添加的包有 commons-fileupload-1.3.1.jar、commons-io-2.2.jar、commons-logging-api-1.1.3.jar、freemarker-2.3.19.jar、javassist-3.11.0.GA.jar、ognl-3.0.6.jar、Struts 2-core-2.3.16.3.jar 和 xwork-core-2.3.16.3.jar。工程的目录结构及引用 jar 包如图 5-3 所示。

注意：由于 Struts 2 版本的差异性，上面提到的包不一定满足所有版本的需求。配置完 Struts 2 后，需要部署运行一下，根据运行时的错误提示来添加 jar 包解决问题。

图 5-3　工程目录结构

2. 编写 struts. xml 文件

在 MyEclipse 项目中的 src 根目录下建立一个 struts. xml 文件。可以打开下载的 Struts 2 安装包里的 apps 目录下的任意一个 jar 包，在里面的 WEB_INFR/src 目录下，查找 struts. xml 文件，将该文件复制到项目的 src 根目录下，然后将里面的内容清空(只留下<struts>标签和头部标签即可)。

3. 在 web. xml 中配置 Struts 2

Struts 2 的入口点是一个过滤器(Filter)。因此，Struts 2 要按过滤器的方式配置，和 struts. xml 文件的创建生成过程类似，在 Struts 2 安装包里找到 web. xml 文件，将里面的<filter>和<filter-mapping>标签及其内容复制到项目中的 web. config 文件即可，然后进行配置。

```
<? xml version = "1.0" encoding = "UTF-8"? >
<web-app version = "3.0"
    xmlns = "http://java.sun.com/xml/ns/javaee"
    xmlns:xsi = "http://www.w3.org/2001/XMLSchema-instance"
    xsi:schemaLocation = "http://java.sun.com/xml/ns/javaee
    http://java.sun.com/xml/ns/javaee/web-app_3_0.xsd">
<filter>
        <filter-name>struts2</filter-name>
<filter-class>org.apache.struts2.dispatcher.ng.filter.StrutsPrepareAndExecuteFilter</
filter-class>
      </filter>

      <filter-mapping>
          <filter-name>struts2</filter-name>
          <url-pattern>/*</url-pattern>
      </filter-mapping>

  <welcome-file-list>
      <welcome-file>index.jsp</welcome-file>
  </welcome-file-list>
</web-app>
```

4. 编写 Action 类

Struts 2. x 的动作类需要从 com. opensymphony. xwork2. ActionSupport 类继承。

```
import com.opensymphony.xwork2.ActionSupport;

public class HelloWorldAction extends ActionSupport{
    private String name ;

    /**
     * @return the name
     */
    public String getName() {
          return name;
    }

    /**
     * @param name the name to set
     */
```

```
    public void setName(String name) {
            this.name = name ;
    }

    @Override
    public String execute() throws Exception {
            name = "Hello, " + name ;
            return SUCCESS;
    }

}
```

动作类的一个特征就是要覆盖 execute 方法，execute 方法没有参数，返回一个 String，用于表述执行结果（逻辑名称）。

5. 配置 Action 类

在 Struts 2.x 中的配置文件一般为 struts.xml，放到 WEB-INF 的 classes 目录中。

```
<?xml version = "1.0" encoding = "UTF-8" ? >
  <!DOCTYPE struts PUBLIC
      "-//Apache Software Foundation//DTD Struts Configuration 2.0//EN"
      "http://struts.apache.org/dtds/struts-2.0.dtd">

  <struts>
      <!--
      <include file = "struts-default.xml" />
       -->
      <package name = "struts2" extends = "struts-default">
          <action name = "hello" class = "com.bcpl.action.HelloWorldAction">
              <result>/success.jsp</result>

          </action>
      </package>
  </struts>
```

主要属性说明如下所示。

package-name：用于区别不同的 package；必须是唯一的、可用的变量名；用于其他 package 来继承。

package-namespace：用于减少重复代码（和 Struts 1 比较），是调用 Action 时输入路径的组成部分。

package-extends：用于继承其他 package 以使用里面的过滤器等。

action-name：用于在一个 package 里区别不同的 Action；必须是唯一的、可用的变量名；是调用 Action 时输入路径的组成部分。

action-class：Action 所在的路径（包名+类名）。

action-method：Action 所调用的方法名。

关于其他的属性，可以查阅相关文档。

在<struts>标签中可以有多个<package>，extends 属性继承一个默认的配置文件“struts-default”，一般都继承于它。

<action>标签中的 name 属性表示动作名，class 表示动作类名。

<result>标签的 name 实际上就是 execute 方法返回的字符串，根据返回字符串跳转到某个页面。

在<struts>中可以有多个<package>，在<package>中可以有多个<action>。

可以用如下的 URL 来访问这个动作，链接 URL：http://localhost:8080/Struts 2Hello/hello. action? name=wjj。

注意：Struts 2 是以. action 结尾的。

6. 编写 JSP 页面

在 JSP 页面中使用 Struts 2 的标签。在 Struts 2 中已经将 Struts 1. x 的好几个标签库都统一了，在 Struts 2 中只有一个标签库/struts-tags，其包含了所有的 Struts 2 标签，创建页面 hello. jsp 代码如下：

```
<%@ page language="java" %>
<%@ taglib prefix="s" uri="/struts-tags" %>

  <html>
      <body>
          <s:form action="hello">
                      name:<s:textfield  name="name" />
                          <s:submit/>
          </s:form>
      </body>
  </html>
```

success. jsp 页面主要代码如下：

```
<%@ page language="java" import="java.util.*" pageEncoding="UTF-8"%>
<%@ taglib prefix="s" uri="/struts-tags" %>
<html>
  <head>

    <title>My JSP 页面</title>

  </head>

  <body>
  <s:property value="name"/>
  </body>
</html>
```

7. 启动 Tomcat

在网页地址栏里输入 http://localhost:8080/项目的名称/hello.jsp，打开页面，如图 5-4 所示。

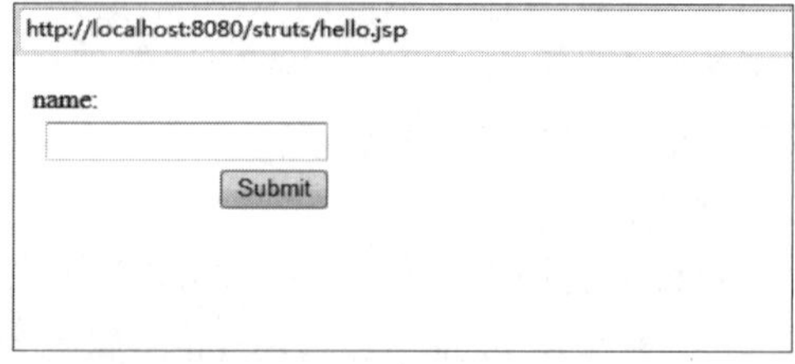

图 5-4　hello. jsp 页面

注意:本例没有使用工程名作为 URL 路径。

8. 测试

输入 name:wjj,单击 submit 按钮,就来到了 success. jsp 页面,显示页面信息如图 5-5 所示。

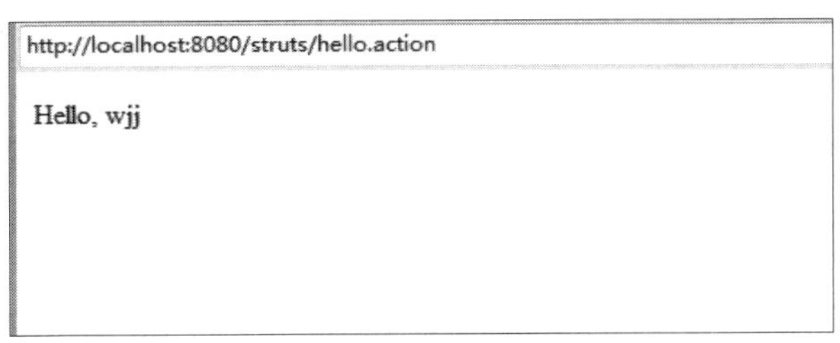

图 5-5　提交成功信息显示页面

5.2.4　Struts 2 常用配置

在 Struts 2 中提供了一个默认的 struts. xml 文件,但如果 Package、Action 和 Interceptors 等配置比较多时,都放到一个 struts. xml 文件不太容易维护。因此,就需要将 struts. xml 文件分成多个配置文件,然后在 struts. xml 文件中使用<include>标签引用这些配置文件。其目的是使配置文件结构更清晰,更容易维护配置信息,使配置文件可以复用。如果在多个 Web 程序中都使用类似或相同的配置文件,那么可以使用<include>标签来引用这些配置文件,这样可以减少工作量。通常,Struts 2 的配置文件有两类,分别为包括配置 Action 的 struts. xml 文件和配置 Struts 2 全局属性的 struts. properties 文件。接下来分别对它们进行详细介绍。

1. struts. xml 配置文件

在 Struts 2 框架所包含的 jar 包中,通常会包含一个名称为 struts-default. xml 的文件,它是 Struts 2 框架的默认配置文件,Struts 2 框架每次都会自动加载该文件。在 struts-default. xml 文件中定义了一个名字为 struts-default 的包空间,该包空间里定义了 Struts 2 内建的 Result 类型,配置了大量的核心组件,以及 Struts 2 内建的系列拦截器、由不同拦截器组成的拦截器栈和默认的拦截器引用等。另外,Struts 2 框架允许以一种"可插拔"的方式来安装插件,它们都提供了一个类似 Struts 2-xxx-plugin. jar 的文件,例如,Spring 插件,它提供了 Struts 2-spring-plugin2. 3. 16. 3. jar 文件,只要将该文件放在 Web 应用的 WEB-INF/lib 路径下,Struts 2 框架将自动加载该框架。对于大部分 Struts 2 的应用而言,用户并不需要重新定义上面这些配置文件,只需应用即可。

Struts 框架的核心配置文件就是 struts. xml 配置文件。在默认情况下,Struts 2 框架将自动加载放在 WEB-INF/classes 路径下的 struts. xml 文件,该文件主要负责管理应用中的 Action 映射、该 Action 包含的 result 定义等和一些其他相关配置。

Struts 2 框架中核心组件包含 Action、拦截器等,struts. xml 的主要包含内容如下所示。

在 strust. xml 文件中,package 元素用于定义包配置,每个 package 元素定义了一个包配置。Struts 2 框架使用包(package)来管理 Action 和拦截器等,package 是多个 Action、多个拦截器和多个拦截器引用的集合。定义 package 元素时,常用的属性有 name、extends、namespace 和 abstract,其具体含义描述如表 5-1 所示。

表 5-1　package 属性描述

属　性	描　述
name	必填属性，该属性指定该包的名字，该名字是该包被其他包引用的 key
extends	可选属性，该属性指定该包继承其他包。继承其他包，可以继承其他包中的 Action 定义和拦截器定义等
namespace	可选属性，该属性定义该包的命名空间
abstract	可选属性，指定该包是否为一个抽象包，抽象包中不能包含 Action 定义

在同一个 Web 应用中如有同名的 Action，Struts 2 采用命名空间的方式来管理 Action。同一个命名空间里不能有同名的 Action，不同的命名空间里可以有同名的 Action。Struts 2 的命名空间的作用如同程序中包或代码模块的作用，它允许以模块化的方式来组织 Action。Struts 2 不支持单独配置命名空间，而是通过为包指定 namespace 属性来为包里所有 Action 指定共同的命名空间，如 package 的配置代码：

```
<package name = "bcpl" extends = "struts-default" namespace = "/xxjsx">
        <action name = "getUserName" class = "com.bcpl.action.GetNameAction">
                <result name = "login">/login.jsp</result>
                <result name = "success">/listName.jsp</result>
        </action>
</package>
```

当包指定了命名空间后，该包下所有的 Action 处理的 URL 应该是“命名空间＋Action 名”。上述名为 xxjsx 的包，该包下名为 getUserName 的 Action，则该 Action 处理的 URL 为 http://locahost:8080/Demo/xxjsx/getUserName.action。

如果某个包没有指定 namespace 属性，那么该包使用默认的命名空间，默认的命名空间总是“”。默认命名空间里的 Action 可以处理任何模块下的 Action 请求，也就是说，如果存在 URL 为/xxjsx/hello.action 的请求，即使/xxjsx 的命名空间下没有名为 hello 的 Action，则默认命名空间下名为 hello 的也会接受并处理用户请求。此外，Struts 2 可以指定根命名空间，即通过设置某个包的 namespace＝“/”来指定根命名空间。如果请求为/hello.action，系统会在根命名空间（“/”）中查找名为 hello 的 Action，如果在根命名空间中找到了名为 hello 的 Action，则由该 Action 处理用户请求。否则，系统将转入默认命名空间中查找名为 hello 的 Action，如果默认的命名空间里有名为 hello 的 Action，则由该 Action 处理用户请求；如果两个命名空间里都找不到名为 hello 的 Action，则系统出现错误。

（1）Action 配置

Struts 2 使用包来组织 Action，Action 是 struts.xml 中包下的基本配置，其配置方式包含基本配置、通配符配置和处理结果配置等。

① Action 基本配置

Action 配置时，其定义是放在包定义下完成的，定义 Action 通过使用 package 下的 Action 子元素来完成。定义 Action 时，需要指定它的 name 属性，该 name 属性既是该 Action 的名字，也是它需要处理的 URL 的一部分。同时，并定义 Action 处理结果与视图资源之间的映射关系。另外，通常还需要为 Action 元素指定一个 class 属性，指定该 Action 的实现类，如<action name=" LoginAction" class="com.bcpl.action. LoginAction">，具体代码如下：

```
<package name = "xxjsx" extends = "struts-default">
...
<action name = "LoginAction" class = "com.bcpl.action. LoginAction">
            <result>/index.html</result>
            <result name = "welcome">/welcome.jsp</result>
            <result name = "success">/users.jsp</result>
            <result name = "error">/recovery.jsp</result>
            <result name = "input">login.jsp</result>
</action>
...
</package>
```

如上所述，Action 充当业务控制器的角色，它在处理完用户请求后，需要调用指定的视图资源返回给用户。因此，在配置 Action 时，需要配置逻辑视图和物理视图资源之间的映射。它们之间的映射是通过标签元素<result…/>来定义的，每个<result…/>元素定义逻辑视图和物理视图之间的一次映射。

Action 元素除了 name 属性之外，还有 method 属性可以进行设定。Struts 框架允许一个页面表单元素里包含多个按钮，分别提交给不同的处理逻辑。Struts 2 提供了一种处理方法，即将一个 Action 处理类定义成多个逻辑 Action。如果在配置<action…/>元素时，指定 Action 的 method 属性，则可以让 Action 类调用指定方法，而不是让 execute 方法来处理用户请求，如下所示：

```
<package name = "xxjsx" extends = "struts-default">
...
<action name = "login" class = "com.bcpl.LoginAction" method = "login" />
        ...
</action>

<action name = "register" class = " com.bcpl.LoginAction" />
        ...
</action>
...
</package>
```

上述示例代码定义了两个 Action，分别为 login 和 register，它们对应的实现处理类为同一的类 com. bcpl. LoginAction。虽然 login 和 register 两个 Action 有相同的实现处理类，但其处理逻辑并不相同，通过 method 方法指定，Action 名为 register 的 Action 对应的处理逻辑为默认的 execute 方法，Action 名为 login 的 Action 对应的逻辑处理为指定的 login 方法。

② 通配符配置

Struts 框架为了解决在 struts. xml 文件中 Action 重复定义和代码冗余的问题，Struts 2 采用了动态方法调用的方式，即使用通配符配置的方式。Action 标签元素配置时，name、class 和 method 属性，都支持通配符的应用及设置，通配符的方式是动态方法调用的一种形式。当我们使用通配符定义 Action 的 name 属性时，相当于给一个 Action 标签元素定义多个逻辑 Action，代码如下所示。

```
<action name = " * Action"  class = "packageName.MyAction" method = "{1}">
        <result>…</result>
    </action>
```

当在 URL 里访问该 Action 时以 * Action. action 形式访问，都可以通过该 Action 进行处理。但该 Action 定义了一个表达式{1}，该表达式就是 name 属性值中的 * 的值。如果用户请求的 URL 是 loginAction. action，则调用该 Action 的 login 方法；如果用户请求的 URL 是 registerAction. action，则调用该 Action 的 register 方法。如果定义了通配符配置，则该 Action 类里将不再包含 execute 方法，而包含了 register 和 login 两个方法，这两个方法与 execute方法除了方法名不同外，其他的完全相同。

此外，表达式也可出现在<action…/>元素的 class 属性中，即 Struts 2 允许将一系列的 Action 类配置成一个<action…/>元素。如：

```
<action name = " * Action" class = "com.bcpl.action.{1}Action">
    …
  </action>
```

上面的<action…/>定义了一系列的 Action，这些 Action 名字应该匹配 * Action 模式，没有指定 method 属性，所以总是使用 exeute 方法来处理用户请求。但 class 属性值使用了表达式，即如果有 URL 为 registerAction. action 的请求，将匹配 * Action 模式，而交给该 Action 处理，其第一个 * 的值为 register，该 register 传入 class 属性值，即该 Action 的处理类为 com. bcpl. action. registerAction。还有另外一种情况是，在 Struts 2 中可以在 class 属性和 method 属性中同时使用表达式，配置如下：

```
<action name = " * _ * "class = "com.bcpl.action.{1}Action" method = "{2}">
```

当一个 Action 为 computer_update. action 的时候将调用 computerAction 的 update 方法来处理用户请求。当用户请求的 URL 同时匹配多个 Action 时，Struts 2 会根据优先匹配的原则处理请求，即按规则先找到 Action，则会由找到的 Action 来处理其请求。

其匹配查找的顺序：如果用户发起的 URL 为 loginAction. action 的请求，在 struts. xml 文件配置了名为 loginAction 的 Action，则一定由该 Action 来处理用户请求；如果 struts. xml 文件没有名为 loginAction 的 Action，则搜索 name 属性值匹配 loginAction 的 Action，如 name 为 * Action 或 * ，* Action 并不会比 * 更优先匹配 loginAction 的请求，查找的顺序为先找到哪一个 Action，则由其来处理用户的请求。通常，在 struts. xml 文件中将 * 的 Action 配置放置所有配置的后面，以免 Struts 2 使用该 Action 来处理所有希望使用模式匹配的请求，后面的应用永远无法进行匹配。

③ Action 指定参数

在 Struts 2 中还可以为 Action 指定一个或多个参数，在 Struts 2 中可以通过<param>标签指定任意多个参数。

```
<action name = "submit"  class = "packageName.MyAction">
<param name = "param1">value1</param>
<param name = "param2">value2</param>
      <result name = "ok">/ok.jsp</result>
</action>
```

当然，在 Action 中读这些参数也非常简单，只需要如获取请求参数一样在 Action 类中定义相应的 setter 方法即可（一般不用定义 getter 方法）。如下面的代码将读取 param1 和 param2 参数的值：

```
package www.bcpl.action;

import com.opensymphony.xwork2.ActionSupport;
public class MyAction extends ActionSupport
{
    private String param1;
    private String param2;

    public String execute() throws Exception
    {
        System.out.println(param1 + param2);
        return "ok";
    }
    public void setParam1(String param1)
    {
        this.param1 = param1;
    }
    public void setParam2(String param2)
    {
        this.param2 = param2;
    }
}
```

当 Struts 2 在调用 execute 之前，param1 和 param2 的值就已经是相应参数的值了，因此，在 execute 方法中可以直接使用 param1 和 param2。

④ result 类型（结果处理）

<result>标签的 type 属性值，默认是“dispatcher”，开发人员可以根据自己的需要指定不同的类型，如 redirect 等。

```
<result name = "save" type = "redirect">/result.jsp</result>
```

result-type 可以在 Struts 2-core-2.3.16.3.jar 包或 Struts 2 源代码中的 struts-default.xml 文件中找到<result-types>标签，所有的 result-type 都在里面定义了。

```
<result-types>
        <result-type name = "chain" class = "com.opensymphony.xwork2.ActionChainResult"/>
        <result-type name = "dispatcher" class = "org.apache.struts2.dispatcher.Servlet-
        DispatcherResult" default = "true"/>
        <result-type name = "freemarker" class = "org.apache.struts2.views.freemarker.
        FreemarkerResult"/>
        <result-type name = "httpheader" class = "org.apache.struts2.dispatcher.HttpHead-
        erResult"/>
        <result-type name = "redirect" class = "org.apache.struts2.dispatcher.ServletRedi-
        rectResult"/>
        <result-type name = "redirectAction" class = "org.apache.struts2.dispatcher.Serv-
        letActionRedirectResult"/>
```

```
        <result-type name = "stream" class = "org.apache.struts2.dispatcher.StreamResult"/>
        <result-type name = "velocity" class = "org.apache.struts2.dispatcher.VelocityResult"/>
        <result-type name = "xslt" class = "org.apache.struts2.views.xslt.XSLTResult"/>
        <result-type name = "plainText" class = "org.apache.struts2.dispatcher.PlainTex-
        tResult" />
        <! -- Deprecated name form scheduled for removal in Struts 2.1.0. The camelCase ver-
        sions are preferred. See ww-1707 -->
        <result-type name = "redirect-action" class = "org.apache.struts2.dispatcher.Serv-
        letActionRedirectResult"/>
        <result-type name = "plaintext" class = "org.apache.struts2.dispatcher.PlainTex-
        tResult" />
</result-types>
```

Action 仅负责处理用户请求，它只是一个控制器，它并不直接提供对浏览者的响应。当 Action 处理完用户请求后，处理结果通过视图资源来实现，而控制器通过控制将对应的视图资源呈现给客户(浏览者)。

Action 处理完用户请求后，将返回一个普通字符串，普通字符串表示一个逻辑视图名。struts. xml 中包含逻辑视图名和物理视图之间的映射关系，一旦收到 Action 返回的某个逻辑视图名，系统就会调用对应的物理视图给浏览者。

Struts 2 的逻辑视图通过映射，返回一个字符串，实现了 Action 类与 Struts 2 框架的分离，为代码复用性提供了条件。

除此之外，Struts 2 还支持多种结果映射，实际资源不仅可以是 JSP 视图资源，也可以是 FreeMaker 或 Velocity 等视图资源，甚至可以将请求转给下一个 Action 处理，形成 Action 的链式处理。

⑤ result 处理结果配置

Struts 2 通过在 struts. xml 文件中使用<result…/>元素来配置结果。根据<result…/>元素所在位置的不同，Struts 2 提供了两种处理方式，分别为局部结果处理和全局结果处理。

- 局部结果：将<result…/>作为<action…/>元素的子元素配置。

局部结果，它的作用范围是对特定的某个 Action 有效。局部结果是通过在<action…/>元素中指定<result…/>元素来配置的，一个<action…/>元素可以有多个<result…/>元素，这表示一个 Action 可以对应多个结果，如上代码所示。

在此过程中，还可以使用<param…/>子元素配置结果，其中<param…/>元素的 name 属性可以为如下两个值。

location：该参数指定了该逻辑视图对应的实际视图资源。

parse：该参数指定是否允许在实际视图名字中使用 OGNL 表达式，该参数值默认为 true。如果设置该参数值为 false，则不允许在实际视图名中使用表达式。通常无须修改该属性值。

- 全局结果：将<result…/>作为<global-result…/>元素的子元素配置。

Struts 2 的<result…/>元素配置，也可放在<global-results…/>元素中配置，当在<global-results…/>元素中配置<result…/>元素时，该<result…/>元素配置了一个全局结果，全局结果的作用范围是对所有的 Action 都有效。

```
<struts>
    <package name = "demo" extends = "struts-default">
        <global-results>
            <result name = "bye">/bye.jsp</result>
        </global-results>

        <action name = "my" class = "packageName.MyAction"/>
        <action name = "your" class = "packageName.YourAction" />
    </package>
</struts>
```

如果<action>中没有相应的<result>,Struts 2 就会使用全局的<result>。

如果一个 Action 里包含了与全局结果里同名的结果,则 Action 里的局部 Action 会覆盖全局 Action。也就是说,当 Action 处理用户请求结束后,会首先在本 Action 里的局部结果里搜索逻辑视图对应的结果,只有在 Action 里的局部结果里找不到逻辑视图对应的结果,才会到全局结果里搜索。

Struts 2 支持使用多种视图技术,如 JSP、Velocity 和 FreeMarker 等。当一个 Action 处理用户请求结束后,仅仅返回一个字符串,这个字符串就是逻辑视图名,但该逻辑视图并未与任何的视图技术及任何的视图资源关联。实际上,结果类型决定了 Action 处理结束后,下一步将执行的类型的动作。

Struts 2 的结果类型要求实现 com. opensymphony. xwork. Result,这个结果是所有 Action 执行结果的通用接口。如果需要自己的结果类型,则应该提供一个实现该接口的类,并且在 struts. xml 文件中配置该结果类型。

Struts 2 的 struts-default. xml 和各个插件中 struts-plugin. xml 文件中提供了一系列的结果类型,Struts 2 支持的结果类型,如表 5-2 所示。

表 5-2 Struts 2 支持的结果类型

结果类型	描 述
Chain	Action 链式处理的结果类型
Chart	用于整合 JFreeChart 的结果类型
dispatcher	用于 JSP 整合的结果类型
Free Marker	用于 FreeMarker 整合的结果类型
httpheader	用于控制特殊的 HTTP 行为的结果类型
Jasper	用于 JasperReports 整合的结果类型
JSF	用于与 JSF 整合的结果类型
redirect	用于直接重定向到其他 URL 的结果类型
redirect-action	用于直接重定向到 Action 的结果类型
Stream	用于向浏览器返回一个 InputStream(一般用于文件下载)
Tiles	用于与 Tiles 整合的结果类型
Velocity	用于与 Velocity 整合的结果类型
XSLT	用于与 XML/XSLT 整合的结果类型
plaintext	用于显示某个页面的源代码的结果类型

其中，dispatcher 结果类型是默认的类型，也就是说如果省略了 type 属性，默认 type 属性为 dispatcher，它主要用于与 JSP 页面整合。下面就 plaintext、redirect 和 redirect-action 3 种结果类型进行介绍。

plaintext 结果类型：主要用于显示实际视图资源的源代码，在 struts. xml 文件中采用如下配置。

```
        <result type = "plaintext">
        <param name = "location">/welcome. jsp</param>
        <! —设置字符集编码-->
        <param name = "charset">gb2312</param>
    </result>
```

使用 plaintext 结果类型，系统将把视图资源的源代码呈现给用户。如果在 welcome. jsp 页面的代码中包含了中文字符，使用 plaintext 结果将会看到乱码。Struts 2 通过<param name="charset">gb2312</param>元素设置使用特定的编码解析页面代码，防止出现乱码。

redirect 结果类型：与 dispatcher 结果类型相对，dispatcher 结果类型是将请求 forward(转发)到指定的 JSP 资源；而 redirect 结果类型，则意味着将请求 redirect(重定向)到指定的视图资源。

dispatcher 结果类型与 redirect 结果类型的差别主要就是转发和重定向的差别，重定向的效果就是重新产生一个请求，因此所有的请求参数、请求属性、Action 实例和 Action 中封装的属性全部丢失。

完整地配置一个 redirect 的 result，可以指定如下两个参数。

- location：该参数指定 Action 处理完用户请求后跳转的地址。
- parse：该参数指定是否允许在 location 参数值中使用表达式，该参数默认值为 true。

redirect-action 结果类型：当需要让一个 Action 处理结束后，直接将请求重定向(是重定向，不是转发)到另一个 Action 时，即可使用该结果类型。配置 redirect-action 结果类型时，可以指定如下两个参数。

- actionName：该参数指定重定向的 Action 名。
- namespace：该参数指定需要重定向的 Action 所在的命名空间。

下面是一个使用 redirect-action 结果类型的配置实例代码。

```
<result type = "redirect-action">
        <! --指定 Action 的命名空间-->
        <param name = "namespace">/ss</param>
<! --指定 Action 的名字-->
        <param name = "actionName">login </param>
</result>
```

⑥ 动态结果

动态结果的意思是在指定实际视图资源时使用了表达式语法，通过这种语法可以允许 Action 处理完用户请求后，动态转入实际的视图资源。

实际上，Struts 2 不仅允许在 class 属性和 name 属性中使用表达式，还可以在<action…/>

元素的<result…/>子元素中使用表达式。下面提供了一个通用 Action，该 Action 可以配置成如下形式：

```
<action name = "*">
    <result>/{1}.jsp</result>
</action>
```

在上面的 Action 定义中，Action 的名字是一个“*”，即它可以匹配任意的 Action，即所有的用户请求都可通过该 Action 来处理。因为没有为该 Action 指定 class 属性，即该 Action 使用 ActionSupport 来作为处理类，而且因为该 ActionSupport 类的 execute 方法返回 success 的字符串，即该 Action 总是直接返回 result 中指定的 JSP 资源，JSP 资源使用了表达式来生成资源名。上面 Action 定义的含义是：如果请求 a.action，则进入 a.jsp；如果请求 b.action，则进入 b.jsp 页面，依此类推。

另外，在配置<result…/>元素时，还允许使用 OGNL 表达式，这种用法允许让请求参数来决定结果。在我们配置<result…/>元素时，不仅可以使用 ${0}表达式形式来指定视图资源，还可以使用 ${属性名}的方式来指定视图资源。在后面这种配置方式下，${属性名}里的属性名就是对应 Action 实例里的属性。如：

```
<result type = "redirect">edit.action? comName = ${myCom.name}</result>
```

对于上面的表达式语法，要求 Action 中必须包含 myCom 属性，并且 myCom 属性必须包含 name 属性，否则 ${myCom.name}表达式值为 null。

(2) <include>标签重用配置文件

在大部分应用里，随着应用规模的增加，系统中 Action 数量也大量增加，导致 struts.xml 配置文件变得非常臃肿。为了避免这种情况，可以将一个 struts.xml 配置文件分解成多个配置文件，然后在 struts.xml 文件中包含其他配置文件。通过这种方式，Struts 2 提供了一种模块化的方式来管理 struts.xml 配置文件，然后在 struts.xml 文件中使用<include>标签引用这些配置文件，这样做可以使结构更清晰，更容易维护配置信息，并且配置文件可以复用。如果在多个 Web 程序中都使用类似或相同的配置文件，那么可以使用<include>标签来引用这些配置文件，这样可以减少工作量。

Struts 2 默认只加载 WEB-INF/class 下的 struts.xml 文件，所以我们就必须通过 struts.xml 文件来包含其他配置文件。

在 struts.xml 文件中包含其他配置文件通过<include…/>元素完成，配置<include…/>元素需要指定一个必需的属性，该属性指定了被包含配置文件的文件名。被包含的 Struts 配置文件，也是标准的 Struts 2 配置文件，一样包含了 DTD 信息和 Struts 2 配置文件的根元素等信息。通常，将 Struts 2 的所有配置文件都放在 Web 应用的 WEB-INF/classes 路径下，strust.xml 文件包含了其他的配置文件，Struts 2 框架自动加载 strust.xml 文件，从而完成加载所有配置信息。

假设有一个配置文件，文件名为 newstruts.xml，代码如下。

```
<? xml version = "1.0" encoding = "UTF-8" ? >
<! DOCTYPE struts PUBLIC
    "-//Apache Software Foundation//DTD Struts Configuration 2.0//EN"
    "http://struts.apache.org/dtds/struts-2.0.dtd">
<struts>
    <package name = "demo" extends = "struts-default" >
        <action name = "myAction"  class = "packageName.XXXAction"/>
    </package>
</struts>
```

则 struts.xml 引用 newstruts.xml 文件的代码如下。

```
<? xml version = "1.0" encoding = "UTF-8" ? >
<! DOCTYPE struts PUBLIC
    "-//Apache Software Foundation//DTD Struts Configuration 2.0//EN"
    "http://struts.apache.org/dtds/struts-2.0.dtd">
<struts>
    <include file = "newstruts.xml"/>
    <package name = "test" extends = "struts-default">
        <action name = "yourAtion"  class = "packageName.XXXXAction" />
    </package>
</struts>
```

注意:用<include>引用的 XML 文件也必须是完成的 Struts 2 的配置。

实际上<include>在引用时是单独解析的 XML 文件,而不是将被引用的文件插入到 struts.xml 文件中。

(3) Bean 配置

Struts 2 框架是一个可扩展性的框架。对于框架的大部分核心组件,Struts 2 并不是直接以硬编码的方式写在代码中的,而是以自己的 IoC(控制反转)容器来管理框架的核心组件。

Struts 2 框架以可配置的方式来管理 Struts 的核心组件,从而允许开发者可以很方便地扩展该框架的核心组件。当开发者需要扩展,或者替换 Struts 2 的核心组件时,只需提供自己的组件实现类,并将该组件实现类部署在 Struts 2 的 IoC 容器中即可。

使用<bean/>标签元素在 struts.xml 文件中定义 Bean,Bean 元素的属性如表 5-3 所示。

表 5-3　Bean 元素属性

属　性	描　述
class	这个属性是个必填属性,它指定了 Bean 实例的实现类
type	这个属性是个可选属性,它指定了 Bean 实例实现的 Struts 2 规范,该规范通常是通过某个接口来实现的,因此该属性的值通常是一个 Struts 2 接口。如果需要将 Bean 实例作为 Struts 2 组件来使用,则应该指定该属性值
name	该属性指定了 Bean 实例的名字,对于有相同 type 类型的多个 Bean,则它们 name 属性不能相同,这个属性也是一个可选属性
scope	该属性指定 Bean 实例的作用域,该属性是个可选属性,属性值只能是 default、singleton、request、session 或 thred 其中之一
static	该属性指定 Bean 是否使用静态方法注入,通常而言,当指定了 type 属性时,该属性值不应该指定为 true
optional	该属性指定该 Bean 是否是一个可选 Bean,该属性是一个可选属性

在 struts. xml 文件中定义 Bean 时，通常有两种方法。

- 创建该 Bean 的实例，将该实例作为 Struts 2 框架的核心组件使用。
- Bean 包含的静态方法需要一个值注入。

在第一种用法下，因为 Bean 实例往往是作为一个核心组件使用的，因此需要告诉 Struts 容器该实例的作用就是该实例实现了哪个接口，这个接口往往定义了该组件所必须遵守的规范。

对于第二种用法，则可以很方便地允许不创建某个类的实例，却可以接受框架常量。在这种用法下，通常需要设置 static="true"。

注意：对于绝大部分 Struts 2 应用而言，并不需要重新定义 Struts 2 框架的核心组件，也就无须在 struts. xml 文件中定义 Bean.

（4）常量配置

在 struts. xml 文件中配置常量是一种指定 Struts 2 属性的方式。其与后续说到的 struts. properties 文件中配置 Struts 2 属性的作用基本相似。通常推荐在 struts. xml 文件中定义 Struts 2 属性，而不是在 struts. properties 文件中定义 Struts 2 属性，这主要是为了保持与 WebWork 的向后兼容性。另外，我们还可以在 web. xml 文件中配置 Struts 2 常量。

通常，Struts 2 框架按如下搜索顺序加载 Struts 2 常量。

- struts-default. xml：该文件保存在 Struts 2-2. 0. 6. jar 文件中。
- struts-plugin. xml：该文件保存在 Struts 2-xxx-2. 0. 6. jar 等 Struts 2 插件 jar 文件中。
- struts. xml：该文件是 Web 应用默认的 Struts 2 配置文件。
- struts. properties：该文件是 Web 应用默认的 Struts 2 配置文件。
- web. xml：该文件是 Web 应用的配置文件。

上面指定了 Struts 2 框架搜索 Struts 2 常量顺序，如果在多个文件中配置了同一个 Struts 2 常量，则后一个文件中配置的常量值会覆盖前面文件中配置的常量值。

在不同文件中配置常量的方式是不一样的，但不管在哪个文件中，配置 Struts 2 常量都需要指定两个属性：常量 name 和常量 value。

其中在 struts. xml 文件中通过元素 constant 来配置常量，配置常量需要指定两个必填的属性。

- name：该属性指定了常量 name。
- value：该属性指定了常量 value。

如果需要指定 Struts 2 的国际化资源文件的 baseName 为 mess，则可以在 strust. xml 文件中使用如下的配置代码：

```
<? xml  version = "1.0"  encoding = "UTF-8"? >
< ! --指定 Struts 2 的 DTD 信息-->
<! DOCTYPEStruts  PUBLIC
      "-//Apache  Software  Foundation//DTD  Struts  Configuration  2.0//EN"
      "http://struts.apache.org/dtds/struts-2.0.dtd">
<struts>
      <! --通过 constant 元素配置 Struts 2 的属性-- >
<constant name = "struts.custom.i18n.resources" value = "properties/Messages"/>
      …
</struts>
```

上面代码片段配置了一个常用属性 struts. custom. i18n. resources，该属性指定了应用所需的国际化资源文件 baseName 为 properties/Messages。

对于 struts. properties 文件而言，该文件的内容就是系列的 key-value 对，其中每个 key 对应一个 Struts 2 常量 name，而每个 value 对应一个 Struts 2 常量 value。关于 struts. properties 配置文件,我们稍后详细介绍。

在 web. xml 文件中配置了 Struts 2 常量,可通过<filter>元素的<int-param>子元素指定,每个<int-param>元素配置了一个 Struts 2 常量。

在实际开发中,不推荐将 Struts 2 常量配置在 web. xml 文件中。因采用这种配置方式来配置常量,需要更多的代码量,而且降低了文件的可读性。通常推荐将 Struts 2 常量集中在 strust. xml 文件中进行集中管理。

(5) 拦截器配置

拦截器允许在 Action 处理之前,或者 Action 处理结束之后,插入开发者自定义的代码。

在很多时候,需要在多个 Action 进行相同的操作,如权限控制,就可以使用拦截器来检查用户是否登录,用户的权限是否足够。通常,使用拦截器可以完成如下操作。

- 进行权限控制(检查浏览者是否是登录用户,并且有足够的访问权限)。
- 跟踪日志(记录每个浏览者所请求的每个 Action)。
- 跟踪系统的性能瓶颈(我们可以通过记录每个 Action 开始处理时间和结束时间,从而取得耗时较长的 Action)。

Struts 2 也允许将多个拦截器组合在一起,形成一个拦截器栈。一个拦截器栈可以包含多个拦截器,多个拦截器组成下一个拦截器栈。对于 Struts 2 系统而言,多个拦截器组成的拦截器栈对外也表现成一个拦截器。

定义拦截器之前,必须先定义组成拦截器栈的多个拦截器。Struts 2 把拦截器栈当成拦截器处理,因此拦截器和拦截器栈都放在<interceptors…/>元素中定义。拦截器的定义代码如下。

```
<interceptors>
            <interceptor name = "log" class = "cc.dynasoft.LogInterceptor" />
<interceptor name = "authority" class = "cc.dynasoft. Authority Interceptor" />
<interceptor name = "timer" class = "cc.dynasoft.TimerInterceptor" />
<interceptor-stack name = "default">
          <interceptor-ref name = "authority" />
<interceptor-ref name = "timer" />
</interceptor>
...
  </interceptors>
```

一旦定义了拦截器和拦截器栈之后,在 Action 中使用拦截器或拦截器栈的方式是相同的。

```
<action name = "login" class = "www.bcpl.LoginAction">
      ...
          <interceptor-ref name = "log" />
</action>
```

2. struts. properties 配置文件

除了 struts. xml 核心文件外,Struts 2 框架还包含一个 struts. properties 文件,该文件放在 Web 应用的 WEB-INF/classes 路径下。它定义了 Struts 2 框架的大量属性,开发者可以通过改变这些属性来满足个性化应用的需求。可以在 struts. properties 中定义的 Struts 2 属性,如表 5-4 所示。

表 5-4　struts. properties 中定义的属性

属　性	作　用
struts. configuration	该属性指定加载 Struts 2 配置文件的配置文件管理器，该属性的默认值是 org. apache. Struts 2. config. DdfaultConfiguration，这是 Struts 2 默认的配置文件管理器。如果需要实现自己的配置管理器，开发者则可以实现一个实现的 configuration 接口的类，该类可以自己加载 Struts 2 配置文件
struts. locale	该属性指定 Web 应用的默认 Locale
struts. i18n. encoding	该属性指定 Web 应用的默认编码集，该属性对于处理中文请求参数非常有用，对于获取中文请求参数值，应该将该属性值设置为 GBK 或者 GB2312 提示：当设置该参数为 GBK 时，相当于调用 httpservletrequest 的 setcharacterencoding 方法
struts. objectFactory	该属性指定 Struts 2 默认的 objectFactory bean，该属性默认值是 Spring
struts. objectFactory. spring. autoWire	该属性指定 Spring 框架的自动装配模式，该属性的默认值是 name，即默认根据 Bean 的 name 属性自动装配
struts. objectFactory. spring. useClassCache	该属性指定整合 Spring 框架时，是否缓存 Bean 实例，该属性只允许使用 true 和 false 两个属性值，它的默认值是 true，通常不建议修改该属性值
struts. objecTypeDeterminer	该属性指定 Struts 2 的类型检测机制，通常支持 tiger 和 notiger 两个属性值
struts. multipart. parser	该属性指定处理 multipart/form-data 的 MIME 类型（文件上传）请求的框架，该属性支持 cos、pell 和 jakarta 等属性值，即分别对应使用 cos 的文件上传框架、pell 上传及 common-fileupload 文件上传框架，该属性的默认值为 jakarta 如果需要使用 cos 或者 pell 的文件上传方式，则应该将对应的 jar 文件复制到 Web 应用中，例如，使用 cos 上传方式，则需要自己下载 cos 框架的 jar 文件，并将该文件放在 WEB-INF/lib 路径下
struts. multipart. savedir	该属性指定上传文件的临时保存路径，该属性的默认值是 javax. servlet. context. tempdir
struts. multipart. maxsize	该属性指定 Struts 2 文件上传中整个请求内容允许的最大字节数
sturts. custom. properties	该属性指定 Struts 2 应用加载用户自定义的属性文件，该自定义属性文件指定的属性不会覆盖 struts. properties 文件中指定的属性。如果需要加载多个自定义属性文件，多个自定义属性文件的文件名以英文逗号“,”隔开
struts. mapper. class	该属性指定将 HTTP 请求映射到指定的 Action 映射器，Struts 2 提供了默认的映射器：org. pache. Struts 2. dispatcher. mapper. defaultactionmapper。默认映射器根据请求的前缀与 Action 的 name 属性完成映射
struts. action. extension	该属性指定需要 Struts 2 处理的请求后缀，该属性的默认值是 action，即所有匹配 *. action 的请求都由 Struts 2 处理。如果用户需要指定多个请求后缀，则多个后缀之间以英文逗号“,”隔开
struts. serve. static	该属性设置是否通过 jar 文件提供静态内容服务，该属性只支持 true 和 false 两个属性值，该属性的默认属性值是 true
struts. serve. static. browsercache	该属性设置浏览器是否缓存静态内容。当应用处于开发阶段时，我们希望每次请求都获得服务器的最新响应，则可设置该属性为 false

续表

属　性	作　用
struts. enable. dynamicmethodinvocation	该属性设置 Struts 2 是否支持动态方法调用，该属性的默认值是 true。如果需要关闭动态方法调用，则可设置该属性为 false
struts. enable. slashesinactinanames	该属性设置 Struts 2 是否允许在 Action 名中使用斜线，该属性的默认值是 false。如果开发者希望允许在 Action 名中使用斜线，则可设置该属性为 true
struts. tag. altsyntax	该属性指定是否允许在 Struts 2 标签中使用表达式语法，因为通常都需要在标签中使用表达式语法，故此属性应该设置为 true，该属性的默认值是 true
struts. devmode	该属性设置 Struts 2 应用是否使用开发模式。如果设置该属性为 true，则可以在应用出错时显示更多、更友好的出错提示。该属性只接受 true 和 flase 两个值，该属性的默认值是 false。通常，应用在开发阶段，将该属性设置为 true，当进入产品发布阶段后，则该属性设置为 false
struts. i18n. reload	该属性设置是否每次 HTTP 请求到达时，系统都重新加载资源文件，该属性默认值是 false。在开发阶段将该属性设置为 true 会更有利于开发，但在产品发布阶段应将该属性设置为 false 提示：开发阶段将该属性设置为 true，则可以在每次请求时都重新加载国际化资源文件，从而可以让开发者看到实时开发效果；产品发布阶段应该将该属性设置为 false，是为了提供响应性能，每次请求都重新加载资源文件会大大降低应用的性能
struts. ui. theme	该属性指定视图标签默认的视图主题，该属性的默认值是 xhtml
struts. ui. templateDir	该属性指定视图主题所需要模板文件的位置，该属性的默认值是 template，即默认加载 template 路径下的模板文件
struts. ui. templateSuffix	该属性指定模板文件的后缀，该属性的默认属性值是 ftl，该属性还允许使用 ftl、vm 或 jsp，分别对应 FreeMarker、Velocity 和 JSP 模板
struts. configuration. xml. reload	该属性设置当 struts. xml 文件改变后，系统是否自动重新加载该文件，该属性的默认值是 false
struts. velocity. configfile	该属性指定 Velocity 框架所需的 velocity. properties 文件的位置，该属性的默认值是velocity. properties
struts. velocity. toolboxlocation	该属性指定 Velocity 框架的 Context 位置，如果该框架有多个 Context，则多个 Context 之间以英文逗号“，”隔开
struts. velocity. toolboxlocation	该属性指定 Velocity 框架的 toolbox 的位置
struts. url. http. port	该属性指定 Web 应用所在的监听端口，该属性通常没有太大的用户，只是当 Struts 2 需要生成 URL 时(如 URL 标签)，该属性才提供 Web 应用的默认端口
struts. url. https. port	该属性类似于 Struts. url. http. port 属性的作用，区别是该属性指定的是 Web 应用的加密服务端口
struts. url. includeparams	该属性指定 Struts 2 生成 URL 时是否包含请求参数，该属性接受 none、get 和 all 3 个属性值，分别对应于不包含、仅包含 GET 类型请求参数和包含全部请求参数
sturts. sustom. i18n. resources	该属性指定 Struts 2 应用所需要的国际化资源文件，如果有多份国际化资源文件，则多个资源文件的文件名以英文逗号“，”隔开

续表

属　性	作　用
struts. dispatcher. parametersWorkaround	对于某些 Java EE 服务器，不支持 HttpServletRequest 调用 getparameterMap()方法，此时可以设置该属性值为 true 来解决该问题，该属性的默认值是 false。对于 WebLogic、Orion和 OC4J 服务器，通常设置该属性为 true
struts. freemarker. manager. classname	该属性指定 Struts 2 使用的 FreeMarker 管理器，该属性的默认值是 org. apache. struts2. views. freemakrker. FreemarkerManager，这是 Struts 2 内建的 FreeMarker 管理器
struts. freemarker. wrapper. altMap	该属性只支持 true 和 false 两个属性值，默认值是 true，通常无须修改该属性值
sturst. xslt. nocache	该属性指定 XSLT Result 是否使用样式表缓存。当应用处于开发阶段时，该属性通常被设置为 true；当应用处于产品使用阶段时，该属性通常被设置为 false
struts. configuration. files	该属性指定 Struts 2 框架默认加载的配置文件，如果需要指定的默认加载多个配置文件，则多个配置文件的文件名之间以英文逗号“，”隔开。该属性的默认值为 Struts-default. xml，struts-plugin. xml，struts. xml，看到该属性值，读者应该明白为什么 Struts 2 框架默认加载 struts. xml 文件了

在有些时候，开发者并不喜欢使用额外的 struts. properties 文件。前面提到，Struts 2 允许在 struts. xml 文件中管理 Struts 2 属性，在 struts. xml 文件中管理 Struts 2 属性，在 struts. xml 文件中通过配置 constant 元素，一样可以配置这些属性。前面已经提到，我们建议尽量在 strust. xml 文件中配置 Struts 2 常量。

5.3　创建 Controller 组件

Struts 的核心是 Controller 组件，它是连接 Model 和 View 组件的桥梁，也是理解 Struts 架构的关键。Struts 2 的控制器由 FilterDispatcher 和业务控制器 Action 两个部分组成。

5.3.1　FilterDispatcher

任何 MVC 框架都需要与 Web 应用整合，离不开 web. xml 文件，只有配置在 web. xml 文件中 Filter/Servlet 才会被应用加载。对于 Struts 2 框架而言，需要加载 FilterDispatcher，因为 Struts 2 将核心控制器设计成 Filter，而不是一个 Servlet。故为了让 Web 应用加载 FilterDispatcher，需要在 web. xml 文件中配置 FilterDispatcher。

配置 FilterDispatcher 的代码如下：

```
<!--配置 Struts 2 框架的核心 Filter-->
<filter>
      <!--配置 Struts 2 核心 Filter 的名字-->
<filter-name>struts</ filter-name>
      <!--配置 Struts 2 核心 Filter 的实现类-->
      <filter-class>org.apache. struts2.dispatcher.Filter Dispatcher </filter-class>
      <init-param>
              <!--配置 Struts 2 框架默认加载的 Action 包结构-->
```

```
            <param-name>actionpackages</param-name>
            <param-value> org.apache. struts2.showcase.person</param-value>
    </init-param>
        <!--配置 Struts 2 框架的配置提供者类-->
        <init-param>
            < param- name>configProviders< /param- name>
            <param-value>lee.MyConfigurationProvider</param-value>
        </init-param>
</filter>
```

正如上面看到的，当配置 Struts 2 的 FilterDispatcher 类时，可以指定一系列的初始化参数，为该 Filter 配置初始化参数，其中有 3 个初始化参数有特殊意义。

① Config：该参数的值是一个英文逗号(,)隔开的字符串，每个字符串都有一个 XML 配置文件的位置。Struts 2 框架将自动加载该属性指定的系列配置文件。

② Actionpackages：该参数的值也是一个以英文逗号“,”隔开的字符串，每个字符串都是一个包空间，Struts 2 框架将扫描这些包空间下的 Action 类。

③ Configproviders：如果用户需要实现自己的 Configurationprovider 类，用户可以提供一个或多个实现了 Configurationprovider 接口的类，然后将这些类的类名设置成该属性的值，多个类名之间以英文逗号(,)隔开。

除此之外，还可在此处配置 Struts 2 常量，每个<init-param>元素配置一个 Struts 2 常量，其中，<param-name>子元素指定了常量 name，而<param-value>子元素指定了常量 value。

在 web.xml 文件中配置了该 Filter，还需要配置该 Filter 拦截的 URL。通常让该 Filter 拦截所有的用户请求，因此使用通配符来配置该 Filter 拦截的 URL。

该 Filter 拦截 URL 的配置代码如下：

```
<! --配置 Filter 拦截的 URL-->
<filter-mapping>
<! --配置 Struts2 的核心 FilterDispatcher 拦截所有用户请求-->
<filter-name>struts</ filter-name>
<url-pattern>/ * </url- pattern>
</filter-mapping>
```

配置了 Struts 2 的核心 FilterDispatcher 后，即完成了 Struts 2 在 web.xml 文件中的配置。

5.3.2 Action 的开发

对于 Struts 2 应用而言，Action 是应用系统的核心，也称 Action 为业务控制器。开发者需要提供大量的 Action 类，并在 strust.xml 文件配置 Action。

1. 实现 Action 类

Struts 2 采用了低侵入式的设计，Struts 2 的 Action 类是一个普通的 POJO(通常应该包含一个无参数的 execute 方法)，从而带来很好的代码复用性。对于用户登录模块，LoginAction 类的代码如下。

```
import java.util.ArrayList;
import java.util.List;
import com.opensymphony.xwork2.ActionContext;
@SuppressWarnings("serial")
public class LoginAction extends BaseAction {
    private String username;
    private String password;

    public String getPassword() {
        return password;
    }
    public void setPassword(String password) {
        this.password = password;
    }
    public String getUsername() {
        return username;
    }
    public void setUsername(String username) {
        this.username = username;
    }

    public String execute() throws Exception {
    UserDAO d = new UserDAO();
    User user = d.validate(username, password);
    if(user! = null){
                ActionContextac = ActionContext.getContext();
                Map session = ac.getSession();
                  session.put("user", user);
                 session.put("isLogin", "true");
                 return this.SUCCESS;
               }else{
                   return this.INPUT;
        }
```

上面的 Action 类只是一个普通类，这个 Java 类提供了两个属性：username 和 password，这两个属性分别对应两个 HTTP 请求参数。

Action 类里的属性，不仅可用于封装请求参数，还可用于封装处理结果，如在前面的 Action 代码中看到的，如果希望将服务器提示的登录成功或失败在下一个页面输出，那么我们可以在 Action 类中增加一个属性，并为该属性提供对应的 setter 和 getter 方法。

系统不会严格区分 Action 里哪个属性是用于封装请求参数的，哪个属性是用于封装处理结果的。对系统而言，封装请求参数的属性和封装处理结果的属性是完全平等的。如果用户的 HTTP 请求里包含了名为指定属性的请求参数，系统会调用 Action 类的相应的 set 方法，通过这种方式，名为指定属性的请求参数就可以传给 Action 实例，如果 Action 类里没有包含对应的方法，则名为指定属性的请求参数无法传入该属性。

同样，在 JSP 页面中输出 Action 属性时，它也不会区分该属性是用于封装请求参数的，还是用于封装处理结果的。因此,使用 Struts 2 的标签既可以输出 Action 的处理结果,也可以输出 HTTP 请求参数值。

为了让用户开发的 Action 类更规范,Struts 2 提供了一个 Action 接口,这个接口定义了 Struts 2 的 Action 处理类应该实现的规范。它的里面只定义了一个 execute 方法,该接口的规范规定了 Action 类应该包含这样一个方法,该方法返回了一个字符串。除此之外,该接口还定义了 5 个字符串常量,分别是 error、input、login 、none 和 success,它们的作用是统一 execute 方法的返回值。例如,当 Action 类处理用户请求成功后,有人喜欢返回 welcome 字符串,有人喜欢返回 success 字符串,这样不利于项目的统一管理。Struts 2 的 Action 定义上面的 5 个字符串分别代表了统一的特定含义。

另外,Struts 2 还提供了 Action 类的一个实现类 ActionSupport,该 Action 是一个默认的 Action 类,该类里已经提供了许多默认方法,这些默认方法包括获取国际化信息的方法、数据校验的方法和默认处理用户请求的方法等。实际上,ActionSupport 类是 Struts 2 默认的 Action 处理类,如果让开发者的 Action 类继承该 Action 类,则会大大简化 Action 的开发。

2. Action 访问 Servlet API

Struts 2 的 Action 并未直接与任何 Servlet API 耦合,这是 Struts 2 的一个改进之处,因为这样的 Action 类具有更好的重用性,并且能更轻松地测试该 Action。对于 Web 应用的控制器而言,不访问 Servlet API 几乎是不可能的,如获得 HTTP Request 参数、跟踪 HTTP Session 状态等。

Struts 2 提供了一个 ActionContext 类,Struts 2 的 Action 可以通过该类来访问 Servlet API,包括 HttpServletRequest、HttpSession 和 ServletContext 这 3 个类,它们分别代表 JSP 内置对象中的 request、session 和 appliaction,具体如表 5-5 所示。

表 5-5 ActionContext 类中常用方法

方 法	作 用
Object get(Object key)	该方法类似于调用 HttpServletRequest 的 getAttribute(stringname)方法
Map getApplication	返回一个 Map 对象,该对象模拟了该应用的 ServletContext 实例
Static ActionContext getContext	静态方法,获取系统的 ActionContext 实例
Map getParameters	获取所有的请求参数,类似于调用 HttpServletRequest 对象的 getParameterMap 方法
Map getSession	返回一个对象,该 Map 对象模拟了 HttpSession 实例
Void setApplication(Map application)	直接传入一个 Map 实例,将该 Map 实例里的 key-value 对转换成 session 的属性名和属性值
Void setSession(Map session)	直接传入一个 Map 实例,将该 Map 实例里的 key-value 对转换成 session 的属性名和属性值

虽然 Struts 2 提供了 ActionContext 来访问 Servlet API,但这种访问毕竟不能直接获取 Servlet API 实例,为了在 Action 中直接访问 Servlet API,可使用如表 5-6 所示的系列接口。

表 5-6 Action 中直接访问 Servlet API 的接口

接 口	作 用
ServletContextAware	实现该接口的 Action 可以直接访问应用的 ServletContext 实例
ServletRequestAware	实现该接口的 Action 可以直接访问用户请求的 HttpServletRequest 实例
ServletResponseAware	实现该接口的 Action 可以直接访问服务器响应的 HttpServletResponse 实例

另外，为了直接访问 Servlet API，Struts 2 提供了一个 ServletActionContext 类。借助于这个类的帮助，开发者也能够在 Action 中直接访问 Servlet API，却可以避免 Action 类需要实现上面的接口。这个类包含了如表 5-7 所示的静态方法。

表 5-7　ServletActionContext 类静态方法

方　法	作　用
Static PageContext getPageContext()	取得 Web 应用的 PageContext 对象
Static HttpServletRequest getRequest()	取得 Web 应用的 HttpServletRequest 对象
Static HttpServletResponse getResponse()	取得 Web 应用的 HttpServletResponse 对象
Static ServletContext getServletContext()	取得 Web 应用的 ServletContext 对象

5.3.3　Model 驱动

Struts 2 提供模型驱动模式，这种模式也通过专门的 JavaBean 来封装请求参数。

Struts 2 的 Action 对象封装了更多的信息，它不仅可以封装用户的请求参数，还可以封装 Action 的处理结果。Struts 2 的 Action 既用于封装来回请求的参数，也用于保护控制逻辑。相对而言，这种模式确实不太清晰，出于结构清晰的考虑，采用单独的 Model 实例来封装请求参数和处理结果，这就是所谓的模型驱动，也就是使用单独的 JavaBean 实例来贯穿整个 MVC 流程。与之对应的属性驱动模式，则使用属性 Property 作为贯穿 MVC 流程的信息携带者。简单地说，模型驱动使用单独的 Value Object(值对象)来封装请求参数和处理结果，除了这个 JavaBean 之外，还必须提供一个包含处理逻辑的 Action 类，而属性驱动则使用 Action 实例来封装请求参数和处理结果。

对于采用模型驱动的 Action 而言，该 Action 必须实现 ModelDriven 接口，实现该接口则必须实现 getModel 方法，该方法用于把 Action 和与之对应的 Model 实例关联起来。

配置模型驱动的 Action 与配置属性驱动的 Action 没有任何区别，Struts 2 不要求配置模型对象，即不需要配置 UserBean 实例。模型驱动和属性驱动各有利弊，模型驱动结构清晰，但编程繁琐(需要额外提供一个 JavaBean 来作为模型)；属性驱动则编程简洁，但结构不够清晰。

5.4　Model 组件创建

Struts 中的 Model 指的是业务逻辑组件，它可以使用 JavaBean 实现。通常说来，Model 组件的开发者侧重于创建支持所有功能需求的 JavaBeans 类，通常可以分成以下几种类型。

1. JavaBeans

在一个基于 Web 的应用程序中，JavaBeans 可以被保存在一些不同"属性"的集合中。每一个集合都有集合生存期和所保存的 Beans 可见度的不同的规则。总的说来，定义生存期和可见度的这些规则被称为这些 Beans 的范围。JSP 规范中使用以下术语定义可选的范围(括号中定义的是 Servlet API 中的等价物)。

- page：在一个单独的 JSP 页面中可见的 Beans，生存期限于当前请求。(service()方法中的局部变量)。

- request：在一个单独的JSP页面中可见的Beans，也包括所有包含于这个页面或从这个页面重定向到的页面或Servlet。(Request属性)。
- session：参与一个特定的用户session的所有的JSP和Servlet都可见的Beans，跨越一个或多个请求。(Session属性)。
- application：一个Web应用程序的所有JSP页面和Servlet都可见的Beans。(Servlet Context属性)。

同一个Web应用程序的JSP页面和Servlets共享同样一组Bean集合是很重要的。如一个Bean作为一个request属性保存在一个Servlet中，代码如下：

```
MyCart mycart = new MyCart(…);
request.setAttribute("cart", mycart);
```

将Servlet重定向到的一个JSP页面所使用的标准行为标记，代码如下：

```
<jsp:useBean id = "cart"; scope = "request" class = "com.mycompany.MyApp.MyCart"/>
```

2. 系统状态 Beans

系统的实际状态通常表示为一组一个或多个的JavaBeans类，其属性定义当前状态。例如，一个购物车系统包括一个表示购物车的Bean，这个Bean为每个单独的购物者维护，这个Bean中包括一组购物者当前选择购买的商品。同时，系统也包括保存用户信息(包括他们的信用卡和送货地址)，可提供商品的目录和它们当前库存水平的不同的Beans。

对于小规模的系统，或者对于不需要长时间保存的状态信息，一组系统状态Beans可以包含所有系统曾经经历的特定细节的信息。或者，系统状态Beans表示永久保存在一些外部数据库中的信息(如CustomerBean对象对应于表Customers中的特定的一行)，在需要时从服务器的内存中创建或清除。

3. 商业逻辑 Beans

通常应用程序中的功能逻辑封装成为此目的设计的JavaBeans的方法调用，这些方法可以是用于系统状态Beans的相同的类的一部分，或者可以是在专门执行商业逻辑的独立的类中，在后一种情况下，通常需要将系统状态Beans传递给这些方法作为参数处理。

为了代码最大的可重用性，商业逻辑Beans应该被设计和实现为它们不知道自己被执行于Web应用环境中。如果你发现在你的Bean中必须import一个javax.servlet.*类，你就把这个商业逻辑捆绑在Web应用环境中了。考虑重新组织事物，使Action类把所有HTTP请求处理为对商业逻辑Beans属性set方法调用的信息，然后可以发出一个对execute()的调用，这样的一个商业逻辑类可以被重用在Web应用程序以外的环境中，取决于应用程序的复杂度和范围，商业逻辑Beans可以是与作为参数传递的系统状态Beans交互作用的普通的JavaBeans，或者使用JDBC调用访问数据库的普通的JavaBeans，而对于较大的应用程序，这些Beans经常是有状态或无状态的EJBs。

5.5 View 组件创建

创建应用程序中的View组件，主要是使用JSP技术建立，当然Struts 2也支持其他View技术。在JSP中，我们会大量使用标签。Struts 1.x将标志库按功能分成HTML、Tiles、

Logic和Bean等几部分，而Struts 2.0的标志库(Tag Library)严格上来说没有分类，所有标志都在URI的“/struts-tags”命名空间下，可以从功能上将其分为两大类：一般标志和UI标志。

如果Web应用使用了Servlet 2.3以前的规范，Web应用不会自动加载标签文件，因此必须在web.xml文件中配置加载Struts 2标签库。

配置加载Struts 2标签库的配置代码如下：

```
<!--手动配置 Struts 2 的标签库 -->
<taglib>
     <!--配置 Struts 2 标签库的 URI-->
     <taglib-uri>/s</taglib-uri>
<!--指定 Struts 2 标签库定义文件的位置-->
<taglib-location>/WEB-INF/struts-tags.tld</taglib-location>
     </taglib>
```

在上面配置代码中，指定了Struts 2标签库配置文件物理位置为/WEB-INF/strutstags.tld，因此必须手动复制Struts 2的标签库定义文件，将该文件放置在Web应用的WEB-INF路径下。

如果Web应用使用Servlet 2.4以上的规范，则无须在web.xml文件中配置标签库定义，因为Servlet 2.4规范会自动加载标签库定义文件。加载struts-tag.tld标签库定义文件时，该文件的开始部分包含如下代码片段。

```
<taglib>
      <!--定义标签库的版本 -->
      <tlib--version>2.2.3</tlib-version>
      <!-- 定义标签库所需的 JSP 版 -->
      <jsp-version>1.2</jsp-version>
      <short-name>s</short-name>
      <!--定义 Struts 2 标签库的 URI -->
      <uri>/sturts-tags</uri>
      …
</taglib>
```

因为该文件中已经定义了该标签库的URI：struts-tags。这就避免了在web.xml文件中重新定义Struts 2标签库文件的URI。要在JSP中使用Struts 2.0标志，先要指明标志的引入，通过在JSP代码的顶部加入以下代码。

```
<%@taglib prefix="s" uri="/struts-tags" %>
```

5.6 小　　结

随着Web开发框架技术的快速发展，本章所述的Struts技术，也经历了从Struts 1到Struts 2的发展历程，由于目前关于Struts 2的应用已经成为主流，本书就没有对Struts 1进行详细介绍，关于它们之间在Action的实现、线程模型、封装请求参数、类型转换、表达式语言和Action执行控制等方面的差异，并没有进行比较和分析，而是直接对Struts 2的模型映射、框架及工作流程、简单配置和常用组件进行介绍，需要注意的是开发框架所需的支持包，在开发时，必须全部加载。关于其标签和高级应用，会在后续的部分进行介绍。

第 6 章　Struts 2 标签

6.1　Struts 2 标签简介

为了简化框架开发的难度和规范展示，Struts 2 提供了自己的标签库，供开发者进行 View 层的开发，开发者只需引用标准的标签库中的标签，并进行设置，便可以实现完成各种常用的页面的展示效果，同时，在框架中，实现了展示层 View 和控制层 Action 的有效结合，使用 Action 可以有效地提取 View 层的用户提交的数据和把数据快速地返回并显示给用户。

Struts 2 标签库提供了主题、模板支持，极大地简化了视图页面的编写，而且 Struts 2 的主题、模板都提供了很好的扩展性，实现了更好的代码复用。Struts 2 允许在页面中使用自定义组件，这完全能满足项目中页面显示复杂、多变的需求。

Struts 2 的标签库的优势在于，Struts 2 标签库的标签不依赖于任何表现层技术，也就是说 Strtus2 提供了大部分标签，可以在各种表现技术中使用，包括最常用的 JSP 页面，此外，它也可以在 Velocity 和 FreeMarker 等模板技术中使用。

1. Struts 2 分类

（1）用户界面（User Interface，UI）标签，主要用于生成 HTML 元素标签，UI 标签又可分为表单标签和非表单标签。

（2）非 UI 标签，主要用于数据访问、逻辑控制等的标签。非 UI 标签可分为流程控制标签（包括用于实现分支、循环等流程控制的标签）和数据访问标签（主要包括用户输出 ValueStack 中的值，完成国际化等功能）。

（3）Ajax 标签。

2. Struts 2 标签使用

（1）使用前，需要在 JSP 页面头部引入 Struts 标签库。

```
<%@ taglib uri="/struts-tags" prefix="s" %>
```

（2）在 web.xml 中声明要使用的标签，代码如下所示。

```
<?xml version="1.0" encoding="UTF-8"?>
<web-app version="3.0"
    xmlns="http://java.sun.com/xml/ns/javaee"
    xmlns:xsi="http://www.w3.org/2001/XMLSchema-instance"
    xsi:schemaLocation="http://java.sun.com/xml/ns/javaee
    http://java.sun.com/xml/ns/javaee/web-app_3_0.xsd">
    <display-name></display-name>
    <filter>
```

```
                <filter-name>struts2</filter-name>
<filter-class>org.apache.struts2.dispatcher.ng.filter.StrutsPrepareAndExecuteFilter</filter-class>
        </filter>

        <filter-mapping>
                <filter-name>struts2</filter-name>
                <url-pattern>/*</url-pattern>
        </filter-mapping>

    <welcome-file-list>
        <welcome-file>index.jsp</welcome-file>
    </welcome-file-list>
</web-app>
```

(3) 在 JSP 页面中的编写，代码如下。

```
<%@ page language="java" %>
 <%@ taglib prefix="s" uri="/struts-tags" %>
 <html>
     <body>
         <s:form action="hello">
                 name:<s:textfield  name="name"/>
                     <s:submit/>
         </s:form>
     </body>
 </html>
```

6.2 一般标签(非 UI 标签)

6.2.1 控制标签

1. if、elseif 和 else 标签

用于执行基本的条件流转，条件判断标签有 <s:if>、<s:elseif>和<s:else>，<s:if>和<s:else>拥有 test 属性，返回值为 boolean 类型，返回值为 true 或 false，其表达式的值用来决定标签里的内容是否显示，具体应用代码如下。

```
<s:if test="#request.username=='wjj'">欢迎 wjj</s:if>
<s:if test="%{false}">
    <div>hello,not show</div>
</s:if>
<s:elseif test="%{true}">
        <div>hello,show here</div>
</s:elseif>
```

```
<s:else>
    <div>hello,not show here</div>
</s:else>
```

2. iterator 标签

迭代标签用于遍历集合(java. util. Collection)或者枚举值(java. util. Iterator)类型的对象,value 属性表示集合或枚举对象,status 属性表示当前循环的对象,在循环体内部可以引用该对象的属性。

```
<s:iterator value="userList" status="user">
姓名:<s:property value="user.userName"/>
年龄:<s:property value="user.age"/>
</s:iterator>
```

```
<%@ page contentType="text/html; charset=UTF-8" %>
<%@ page import="java.util.List" %>
<%@ page import="java.util.ArrayList" %>
<%@ taglib prefix="s" uri="/struts-tags" %>
<!DOCTYPE HTML PUBLIC"-//W3C//DTD HTML 4.01 Transitional//EN">
<%
    List list = new ArrayList();
    list.add("zhang");
    list.add("li");
    list.add("xiaoming");
    list.add("test");
    request.setAttribute("names", list);
%>
<html>
    <head>
        <title>Iterator Test</title>
    </head>
    <body>
        <h3>Names: </h3>
        <ol>
            <s:iterator value="#request.names" status="stuts">
                <s:if test="#stuts.odd == true">
                    <li>Hi <s:property /></li>
                </s:if>
                <s:else>
                    <li style="background-color:gray"><s:property /></li>
                </s:else>
            </s:iterator>
        </ol>
    </body>
</html>
```

3. sort 标签

排序标签用于对一组枚举值进行排序,属性 comparator 指向一个继承自 java. util.

comparator的比较器，该比较器可以是对应 Action 页面中的一个比较器变量，source 指定要排序的列表对象，接受集合和比较器作为参数，对集合进行排序。如果声明了 var 属性，排序后的集合会使用 var 作为键名放在 PageContext 中，其常用属性如下。

- comparator：排序使用的比较器。
- source：用来排序的集合。
- var：用来存放排序后集合的键名。

```
<s:sortvar = " mysort"comparator = "myComparator" source = "myList"/>
<%
    Iterator sortedIterator = (Iterator) pageContext.getAttribute("mysort");
    for (Iterator i = sortedIterator; i.hasNext(); ) {
      // do something with each of the sorted elements
    }
%>
</s:sort>
```

6.2.2　数据输出标签

1. append 标签

组合标签用于将多个枚举值对象进行叠加，形成一个新的枚举值列表，例如，将 3 个列表对象进行了组合，形成新的列表对象，代码如下所示。

```
<s:append var = "newIteratorList">
<s:param value = " %{myList1}" />
<s:param value = " %{myList2}" />
<s:param value = " %{myList3}" />
</s:append>
```

输出代码如下所示。

```
<s:iterator value = " %{#newIteratorList}">
<s:property />
</s:iterator>
```

2. generator 标签

分割标签用于将一个字符串进行分隔，产生一个枚举值列表。下面的代码将分隔为 3 个字符串，然后循环输出。

```
<s:generator val = " %{'aaa,bbb,ccc' }">
<s:iterator>
<s:property />
</s:iterator>
</s:generator>
```

3. merge 标签

合并标签用于将多个枚举值按照数组的索引位置进行合并，其代码如下所示。

```
<s:merge var = "newMergeList">
<s:param value = "%{mylist1}">--设 mylist1 列表中有 x,y,z 三个元素
<s:param value = "%{mylist2}">--设 mylist2 列表中有 1,2,3 三个元素
</s:merge>
```

合并后新列表的元素及其顺序为:x,1,y,2,z,3。使用迭代标签进行输出,如下所示。

```
<s:iterator value = "%{#newMergeList}">
<s:property />
</s:iterator>
```

4. set 和 subset 标签

set 标签赋予变量一个特定范围内的值。当给一个变量赋一个复杂的表达式,每次访问该变量而不是复杂的表达式时用到。其在两种情况下非常有用:复杂的表达式很耗时(性能提升)或者很难理解。

subset 子集标签用于取得一个枚举列表的子集,source 用于指定检索的列表对象,start 用于指定起始检索的索引位置,count 用于指定检索的结果数量,decider 属性必须是一个 org. apache. Struts 2. util. SubsetIteratorFilter. Decider 类的子类实例,用以指定检索的条件,如在 Action 中提供了以下方法用来取得一个 Decider 对象。

```
public Decider getDecider()
{
     return new Decider() {
public boolean decide(Object element) throws Exception
   {
     int i = ((Integer)element).intValue();
     return (((i % 2) == 0)? true:false);
       }
      }
}
```

然后引用该对象筛选子集:

```
<s:subset source = "myList" decider = "decider">
</s:subset>
```

或者不使用 Decider 对象:

```
<s:subset source = "myList" count = "10" start = "2">
</s:subset>
```

```
<s:set name = "personName" value = "person.name"/>
Hello, <s:property value = "#personName"/>
```

5. date 标签

日期标签,根据特定日期格式(如“dd/MM/yyyy hh:mm”)对日期对象进行多种形式的格式化。

```
<s:date name = "person.birthday" format = "dd/MM/yyyy" />
<s:date name = "person.birthday" format = " %{getText('some.i18n.key')}" />
<s:date name = "person.birthday" nice = "true" />
<s:date name = "person.birthday" />
```

6. i18n 国际化标签

i18n 标签,加载资源包到值堆栈。它可以允许 text 标志访问任何资源包的信息,而不只当前 Action 相关联的资源包,其属性 value 表示资源包的路径,id 用来标识元素。

```
<%@ page contentType = "text/html; charset = UTF-8" %>
<%@ taglib prefix = "s" uri = "/struts-tags" %>

<! DOCTYPE HTML PUBLIC "-//W3C//DTD HTML 4.01 Transitional//EN">
<html>
  <head>
    <title>国际化</title>
  </head>
  <body>
    <h3>
      <s:i18n name = "myApplicationMessages">
        <s:text name = "hello" />
        </s:i18n>
      </h3>
    </body>
</html>
```

7. include 标签

include 标签用于包含一个 Servlet 的输出(Servlet 或 JSP 的页面)。

```
<%@ page contentType = "text/html; charset = UTF-8" %>
<%@ taglib prefix = "s" uri = "/struts-tags" %>

<! DOCTYPE HTML PUBLIC "-//W3C//DTD HTML 4.01 Transitional//EN">
<html>
    <head>
        <title>Include Test</title>
    </head>
    <body>
        <h3>include page</h3>
        <s:include value = "/test.jsp">
            <s:param name = "name">wjj</s:param>
        </s:include>
        <h3>i18n</h3>
        <s:include value = "/i18n.jsp" />
    </body>
</html>
```

param 参数:为其他标签提供参数,如 include 标签和 bean 标签。参数的 name 属性是可选的,如果提供,会调用 Component 的方法 addParameter(String, Object),如果不提供,则外层嵌套标签必须实现 UnnamedParametric 接口(如 TextTag)。

```
<pre>
<ui:component>
  <ui:param name = "key1"      value = "[0]"/>
  <ui:param name = "key2"    value = "[1]"/>
  <ui:param name = "key3" value = "[2]"/>
</ui:component>
</pre>
```

8. property 标签

property 标签,可以获取得到"value"的属性,用于输出指定值,如果 value 没提供,默认为堆栈顶端的元素。

```
<s:push value = "myBean">
    <! -- Example 1: -->
    <s:property value = "myBeanProperty" />
    <s:property value = "myBeanProperty" default = "a default value" />
</s:push>
```

当 Action 返回到指定页面时,可以通过该标签显示 Action 中的属性信息(注 myBeanProperty 必须在 Action 中存在,并有 get 方法)。

9. text 标签

text 标签是支持国际化信息的标签。国际化信息必须放在一个和当前 Action 同名的 resource bundle 中,如果没有找到相应 message,tag body 将被当作默认 message,如果没有 tag body,message 的 name 会被当作默认 message。

```
<s:i18n name = "struts.action.test.i18n.app">
    <s:text name = "main.title"/>
</s:i18n>
<s:text name = "main.title" />
<s:text name = "i18n.label.greetings">
    <s:param >Hello</s:param>
</s:text>
```

10. url 链接标签

该标签用于创建 URL,可以通过 param 标签提供 request 参数。当 includeParams 的值为 all 或者 get 时,param 标签中定义的参数将有优先权,也就是说其会覆盖其他同名参数的值。

```
<%@ page contentType = "text/html; charset = UTF-8" %>
<%@ taglib prefix = "s" uri = "/struts-tags" %>
<! DOCTYPE HTML PUBLIC "-//W3C//DTD HTML 4.01 Transitional//EN">
<html>
  <head>
    <title>URL</title>
  </head>
  <body>
    <h3>URL</h3>
```

```
        <a href='<s:url value="/i18n.jsp" />'>i18n</a><br />
        <s:url id="url" value="/test.jsp">
          <s:param name="name">wjj</s:param>
        </s:url>
        <s:a href="%{url}">zhang </s:a>
    </body>
</html>
```

11. textfield 标签

textfield 标签用于从页面往 Action 中的对象内传值。

```
<s:textfield name="user.userName" />
```

该标签可以在页面中向 Action 中的实体对象内直接传值(注:在 Action 中要存在 user 对象,并有 set/get 方法)。

6.3　UI 标签

UI 标签分为表单标签和非表单标签两部分。表单 UI 部分是对 HTML 表单元素的包装,包括 form、checkbox、radio、label、file、hidden、select、textfield、textarea 和 submit 等,这里就不再赘述了。在 Struts 2.0 增加的控件标签有 doubleselect 和 optiontransferselect 等。非表单标签部分常用的标签有 actionerror、actionmessage 和 fielderror 等。

6.3.1　表单标签

HTML 表单标签通常都拥有以下 3 个设置样式的属性。

- templateDir:执行模板路径。
- theme:指定主题名称,可选值包括 simple、XHTML 和 Ajax 等。
- template:指定模板名称。

使用 HTML 表单标签会生成大量格式化的 HTML 代码,这些代码是由 Struts 2 的模板生成的,这样的好处是可以让我们的 JSP 代码十分简单,只需要配置使用不同的主题模板,就可以显示不同的页面样式。

Struts 2 默认提供了 5 种主题,如下所示。

- simple 主题:最简单的主题。
- XHTML 主题:默认主题,使用常用的 HTML 技巧。
- CSS XHTML 主题:使用 CSS 实现的 XHTML 主题。
- archive 主题:使用文档生成主题。
- Ajax 主题:基于 Ajax 实现的主题。

通过在 struts. properties 文件中改变 struts. ui. theme、struts. ui. templateDir 和 struts. ui. templateSuffix 3 个标签来自由切实可行换主题。

6.3.2　非表单标签

非表单标签主要包含 Struts 新增加的标签和错误及消息输出标签,用来输出各种错误和消息。

1. 错误标签

错误标签<s:actionerror />用以输出 Action 错误信息，根据特定布局风格提供由 Action 产生的错误。

```
        <s:url id = "url" value = "/test.jsp">
        <s:param name = "name">wjj</s:param>
      </s:url>
       <s:a href = " % {url}">zhang </s:a>
    </body>
</html>
```

2. 消息标签

消息标签<s:atcionmessage />用以输出 Action 普通消息。

```
<s:actionmessage />
   <s:form … >
      …
   </s:form>
```

3. 字段错误标签

字段错误标签<s:fielderror>用以输出 Action 校验中某一个字段的错误信息或所有字段的错误信息。

```
<s:fielderror />--输出所有字段的错误信息
<s:fielderror>
<s:param>username</s:param>--输出字段 username 的错误信息
<s:param>password</s:param>--输出字段 password 的错误信息
<s:fielderror>
```

4. doubleselect 标签

提供两套 HTML 列表框(select)元素，其中第二套元素显示的值会根据第一套元素被选中的值而改变，代码如下。

```
<s:doubleselect label = "doubleselect test1"
name = "menu"
list = "{'fruit','other'}"
doubleName = "dishes"
doubleList = "top == 'fruit' ? {'apple', 'orange'} : {'monkey', 'chicken'}" />
<s:doubleselect label = "doubleselect test2"
name = "menu"
list = "#{'fruit':'Nice Fruits', 'other':'Other Dishes'}"
doubleName = "dishes" doubleList = "top == 'fruit' ? {'apple', 'orange'} : {'monkey', 'chicken'}" />
```

5. optiontransferselect 标签

创建一个可传递选项的列表框组件，该组件是基于在两个<select >标签中间添加按钮，并允许两个列表框之间的选项可以相互移动到对方的选择框中。在包含了表单提交动作基础上可以自动选择所有的列表框选项，具体代码如下。

```
<--minimum configuration-->
<s:optiontransferselect
        label = "Favourite Cartoons Characters"
        name = "leftSideCartoonCharacters"
        list = "{'abc','te','John'}"
        doubleName = "rightSideCartoonCharacters"
        doubleList = "{'number','Mouse','man'}"
/>

<-- possible configuration -->
<s:optiontransferselect
      label = "Favourite Cartoons Characters"
      name = "leftSideCartoonCharacters"
      leftTitle = "Left Title"
      rightTitle = "Right Title"
      list = "{'abc','te','John'}"
      multiple = "true"
      headerKey = "headerKey"
      headerValue = "--- Please Select ---"
      emptyOption = "true"
      doubleList = "{'number','Mouse','man'}"
      doubleName = "rightSideCartoonCharacters"
      doubleHeaderKey = "doubleHeaderKey"
      doubleHeaderValue = "--- Please Select ---"
      doubleEmptyOption = "true"
      doubleMultiple = "true"
/>
```

6.3.3 综合应用

example.jsp 代码如下。

```
<%@ page contentType = "text/html; charset = UTF-8" pageEncoding = "UTF-8" %>
<%@ taglib prefix = "s" uri = "/struts-tags" %>
<html>
      <head>
            <title>UI Tags Example</title>
            <s:head />
      </head>
      <body>
            <s:actionerror />
            <! -- 使用 Struts 的错误验证 -->
            <s:actionmessage />
            <! -- 使用 Struts 的信息验证 -->
            <s:fielderror />
            <! -- 使用 Struts 的域验证 -->
            <s:form action = "exampleSubmit" method = "post"   enctype = "multipart/form-data">
```

```
//单行文本框
            <s:textfield label = "姓名" name = "name" tooltip = " 输入提示信息" />

//文本框区
    <s:textarea tooltip = "Enter your Biography" label = "Biography"
                    name = "bio" cols = "20" rows = "3" />
//下拉列表
    <s:select tooltip = "Choose Your Favourite Color"
                    label = "Favorite Color" list = "{'Red' , 'Blue' , 'Green' }"
                    name = "favoriteColor" emptyOption = "true" headerKey = "None"
                    headerValue = "None" />

//复选框组,对应 Action 中的集合属性

    <s:checkboxlist tooltip = "Choose your Friends" label = "Friends"
                    list = "{'Patrick' , 'Jason' , 'Jay' , 'Toby' , 'Rene' }" name = "friends" />
//单一复选框
    <s:checkbox tooltip = "Confirmed that you are Over 18"
                    label = "Age 18 + " name = "legalAge" />


//单选框组,对应 Action 中的集合属性

    <s:radio label = "请选择你喜欢的书的出版日期" name = "b" labelposition = "top"
                    list = "#{'java' :'2006-10' , 'jsp' :'2008-9' , 'struts2' :'2008-5' }"
                    listKey = "key" listValue = "value" />

//简单单选框
    <s:radio label = "请选择你喜欢的书"  name = "a"  labelposition = "top"
                    list = "{'java' ,'jsp' ,'struts' }">
            </s:radio>

//文件选择组件
    <s:file tooltip = "Upload Your Picture" label = "Picture" name = "picture"></s:file>

            <s:optiontransferselect
                    tooltip = "Select Your Favourite Cartoon Characters"
                    label = "Favourite Cartoons Characters"
                    name = "leftSideCartoonCharacters" leftTitle = "Left Title"
                    rightTitle = "Right Title" list = "{'Popeye' , 'He-Man' , 'Spiderman' }"
                    multiple = "true" headerKey = "headerKey"
                    headerValue = "--- Please Select ---" emptyOption = "true"
                    doubleList = "{'Superman' , 'Mickey Mouse' , 'Donald Duck' }"
                    doubleName = "rightSideCartoonCharacters"
                    doubleHeaderKey = "doubleHeaderKey"
                   doubleHeaderValue = "--- Please Select ---" doubleEmptyOption = "true"
```

```
                doubleMultiple = "true" />
            <s:submit onclick = "alert(' aaaa' );" />
            <s:reset onclick = "alert(' bbbb' );" />
        </s:form>
          <s:form action = "Login" method = "POST" theme = "simple">
                <table>
                      <tr>
                            <td>
                                  User name:
                                  <s:textfield name = "name" label = "User name" />
                            </td>
                            <td>
                                  Password:
                                  <s:password name = "password" label = "Password" />
                            </td>
                      </tr>
                </table>
          </s:form>


      </body>
  </html>
```

6.4　EL 表达式语言

EL(Expression Language)表达式是 Struts 2 中的表达式语言，在 View 层 JSP 页面开发中，我们还会经常使用表达式语言。它是一种简单的语言，基于可用的命名空间(PageContext 属性)、嵌套属性、对集合、操作符(算术型、关系型和逻辑型)的访问符、映射到 Java 类中静态方法的可扩展函数和一组隐式对象。EL 提供了在 JSP 脚本编制元素范围外，使用运行时表达式的功能。脚本编制元素是指页面中能够用于在 JSP 文件中嵌入 Java 代码的元素，它们通常用于对象操作以及执行那些影响所生成内容的计算。JSP 2.0 将 EL 表达式添加为一种脚本编制元素。使用 EL 表达式的优势主要有：

① 避免<% = Var %>、<% = (MyType) request. getAttribute()%>和<% = myBean. getMyProperty()%>之类的语句，使页面更简洁；

② 支持运算符(如+、-、*、/)，比普通的标志具有更高的自由度和更强的功能；

③ 简单明了地表达了代码逻辑，使代码更可读与便于维护。

Struts 2 支持的表达式语言主要有：

① OGNL(Object-Graph Navigation Language)，可以方便地操作对象属性的开源表达式语言；

② JSTL(JSP Standard Tag Library)，JSP 2.0 集成的标准表达式语言；

③ Groovy，基于 Java 平台的动态语言，它具有时下比较流行的动态语言(如 Python、Ruby 和 Smarttalk 等)的一些特性；

④ Velocity,严格来说不是表达式语言,它是一种基于 Java 的模板匹配引擎,据说其性能要比 JSP 好。

6.4.1 EL 基本用法

1. EL 语法结构

EL 语言的语法结构：${expression}。

2. EL 存取数据

EL 提供“.”和“[ ]”两种运算符来存取数据。

当要存取的属性名称中包含一些特殊字符,如“或”等并非字母或数字的符号,就一定要使用“[]”,如 ${user. My-Name}应当改为 ${user["My-Name"] }。

如果要动态取值时,就可以用“[]”来做,而“.”无法做到动态取值,如 ${sessionScope. user[data]}中 data 是一个变量。

3. EL 变量

EL 存取变量数据的方法很简单,如 ${username}。它的意思是取出某一范围中名称为 username 的变量。因为我们并没有指定哪一个范围的 username,所以它会依序从 Page→Request→Session→Application 查找。

4. EL 运算符

EL 运算符有算术运算符、关系运算符、逻辑运算符和其他运算符。

① 算术运算符有 5 个:+、-、* 或 $ 、/或 div、%或 mod。

② 关系运算符有 6 个:==或 eq、!=或 ne、<或 lt、>或 gt、<=或 le、>=或 ge。

③ 逻辑运算符有 3 个:&& 或 and、||或 or、! 或 not。

④ 其他运算符有 3 个:Empty 运算符、条件运算符、()运算符。

最常用的表达式有:

① 是否为空判断:${empty param. name};

② 三元运算:${A? B:C};

③ 算术运算:${A*(B+C)}。

6.4.2 OGNL 表达式

1. OGNL 简介

Struts 2 默认的表达式语言是 OGNL,OGNL(Object-Graph Navigation Language)是一种功能强大的表达式语言,通过它简单一致的表达式语法,可以存取对象的任意属性、调用对象的方法、遍历整个对象的结构图和实现字段类型转化等功能。它使用相同的表达式去存取对象的属性。Struts 2 默认的表达式语言是 OGNL,它的优势主要体现在以下几个方面。

(1) 支持对象方法调用,如 xxx. doSomeSpecial()。

(2) 支持类静态的方法调用和值访问,表达式的格式如下所示。

```
@[类全名(包括包路径)]@[方法名 | 值名]
```

如@java. lang. String@format('te %s', 'b')。

(3) 支持赋值操作和表达式串联,如 price=100, discount=0. 8, calculatePrice(price * discount),这个表达式会返回 80。

(4) 访问 OGNL 上下文(OGNL context)和 ActionContext;

(5) 操作集合对象,可以直接 new 一个对象。

2. OGNL 符号

(1) "#"符号

"#"符号的用途主要 3 种,分别为:

① 访问非根对象属性,如 # session. msg 表达式(# session. msg 表达式相当于 ActionContext. getContext(). getSession(). getAttribute("msg")),由于 Struts 2 中值栈被视为根对象,所以访问其他非根对象时,需要加#前缀,实际上,"#"相当于 ActionContext. getContext();

② 用于过滤和投影(Projecting)集合,如 persons. {? #this. age>25},persons. {? #this. name=='pla1'}. {age}[0];

③ 用来构造 Map,如示例中的#{'foo1':'bar1', 'foo2':'bar2'}。

(2) "%"符号

"%"符号的用途是在标志的属性为字符串类型时,计算 OGNL 表达式的值,这个类似 js 中的 eval。

(3) "$"符号

"$"符号主要有两个方面的用途。

① 在国际化资源文件中,引用 OGNL 表达式,如国际化资源文件中的代码:reg. agerange =国际化资源信息:年龄必须在 ${min}和 ${max}之间。

② 在 Struts 2 框架的配置文件中引用 OGNL 表达式,如下所示。

```
<validators>
    <field name = "test">
            <field-validator type = "int">
            <param name = "min">10</param>
            <param name = "max">100</param>
            <message>BaseAction-test 校验:数字必须为 ${min}和 ${max}之间! </message>

        </field-validator>
    </field>
</validators>
```

3. OGNL 基本用法

OGNL 是通常要结合 Struts 2 的标志一起使用,主要是"#"、"%"和"$"这 3 个符号的使用,使用方法如下所示。

(1) 方法调用。

```
class Test{
    int fun();
}
```

调用方式:t. fun()。

(2) 访问静态方法和变量的格式和示例分别如下所示。

```
@[类全名(包括包路径)]@[方法名 |  值名]
```

```
@some.pkg.SomeClass@CONSTANTS
@some.pkg.SomeClass@someFun()
```

（3）访问 OGNL 上下文(OGNL context)和 ActionContext。

```
ActionContext().getContext().getSession().get("kkk")#session.kkk
ActionContext().getContext().get("person")#person
```

其中，“#”符号相当于 ActionContext。

关于 ActionContext 中常用的属性如表 6-1 所示。

表 6-1　ActionContext 中常用的属性

属性名称	用　途	示　例
parameters	包含当前 HTTP 请求参数的 Map	#parameters.id[0]作用相当于 request.getParameter("id")
request	包含当前 HttpServletRequest 的属性(attribute)的 Map	#request.userName 相当于 request.getAttribute("userName")
session	包含当前 HttpSession 的属性(attribute)的 Map	#session.userName 相当于 session.getAttribute("userName")
application	包含当前应用的 ServletContext 的属性(attribute)的 Map	#application.userName 相当于 application.getAttribute("userName")
attr	用于按 request→session→application 顺序访问其属性(attribute)	#attr.userName 相当于按顺序在以上 3 个范围(scope)内读取 userName 属性，直到找到为止

具体的 OGNL 表达式应用示例代码如下，首先创建 TestAction，然后在 JSP 页面中进行应用。

① 创建 TestAction.java 代码，如下所示。

```
package com.bcpl.cn.action;

import java.util.Date;
import java.util.LinkedList;
import java.util.List;

import javax.servlet.http.HttpServletRequest;

import org.apache.struts2.ServletActionContext;
import org.apache.struts2.convention.annotation.Action;
import org.apache.struts2.convention.annotation.Namespace;
import org.apache.struts2.convention.annotation.ParentPackage;
import org.apache.struts2.convention.annotation.Result;
import org.apache.struts2.convention.annotation.Results;
import org.springframework.stereotype.Controller;

import com.opensymphony.xwork2.ActionContext;
import com.opensymphony.xwork2.ActionSupport;
public classTestAction extends ActionSupport {
    private List<Person> persons;
```

```
    @Action("ognlTest")
    public String ognlTest() throws Exception {
        //获得 ActionContext 实例,以便访问 Servlet API
        ActionContext ctx = ActionContext.getContext();
        //存入 application
        ctx.getApplication().put("msg", "application 信息");
        //保存 session
        ctx.getSession().put("msg", "seesion 信息");
        //保存 request 信息
        HttpServletRequest request = ServletActionContext.getRequest();
        request.setAttribute("msg", "request 信息");
        //为 persons 赋值
        persons = new LinkedList<Person>();
        Person person1 = new Person();
        person1.setName("goft");
        person1.setAge(26);
        person1.setBirthday(new Date());
        persons.add(person1);

        Person person2 = new Person();
        person2.setName("John");
        person2.setAge(12);
        person2.setBirthday(new Date());
        persons.add(person2);

        Person person3 = new Person();
        person3.setName("Quin");
        person3.setAge(16);
        person3.setBirthday(new Date());
        persons.add(person3);

        return SUCCESS;

    }

    public List<Person> getPersons() {
        return persons;
    }

    public void setPersons(List<Person> persons) {
        this.persons = persons;
    }
}
```

② 创建 showognl.jsp 应用页面,代码如下。

```
<%@ page language="java" contentType="text/html; charset=utf-8" pageEncoding="utf-8" %>

<%@ taglib prefix="s" uri="/struts-tags" %>
<!DOCTYPE html PUBLIC "-//W3C//DTD XHTML 1.0 Transitional//EN" "http://www.w3.org/TR/xhtml1/DTD/xhtml1-transitional.dtd">
 <html >
    <head>
       <title> OGNL 表达式</title>
 </head>
<body>
      <h1>获取 OGNL 上下文和 Action 上下文</h1>
      <! -使用 OGNL 访问属性值-->
      <p>parameters: <s:property value="#parameters.msg" /></p>
      <p>request.msg: <s:property value="#request.msg" /></p>
      <p>session.msg: <s:property value="#session.msg" /></p>
      <p>application.msg: <s:property value="#application.msg" /></p>
      <p>attr.msg: <s:property value="#attr.msg" /></p>
        <hr />
<h3>用于过滤和投影(projecting)集合</h3>
        <ul>
          <!--判断年龄-->
           <s:iterator value="persons.{? #this.age>15}">
                  <li><s:property value="name" /> - 年龄:<s:property value="age" /></
                  li>
           </s:iterator>
           </ul>
           <p>姓名为 pla1 的年龄: <s:property value="persons.{? #this.name == goft'}.
           {age}[0]"/></p>
           <hr />

  <h3>构造 Map</h3>
           <s:set name="foobar" value="#{'foo1':'bar1', 'foo2':'bar2'}" />
           <p>The value of key "foo1" is <s:property value="#foobar['foo1']" /></p>
           <hr />
<h4>%符号的用法</h4>
           <s:set name="foobar" value="#{'foo1':'bar1', 'foo2':'bar2'}" />
           <p>The value of key "foo1" is <s:property value="#foobar['foo1']" /></p>
          <p>不使用%:<s:url value="#foobar['foo1']" /></p>
          <p>使用%:<s:url value="%{#foobar['foo1']}" /></p>

<hr />
        <%
           request.setAttribute("req", "request scope");
           request.getSession().setAttribute("sess", "session scope");
           request.getSession().getServletContext().setAttribute("app","aplication scope");
```

```
%>
通过 ognl 表达式获取 属性范围中的值
<br>
<s:property value = "#request.req" />
<br />
<s:property value = "#session.sess" />
<br />
<s:property value = "#application.app" />
<br />
<hr>

通过<span style = "background-color: #fafafa;">OGNL 表达式创建 list 集合 ,并且遍历
出集合中的值
<br>
<s:set name = "list" value = "{'e','d','c','b','a'}"></s:set>
<s:iterator value = "#list" var = "o">
    <! -- ${o }<br/> -->
    <s:property />
    <br />
</s:iterator>
<br />
<hr>
通过 OGNL 表达式创建 Map 集合,并且遍历出集合中的值
<br>
<s:set name = "map"
    value = "#{'1':'e','2':'d','3':'c','4':'b','5':'a'}"></s:set>
<s:iterator value = "#map" var = "o">
    <! -- ${o.key }-> ${o.value }<br/> -->
    <! -- <s:property value = "#o.key"/>-><s:property value = "#o.value"/>
    <br/> -->
    <s:property value = "key" />-><s:property value = "value" />
    <br />
</s:iterator>
<br />
<hr>
通过 OGNL 表达式 进行逻辑判断
<br>
<s:if test = "'aa' in {'aaa','bbb'}">
    aa 在集合{'aaa','bbb'}中;
</s:if>
<s:else>
    aa 不在集合{'aaa','bbb'}中;
</s:else>
<br />
<s:if test = "#request.req not in #list">
        不在集合 list 中;
```

```
        </s:if>
        <s:else>
                在集合 list 中;
        </s:else>
        <br />
        <hr>

    通过 OGNL 表达式的投影功能进行数据筛选
        <br>
        <s:set name = "list1" value = "{1,2,3,4,5}"></s:set>
        <s:iterator value = "#list1.{? #this>2}" var = "o">
            <! -- #list.{? #this>2}:在 list1 集合迭代的时候,从中筛选出当前迭代对象
            >2 的集合进行显示 -->
            ${o }<br />
        </s:iterator>
        <br />
        <hr>
     通过 OGNL 表达式 访问某个类的静态方法和值
        <br>
        <s:property value = "@java.lang.Math@floor(32.56)" />

        <s:property value = "@com.rao.struts2.action.OGNL1Action@aa" />
        <br />
        <br />
        <hr>
</body>
</html>
```

6.5 小　　结

Struts 2 标签技术,是为了简化 Web 前端开发的繁琐、复杂及框架开发的难度,同时,也规范了开发者对框架 View 层的开发和统一应用展示。基于此,Struts 2 框架提供了自己的标签库,供开发者进行 View 层的开发,开发者只需引用标准的标签库中的标签,并进行设置,便可以实现完成各种常用的页面的展示效果,同时,也可以在框架结构中,有效地实现展示层 View 和控制层 Action 的完美结合,使用用户创建的 Action 处理类,能够可以有效地提取 View 层的用户提交的数据,并根据数据的需求,把数据分发给不同的处理者,并最终把处理结果数据,通过视图层标签快速的返回,在客户端显示给终端用户。

关于 Struts 2 的标签,本书只是介绍了部分,并根据不同的应用目的进行了分类,同时,也对 Struts 2 的表达式语言 EL 和 OGNL 进行了重点介绍和应用分析,实际开发中,还需要更为深入地拓展了解其他标签的通常用法。

第 7 章　Struts 高级技术

7.1　Struts 2 国际化

以前，应用程序开发者能够考虑到仅仅支持他们本国的只使用一种语言(或者有时候是两种)和通常只有一种数量表现方式(如日期、数字和货币值)的应用。然而，基于 Web 技术应用程序的爆炸性增长，以及将这些应用程序部署在 Internet 或其他被广泛访问的网络之上，已经在很多情况下使得国家的边界淡化到不可见，这种情况转变成为一种对于应用程序支持国际化(internationalization，经常被称为 i18n，因为 18 是字母 i 和字母 n 之间的字母个数)和本地化的需求。国际化是商业系统中不可或缺的一部分，国际化的作用就是根据不同国家的用户在访问 Web 或其他类型的程序时，将各种信息以本地的常用形式显示出来，如界面信息在中国，就会显示中文信息，在以英文为主的国家里，就会显示英文信息，还有就是一些信息的格式，如日期格式等。

例如，如果您要输出一条国际化的信息，首先只需在代码包中加入 FILE-NAME_语种_时区. properties(其中 FILE-NAME 为默认资源文件的文件名)，然后在 struts-config. xml 中指明其路径，最后在页面用<bean:message>标志输出即可。不过 Struts 2.0 并没有在这部分止步，而是在原有的简单易用的基础上，将其做得更灵活、更强大。所以无论您学习的是什么 Web 框架，它都是必须掌握的技能。

Struts 1.x 对国际化有很好的支持，它极大地简化了我们程序员在做国际化时所需的工作，例如，如果您要输出一条国际化的信息，只需在代码包中加入 FILE-NAME_xx_XX. properties(其中 FILE-NAME 为默认资源文件的文件名)，然后在 struts-config. xml 中指明其路径，再在页面用<bean:message>标志输出即可。而 Struts 2.0 在原有的 Struts 1 简单易用的基础上，将其做得更灵活、更强大。

7.1.1　Struts 2 国际化方式

Struts 2 国际化是建立在 Java 国际化的基础之上，一样也是通过提供不同国家/语言环境的消息资源，然后通过 ResourceBundle 加载指定 Locale 对应的资源文件，再取得该资源文件中指定 key 对应的消息，整个过程与 Java 程序的国际化完全相同，只是 Struts 2 框架对 Java 程序国际化做了进一步封装，从而简化了应用程序的国际化。

从属性文件中获得字符串信息是国际化的基本应用，在 Struts 2 中使用的属性文件扩展名为. properties。

1. 全局资源文件加载

Struts 2 支持 4 种配置和访问资源文件的方法，包括：

- 使用全局的资源文件；
- 使用包范围内的资源文件；
- 使用 Action 范围的资源文件；
- 使用<s:i18n>标志访问特定路径的 properties 文件。

其中，最常用的就是加载全局的国际化资源文件，至于其他几种方式，本书后续会进行详细论述。加载全局的国际化资源文件的方式通过配置常量来实现，不管在 struts. custom. xml 文件中配置常量，还是在 struts. properties 文件中配置常量，只需要配置 struts. custom. il8n. resources 常量即可。配置 struts. custom. il8n. resources 常量时，该常量的值为全局国际化资源文件的 baseName。假如系统需要加载的国际化资源文件的 baseName 为 properties/messageResource，则我们可以在 struts. properties 文件中指定，如下所示。

```
Struts.custom.il8n.resources = properties.messageResource
```

或者更好的做法是在 struts. xml 文件中配置如下的一个常量：

```
<constant name = "struts.custom.i18n.resources" value = "properties/messageResource "/>
```

通过上述方式加载国际化资源文件后，Struts 2 应用就可以在所有地方获得国际化资源文件，其中包括 JSP 页面和 Action 的应用。

2. 实现国际化资源访问

Struts 2 既可以在 JSP 页面中通过标签输出国际化消息，也可以在 Action 类中输出国际化消息，不管采用哪种方式，Struts 2 都提供了支持，使用非常方便。

(1) Struts 2 访问国际化消息主要有 3 种方式，分别如下所示。

① JSP 页面国际化：在 JSP 页面中输出国际化消息，可以使用 Struts 2 的<s:text…/>标签，该标签可以指定一个 name 属性，该属性指定了国际化资源文件中的 key。

② 表单元素国际化：在表单元素里输出国际化信息，可以在该表单标签中指定一个 key 属性，该 key 指定了国际化资源文件中的 key。

③ Action 国际化：在 Action 类中访问国际化消息，可以使用 ActionSupport 类的 getText 方法，该方法可以接受一个 name 参数，该参数指定了国际化资源文件中的 key。

(2) Struts 2 访问国际化资源消息的具体应用如下所示。

① 创建资源文件，先创建 struts. properties 资源文件的内容，如下所示。

```
struts.custom.i18n.resources = globalMessages
```

② 创建以 globalMessage 为前缀的国际化资源文件，命名为 globalMessage_en_US. properites，并将其保存在 src 目录下，或保存在 WEB-INF/classes/properties 路径下。

```
HelloWorld = hello,world!
loginPage = loginPage
login.username = username
login.password = password
login.submit = login
msg = welcome {0}, time now:{1}
login.success = welcome {0},login success! time is now:{1}
login.failure = {0},login failure \!
username.required = username is required.
password.required = password is required.
```

③ 创建 globalMessage_zh_CN. properties 资源文件，其与上述英文资源文件对应的文件内容为 key-value 键值对。

```
HelloWorld = 您好,世界!
loginPage = 登录页面
login.username = 用户名
login.password = 密码
login.submit = 登录
msg = 欢迎{0}, 今天日期:{1}
login.success = 欢迎 {0},登录成功! 时间:{1}
login.failure = {0},登录失败!
username.required = 用户名不能为空
password.required = 密码不能为空
```

将上面的资源文件以 globalMessage_zh_CN. properties 文件名保存，可以使用 native2ascii 进行处理，保存在 WEB-INF/classes/properties 路径下，其对应的十六进制文件具体内容如下。

```
HelloWorld = \u4F60\u597D\uFF0C\u4E16\u754C\uFF01
loginPage = \u767B\u5F55\u9875\u9762
login.username = \u7528\u6237\u540D
login.password = \u5BC6\u7801
login.submit = \u767B\u9646
msg = \u6B22\u8FCE{0},\u4ECA\u5929\u65E5\u671F\uFF1A{1}
login.success = \u6B22\u8FCE{0},\u767B\u9646\u6210\u529F\uFF01 \u65F6\u95F4\u662F\uFF1A{1}
login.failure = {0},\u767B\u9646\u5931\u8D25\uFF01
password.required = \u5bc6\u7801\u4e0d\u80fd\u4e3a\u7a7a
username.required = \u7528\u6237\u540d\u4e0d\u80fd\u4e3a\u7a7a
password.required = \u5bc6\u7801\u4e0d\u80fd\u4e3a\u7a7a
```

在创建好上述资源文件后，系统会根据浏览者所在的 Locale 来加载对应的语言资源文件，下面是登录页面代码。

```
<%@ page  contentType="text/html; charset=UTF-8" %>
    <%@taglib prefix="s" uri="/struts-tags" %>
    <html>
        <head>
            <title><s:text name="loginPage"/></title>
        </head>
        <body>
            <h2>
            <s:text name="msg">
                <s:param>wjj</s:param>
                <s:param>2010-04-22</s:param>
            </s:text>
            </h2>
            <h2><s:property value="%{getText('HelloWorld')}"/></h2>

            <s:form action="">
                <s:textfield name="username" label="%{getText('login.username')}"></s:textfield>
                <s:password name="password" key="login.password"></s:password>
                <s:submit key="login.submit"></s:submit>
            </s:form>
        </body>
</html>
```

上面的 JSP 页面中使用了<s:text…/>标签来直接输出国际化信息，也通过在表单元素中指定 key 属性来输出国际化信息，通过这种方式，就可以完成 JSP 页面中普通文本和表单元素标签的国际化。

如果在简体中文环境下，浏览该页面将看到如图 7-1 所示的页面。

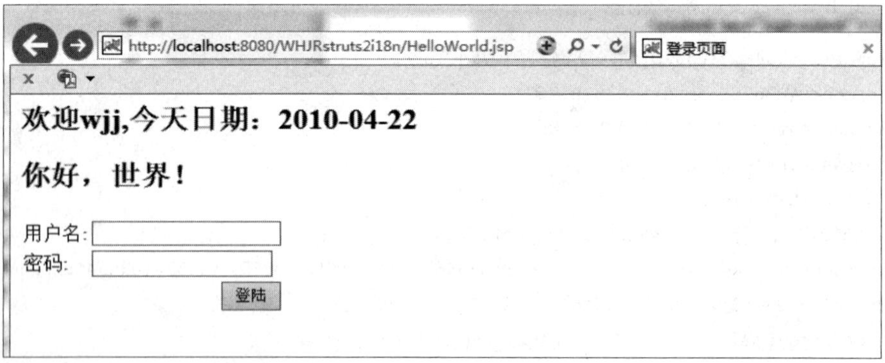

图 7-1　中文登录页面

如果在浏览器中修改语言，将机器的语言环境修改成美国英语(en-US)环境，再次浏览该页面，将看到如图 7-2 所示的页面。

如果为了在 Action 中访问国际化消息，则可以在 Action 类中调用 ActionSupport 类的 getText 方法，就能够取得国际化资源文件中的国际化消息。通过这种方式，即使 Action 需要设置在下一个页面显示的信息，也无须直接设置字符串常量，而是使用国际化消息的 key 来输

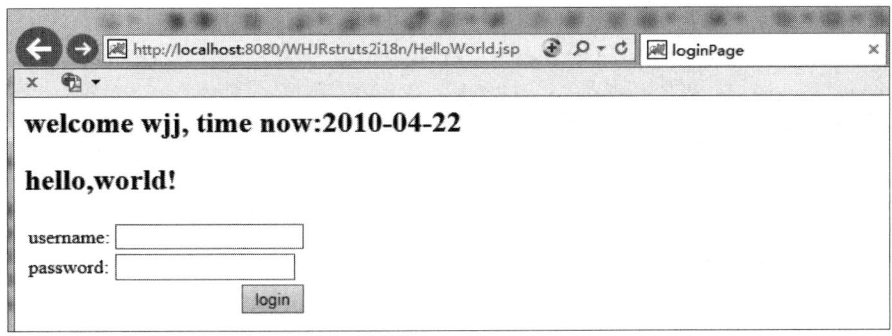

图 7-2　英文登录页面

出，从而实现程序的国际化，具体如下例应用中 Action 类的代码。

```
package com.bcpl.cn;

import java.util.Date;
import java.util.Map;

import com.opensymphony.xwork2.ActionContext;
import com.opensymphony.xwork2.ActionSupport;

/**
 * @author wjj
 *
 */
public class LoginAction extends ActionSupport {
    private String uname ;
    private String upass ;

    @Override
    public String execute() throws Exception {
        ActionContext ac = ActionContext.getContext() ;
        Map session = ac.getSession();
        if(uname.equals("wjj") && upass.equals("123456")){
            session.put("tip", this.getText("login.success", new String[]{this.uname,new
            Date().toString()}));
            return this.SUCCESS ;
        }else{
            session.put("tip", this.getText("login.failure", new String[]{this.uname}));
            return this.INPUT ;
        }
    }
    public void validate(){
//调用 get Text 方法取出国际化信息
        if(getUsername() == null||"".equals(this.getUsername().trim())){
            this.addFieldError("username", this.getText("username.required"));
```

```
            }
            if(this.getPassword() == null||"".equals(this.getPassword().trim())){
                this.addFieldError("password", this.getText("password.required"));
            }
        }
    }
    /**
         * @return the uname
         */
        public String getUname() {
            return uname;
        }
        /**
         * @param uname the uname to set
         */
        public void setUname(String uname) {
            this.uname = uname;
        }
        /**
         * @return the upass
         */
        public String getUpass() {
            return upass;
        }
        /**
         * @param upass the upass to set
         */
        public void setUpass(String upass) {
            this.upass = upass;
        }
    }
```

注意:上述代码中直接使用了数据验证技术,也可以重写 validate 方法。如果用户登录时没有填写用户名或密码,系统就会转到登录失败页面。在简体中文环境下,提示消息分别为用户名不能为空和密码不能为空。如果在浏览器中修改语言,将机器的语言环境修改成美国英语环境,那么提示消息分别变为"username is required;password is required",即可提供国际化的支持。

关于数据验证的详细内容,会在后续部分进行论述,在这里就不再赘述了。

7.1.2　参数化国际化字符串

许多情况下,需要动态地为国际化字符插入一些参数,在 Struts 2.0 中可以方便地做到这点,例如,在输入验证提示信息的时候,在 Struts 2.0 中,我们通过以下两种方法实现。

(1) 在资源文件的国际化字符串中使用 OGNL,格式和示例分别如下所示。

```
${表达式}
```

```
validation.require = ${getText(fileName)} is required
```

(2) 使用 java. text. MessageFormat 中的字符串格式,格式和示例分别如下所示。

```
{ 参数序号(从 0 开始),
格式类型(number | date | time | choice), 格式样式}
```

```
validation.between = Date must between {0, date, short} and {1, date, short}
```

在显示这些国际化字符时,同样有两种方法设置参数的值。

(1) 使用标志的 value0、value1、…、valueN 的属性,如下所示。

```
<s:text name = "validation.required" value0 = "User Name"/>
```

(2) 使用 param 子元素,这些 param 将按先后顺序,代入到国际化字符串的参数中,如下所示。

```
<s:text name = "msg">
    <s:param value = "admin"/>
    <s:param value = "2009-03-19"/>
</s:text>
```

在 Action 类中输出占位符的消息,ActionSupport 提供了 getText 方法,调用该方法时,传入用于填充占位符的参数值。

```
ActionContext ctx = ActionContext.getContext();
ctx.put("tip",this.getText("msg",new String[]{this.uname,d}));
```

如果需要在 JSP 页面中填充国际化消息里的占位符,则可以通过在<s:text…/>标签中使用多个<s:param…/>标签来填充消息中的占位符。第一个<s:param…/>标签指定第一个占位符值,第二个<s:param…/>标签指定第二个占位符值,依此类推。

如果需要在 Action 中填充国际化消息里的占位符,则可以通过在调用 getText 方法时使用 getText(String aTextName,List args)或 getText(String key, String[] args)方法来填充占位符。该方法的第二个参数既可以是一个字符串数组,也可以是字符串组成的 List 对象。

在上述实例的资源文件中有如下国际化消息。

```
#带占位符的国际化信息
welcomeHint = 欢迎,{0},您已经登录成功!
```

为了在 Action 类中输出占位符的消息,我们在 Action 类中调用 ActionSupport 类的 getText 方法,调用该方法时,传入用于填充占位符的参数值。访问该带占位符消息的 Action 类如下所示。

```
package com.bcpl.action;
import com.opensymphony.xwork2.ActionContext;
import com.opensymphony.xwork2.ActionSupport;
@SuppressWarnings("serial")
public class LoginAction extends ActionSupport{
    private String username;
    private String password;
    public String getPassword() {
        return password;
    }
    public void setPassword(String password) {
        this.password = password;
```

```
        }
        public String getUsername() {
            return username;
        }
        public void setUsername(String username) {
            this.username = username;
        }
        @SuppressWarnings("unchecked")
        public String execute() throws Exception {
            ActionContext ac = ActionContext.getContext() ;
            Map session = ac.getSession();
            if(uname.equals("wjj") && upass.equals("123456")){
                session.put("tip", this.getText("login.success", new String[]{this.uname,
                new Date().toString()}));
                return this.SUCCESS ;
            }else{
                session.put("tip", this.getText("login.failure", new String[]{this.uname}));
                return this.INPUT ;
            }

        }
    }
```

通过上面的带参数的 getText 方法,就可以为国际化消息的占位符传入参数了。

在 JSP 页面中输出带两个占位符的国际化消息,只需在<s:text…/>标签中指定<s:param…/>子标签即可,下面是 index.jsp 页面的代码。

```
<%@ page language="java" import="java.util.*" pageEncoding="GB18030" %>
<%@ taglib prefix="s" uri="/struts-tags" %>
<html>
    <head>
        <title>My JSP 'login.jsp' starting page</title>
    </head>
    <body>
        <s:form action="login">
            <s:textfield name="uname" label="username"></s:textfield>
            <s:textfield name="upass" label="password"></s:textfield>
            <s:submit></s:submit>
        </s:form>
    </body>
</html>
```

在 success.jsp 页面使用 ${sessionScope.tip}输出的国际化消息来自 Action 类。

```
<body>
    ${sessionScope.tip }
    </body>
```

当我们以 wjj 用户名登录成功后,结果如图 7-3 所示。

图 7-3　成功登录后的结果显示

如果美国英语语言环境下用户通过登录页面登录成功，进入 success. jsp 页面，将看到如图 7-4 所示的页面。

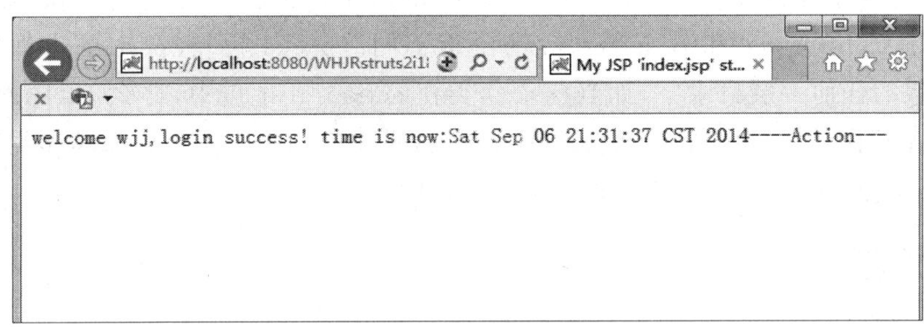

图 7-4　美国英语语言环境下成功登录后的结果显示

7.1.3　Struts 2 定位资源属性文件顺序

（1）与动作类同名：XXXAction. properties。Struts 2 会首先查询与当前访问的动作类同名，并且和 XXXAction. class 在同一个目录下的属性文件，如 LoginAction_en_US. properties 和 LoginAction_zh_CN. properties。

（2）与动作类的基类相同：BaseAction. properties。BaseAction 表示动作类的基类，其属性文件为 BaseAction. properties。所有动作类都会查找 Object. properties 文件（因为 Object 是所有 Java 类的基类），但要注意的是 Object. properties 文件不能放到当前动作类的目录中，由于 Object 在 java. lang 包中，因此，Object. properties 要放到 jdk 包的 java. lang 目录中。

对于 ActionSupport. properties 文件，不能放到动作类的当前目录中，由于 ActionSupport 类位于 com. opensymphony. xwork2 中，因此，需要将 ActionSupport. properties 文件放到 xwork2. jar 包中的 com\opensymphony\xwork2 目录中，由于放到 jar 文件中并不方便，因此，可以创建一个和当前动作类在一个目录的类来继承 ActionSupport，然后所有的动作类都继承于这个类，如下所示。

```
public class BaseAction extends ActionSupport{
        ...
        }

        public class XXXAction extends BaseAction{
            ...
        }
```

这样只要存在一个 BaseAction. properties，在当前目录下的所有动作类都会读取这个文件。

（3）与接口名相同：XXInterface. properties。这类文件和 BaseAction. properties 类似，XXInterface 表示动作类实现的接口。

（4）与包名相同：package. properties。如 package_en_US. properties 和 package_zh_CN. properties，与 XXInterface 和 BaseAction 的泛指不同。package. properties 文件可以放到当前动作类的包的任何一层目录下，如当前动作类在 action. test 包中，那么 package. properties 可以放到 action 目录中，也可以放到 action/test 目录中。Struts 2 会从离动作类最近的位置开始查找 package. properties 文件，如下是一个动作类。

```
package action.test;

import org.apache.struts2.*;
import com.opensymphony.xwork2.ActionSupport;

public class MyAction extends ActionSupport{
        public String execute() {
                return "forward";
        }
}
```

在 action\test 目录下有一个 MyAction. properties 文件，内容如下。

```
hello = hello, how are you
say = bye, bye!!!
```

然后可以在 jsp 文件中使用如下几种方法取出资源信息。

```
<s:property value = "getText('hello')"/>
<s:text name = "say" />
```

（5）使用<s:il8n>标签搜索 il8n 资源信息。

<s:il8n>标签可以直接定位属性文件，例如，wjj. properties 在 WEB-INF\classes\test 目录下，wjj. properties 内容和 MyAction. properties 一样，则可以使用如下代码读取 wjj. properties 的内容。

```
<%@ taglib prefix="s" uri="/struts-tags" %>
<s:i18n name="test.abc">
        <s:text name="hello" />
        <s:text name="say" />
</s:i18n>
```

(6) 查找全局资源属性文件。

在 WEB-INF 的 classes 目录下建立一个 struts. properties 文件，内容如下。

```
globalMessages.properties
struts.custom.i18n.resources=action/test/globalMessages
```

或者在 struts. xml 文件里加入：

```
<constant name="struts.custom.i18n.resources" value="action/test/globalMessages"/>
```

在 WEB-INF\classes\action\test 目录下建立一个 globalMessages. properties 文件，当 Struts 2 按着上述的顺序没有找到相应的属性文件时，最后就会考虑寻找全局的属性文件，因此，就会找到 globalMessages. properties。

Struts 支持 4 种配置和访问资源的方法，其依次查找顺序分别如下：

① 使用全局的资源文件；

② 使用包范围内的资源文件；

③ 使用 Action 范围的资源文件；

④ 使用<s:i18n>标志访问特定路径的 properties 文件。

它们的范围分别是从大到小，而 Struts 2.0 在查找国际化字符串所遵循的是特定的顺序，如图 7-5 所示。

Struts 2.0 在查找国际化字符串所遵循的是特定的顺序，假设在某个 LoginAction 中调用了 getText("user. name")，Struts 2.0 将会执行以下的操作：

① 查找 LoginAction_xx_XX. properties 文件或 LoginAction. properties；

② 查找 LoginAction 实现的接口，查找与接口同名的资源文件 TestInterface. properties；

③ 查找 LoginAction 的父类 RegisterAction 的 properties 文件，文件名为 RegisterAction. properties；

④ 判断当前 LoginAction 是否实现接口 ModelDriven，如果是，调用 getModel()获得对象，查找与其同名的资源文件；

⑤ 查找当前包下的 package. properties 文件；

⑥ 查找当前包的父包，直到最顶层包；

⑦ 在值栈(Value Stack)中，查找名为 user 的属性，转到 user 类型同名的资源文件，查找键为 name 的资源；

⑧ 查找在 struts. properties 配置的默认资源文件；

⑨ 输出 user. name。

7.1.4　其他加载国际化资源文件的方式

前面介绍了 Struts 2 中加载国际化资源的最常用方式，除此之外，Struts 2 还提供了其他

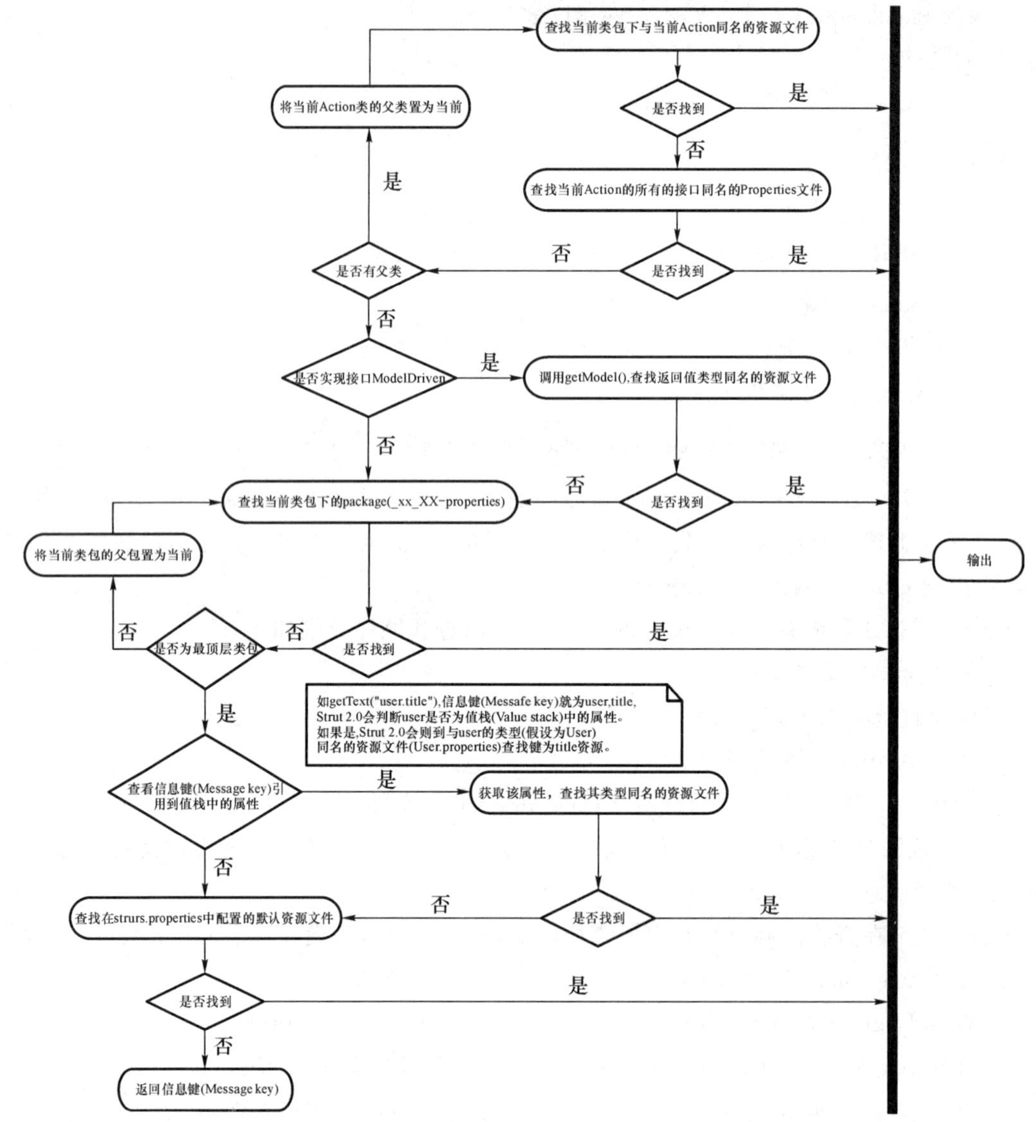

图 7-5　查找资源文件查找顺序图

方式来加载国际资源文件,包括上述介绍的指定包范围资源文件、类范围资源文件和临时指定资源文件等。

1. 包范围内的资源文件

对于一个大型商业应用而言,国际化资源文件的管理是一个复杂的工程,因为整个应用中有大量内容需要实现国际化,如果我们把国际化资源都放在同一个全局文件里,这将是不可想象的事情。为了更好地体现软件工程里"分而治之"的原则,Struts 2 允许针对不同模块和不同 Action 来组织国际化资源文件。

Struts 2 应用指定包范围资源文件的方法是:在包的根路径下建立多个文件名为 package_language_country. properties 的文件,按要求建立了这个系列国际化资源文件,应用中处于其包下的所有 Action 都可以访问该资源文件,如下例的 Action 类。

```
package com.bcpl.struts2.action;
import com.opensymphony.xwork2.ActionContext;
import com.opensymphony.xwork2.ActionSupport;
@SuppressWarnings("serial")
public class LoginAction extends ActionSupport{

    private String username;
    private String password;
    public String getPassword() {
        return password;
    }
    public void setPassword(String password) {
        this.password = password;
    }
    public String getUsername() {
        return username;
    }
    public void setUsername(String username) {
        this.username = username;
    }
    @SuppressWarnings("unchecked")
    public String execute(){
        if(getUsername().equals("wjj")&& getPassword().equals("123456")){

            //调包范围资源文件
            ActionContext.getContext().put("tip", this.getText("login.success"));
            return SUCCESS;
        }
        ActionContext.getContext().put("tip", this.getText("login.failed"));
        return ERROR;
    }
}
```

接着我们创建中英文对照的资源文件，中文资源文件为 package_zh_CN.properties，文件内容为：

```
flogin.success = 欢迎{0},登录成功:{1}
login.failure = {0},登录失败!
```

可以使用 myeclipse 自带的工具或 native2ascii 工具把上述内容转为十六进制内容，如下所示。

```
login.success = \u6B22\u8FCE{0},\u767B\u9646\u6210\u529F\uFF01 \u65F6\u95F4\u662F\uFF1A{1}
login.failure = {0},\u767B\u9646\u5931\u8D25\uFF01
```

对应的英文资源文件为 package_en_US.properties，文件内容为：

```
login.success = welcome {0},login success! time is now:{1}
login.failure = {0},login failure \!
```

然后将中英文资源文件保存在 WEB-INF/classes/com/bcpl/Struts 2 路径下，该资源文件就可以被位于 Struts 2 包和 Struts 2 子包下的所有 Action 访问了。

因此，当我们在简体中文语言环境下成功登录时，将看到如图 7-3 所示的页面。

2. Action 范围内资源文件

Struts 2 允许为 Action 单独指定一份国际化资源文件，为 Action 单独指定国际化资源文件的方法是：在 Action 类文件所在的路径建立多个文件名为 ActionName_language_country. properties 的文件，建立好这些国际化资源文件后，该 Action 就可以访问其范围内的资源文件。同上述过程一样，创建与 Action 同名的资源文件，命名如 LoginAction_zh_CN. properties、LoginAction_en_US. properties，并将其放置到 WEB-INF/classes/ com/bcpl/Struts 2/action 路径下。

通过在 Action 范围内创建的资源文件，就可以在不同的 Action 里使用相同的 key 名来表示不同的字符串值。

3. 指定临时资源文件

除了上述的资源放置之外，还可以采用指定临时资源文件的方式，在 JSP 页面中输出国际化消息。Struts 2 的标签<s:il8n>可以实现此功能。

如果把<s:il8n>标签作为<s:text…/>标签的父标签，则<s:text…/>标签将会直接加载<s:il8n>标签里指定的国际化资源文件；如果把<s:il8n>标签当成表单标签的父标签，则表单标签的 key 属性将会从国际化资源文件中加载其消息。

同前述一样，创建中英文对照的资源文件，并将其保存在 WEB-INF/classes 路径下，然后在 login. jsp 登录页面中通过<s:il8n>标签来使用上述创建资源文件，资源文件命名为 login_ zh_CN. properties 或 login_en_US. properties，文件内容分别如下所示。

login_ zh_CN. properties 文件内容：

```
loginPage = 登录页面
loginTip = 用户登录
userName = 用户名
password = 用户密码
submit = 提交
```

login_en_US. properties 文件内容：

```
loginPage = loginPage
loginTip = Login
userName = username
password = password
submit = submit
```

使用方法如下：

```
<%@ page language = "java" contentType = "text/html; charset = utf-8" %>
<%@ taglib uri = "/struts-tags" prefix = "s" %>
<html>
<head>
```

```
<title>
<s:i18n name = "login">
<s:text name = "loginPage"/>
</s:i18n>
</title>
</head>
<body>
<h3>
<! --使用 i18n 作为 s:text 标签的父标签,指定临时的国际化资源文件的 baseName 为 login-->
<s:i18n name = "login">
<! --输出国际化信息-->
<s:text name = "loginTip"/>
</s:i18n>
</h3>
<! --使用 i18n 作为 s:text 标签的父标签,临时指定国际化资源文件的 baseName 为 login-->
<s:i18n name = "login">
<s:form action = "Login" method = "post">
<s:textfield name = "username" key = "user"/>
<s:password name = "password" key = "password"/>
<s:submit name = "submit" key = "submit" />
</s:form>
</s:i18n>
</body>
</html>
```

7.1.5　国际化应用实例

创建国际化 Hello World 涉及的要点及关键步骤如下所示。

(1) 在 Eclipse 创建工程(Struts 2i18n)配置开发和运行环境。

(2) 在 src 文件夹中加入 struts. properties 文件,内容如下:

```
struts.custom.i18n.resources = globalMessages
```

或者在 struts. xml 文件中添加:

```
<constant name = "struts.custom.i18n.resources" value = "globalMessages"/>
```

(3) 在 src 文件夹中加入 globalMessages_en_US. properties 文件,内容如下:

```
HelloWorld = Hello World!
```

在 src 文件夹中加入 globalMessages_zh_CN. properties 文件,内容如下:

```
HelloWorld = 你好,世界!
```

转换为十六进制如下:

```
HelloWorld = \u4f60\u597d\uff0c\u4e16\u754c\uff01
```

(4) 在 WebContent 文件夹下加入 HelloWorld. jsp 文件,内容如下:

```
<%@ page   contentType="text/html; charset=UTF-8" %>
<%@taglib prefix="s" uri="/struts-tags" %>
<html>
    <head>
        <title>Hello World</title>
    </head>
    <body>
        <h2><s:text name="HelloWorld"/></h2>
        <h2><s:property value="%{getText('HelloWorld')}"/></h2>
    </body>
</html>
```

(5) 发布运行应用程序，在浏览器地址栏中输入 http://localhost:8080/Struts 2i18n/HelloWorld.jsp，中文环境下运行结果如图 7-6 所示。

图 7-6　中文环境运行结果

(6) 在使用 UI 表单标志时，getText 可以用来设置 label 属性，如下所示。

```
<s:textfield name="name" label="%{getText('HelloWorld')}"/>
```

或者

```
<s:textfield name="name" key="HelloWorld"/>
```

7.1.6　数据库中文问题的处理

除了读取资源文件外，还需要处理页面和数据库之间数据存取的中文问题。可以通过开发 SetCharacterEncodingFilter 实现。

(1) 在 struts.xml 中<struts>标签下加入：

```
<constant name="struts.i18n.encoding" value="GBK"/>
```

(2) 在 WebRoot/WEB-INF/web.xml 加入编码的 filter 声明。

```
<filter>
<filter-name>encodeFilter</filter-name>
<filter-class>packageName.SetCharacterEncodingFilter</filter-class>
        </filter>
        <filter-mapping>
                <filter-name>encodeFilter</filter-name>
                <url-pattern>/*</url-pattern>
        </filter-mapping>
```

(3) 实现 filter 类 SetCharacterEncodingFilter，方法如下所示。

```
package com.bcpl.util;

import java.io.IOException;
import javax.servlet.Filter;
import javax.servlet.FilterChain;
import javax.servlet.FilterConfig;
import javax.servlet.ServletException;
import javax.servlet.ServletRequest;
import javax.servlet.ServletResponse;
import javax.servlet.UnavailableException;
    public class SetCharacterEncodingFilter implements Filter{

            public void init(FilterConfig config) throws ServletException {   }

            public void doFilter(ServletRequest request, ServletResponse response,FilterChain chain)
                                throws IOException, ServletException {

            request.setCharacterEncoding("UTF-8");  //!!!!!!!!!
            chain.doFilter(request, response);
            }

          public void destroy() {   }
        }
```

(4) 在所有 JSP 页面里设定编码方式。

```
<%@ page language = "java" contentType = "text/html;charset = UTF-8" %>
```

(5) 在 JSP 页面使用 Struts 标签。

```
<s:property value = "getText('hello')"/>
<s:text name = "bye" />
```

(6) 写一个临时文件 XXXX_temp.properties。

```
error.login = 用户名或密码错误
username.length.short = 用户名至少 4 位
password.short = 密码至少 6 位
```

(7) Encode(编码)，把临时文件所对应的中文内容编码，把 XXXX_temp.properties 编码后转换成 XXXX_zh_CN.properties。

```
输入命令 native2ascii (中文转化成 UTF-8)
\>native2ascii-encoding  gb2312  XXXX_temp.properties  XXXX_zh_CN.properties
输入命令 native2ascii-reverse (UTF-8 转化成中文)
\>native2ascii-reverse -encoding  gb2312  XXXX_zh_CN.properties  XXXX_temp2.properties
```

(8) 浏览器的语言设定选中文 zh-cn。

通过上述应用可以看出，Struts 国际化应用过程中，实现国际化应该遵循一些基本的原

则,其主要有:

① 尽量不要在 Servlet/JSP 中使用含非英文字符的常量字符串;

② 对于 JSP 文件,应该对 page 指令中的 charset 属性进行相应的设置;

③ 不要在 JSP 文件中直接包含本地化的消息资源,而是应该把消息资源存放在资源文件里;

④ 不必在每个 JSP 中设置 HTTP 请求的字符编码,可以在 Servlet 过滤器中设置 HTTP 请求的字符编码;

⑤ 尽量使用 UTF-8 作为 HTTP 请求和响应的字符编码,而不是 GBK 或 GK2312;

⑥ 充分考虑底层数据库所使用的编码,它可能会给应用程序的移植带来麻烦。

7.2　Struts 2 下快捷地选择或切换语言

开发国际化的应用程序时,让用户快捷地选择或切换语言是网站首页 Web 页面常用的功能。在 Struts 2.0 中,可以通过 ActionContext. getContext(). setLocale(Locale arg)设置用户的默认语言。Struts 2.0 针对此应用,提供了 i18n 的拦截器(Interceptor),并在默认情况下将其注册到拦截器链(Interceptor Chain)中。其原理:在执行 Action 方法前,i18n 拦截器会在请求中查找名为 request_locale 的参数,如果其存在,拦截器就将其作为参数实例化 Locale 对象,并将其设为用户默认的区域(Locale),然后,将 Locale 对象保存在 session 的 WW_TRANS_I18N_LOCALE属性中,具体应用如下例所示。

(1) 建一个 JavaBean,命名为 Locales. java。

```
package com.bcpl.action;
import java.util.Hashtable;
import java.util.Locale;
import java.util.Map;
public class Locales {
    private Locale current;

    //定义当前用户的语言种类
    public void setCurrent(Locale current) {
        this.current = current;
    }

    public Map<String, Locale> getLocales() {
        Map<String, Locale> locales = new Hashtable<String, Locale>();
        locales.put("American English", Locale.US);
        locales.put("Simplified Chinese", Locale.CHINA);
        //…添加其他所支持的语言版本
        return locales;
    }
}
```

(2) 创建 languageSelector.jsp 控件页面,进行不同语言的选取。

```
<%@ page language="java" contentType="text/html; charset=gb2312" %>
<%@ taglib prefix="s" uri="/struts-tags" %>
<script type="text/javascript">
function langSelecter_onChanged()
{
    document.getElementById("langForm").submit();
}
</script>
<s:set name="SESSION_LOCALE" value="#session['WW_TRANS_I18N_LOCALE']" />

<s:bean id="locales" name="com.ascent.action.Locales">
    <s:param name="current" value="#SESSION_LOCALE == null ? locale : #SESSION_LOCALE" />
</s:bean>

<form action="<s:url/>" id="langForm" style="background-color: #bbbbbb; padding-top: 4px; padding-bottom: 4px;">
    <s:text name="language" />
    <s:select label="Language" list="#locales.locales" listKey="value"
        listValue="key" value="#SESSION_LOCALE == null ? locale : #SESSION_LOCALE"
        name="request_locale" id="langSelecter"
        onchange="langSelecter_onChanged()" theme="simple" />
</form>
```

上述代码中 languageSelector.jsp 先实例化一个 Locales 对象,并把对象的 Map 类型的属性 Locales 赋予下拉列表,然后下拉列表就获得可用语言的列表。languageSelector 有<s:form>标志和一段 JavaScript 脚本,其作用是在用户在下拉列表中选择了后,提交包含 reqeust_locale 变量的表单到 Action。在打开页面时,为了下拉列表的选中的当前区域,我们需要到 session 取得当前区域(键为 WW_TRANS_I18N_LOCALE 的属性),而该属性在没有设置语言前是空的,所以通过值栈中 locale 属性来取得当前区域(用户浏览器所设置的语言)。

(3) 在资源文件里加上:

```
language = language is :
```

(4) 可以把 languageSelector.jsp 作为一个控件使用,方法是在 JSP 页面中把它包含进来,代码如下所示。

```
<s:include value="languageSelector.jsp"/>
```

(5) 需要通过访问 Action 到达包含 languageSelector.jsp 的页面,即可实现国际化的功能。

7.3　Struts 2 类型转换

在面向对象的设计模式 MVC 中,Web 应用程序实际上是分布在不同主机上的两个进程之间的交互。交互之间的通讯是通过 HTTP 协议,它们通讯传递的为字符串。服务器可以接收到的来自用户的数据只能是字符串或字符数组，而在服务器端程序的 Java 语言要求提交的数据必须有指定的数据类型，如日期(Date)、整数(int)、浮点数(float)或自定义类型(UDT)等。要将两种不同环境下的数据进行映射，必须通过类型转换机制，将 HTML 客户端提交的数据映射为 Java 的各种数据类型。

在 Struts 开发中，客户端和服务端之间的数据传输流是双向的，浏览器通常充当客户端，也即表现层 View，表现层通常用来与用户交互和收集用户输入的各种数据，并在用户输入数据后，触发事件进行提交，提交给服务器，在此数据传输过程中，需要进行数据类型的转换，因客户端提交的数据为字符串类型，而服务器需要 Java 类型的数据。在服务器端处理完数据后，把数据传输到客户端并不需要进行数据类型的转换。

关于 Struts 2，提供了专门的转换器来实现数据类型的转换，也可以采用传统的创建 JavaBean 类和 Servlet 类的方式完成数据的转换。Struts 2 内置的类型转换器是基于 OGNL 表达式的，需要把 HTML 输入项(表单元素和其他 GET/POET 的参数)命名为合法的 OGNL 表达式，即可利用 Struts 2 的转换机制。在 Struts 2 中数据校验流程通常包含以下几个方面：首先通过转换器将 HTML 请求参数转换成相应的 Bean 属性；然后判断转换过程中是否出现了异常，如果有，则将其保存到 ActionContext 上下文中，通过 conversionError 拦截器再封装为 fieldError，如果没有，则转到下一步；通过反射来调用 validateXxx()方法(Xxx 表示相应的 Action 的方法名)；接着调用其中实现的 validate()方法。在此过程中，如果经过上述过程没有发生 fieldError，则调用 Action 方法；如果有，则会跳过 Action 方法，通过前面实现的国际化资源文件将 fieldError 输出到对应的页面。

1. Struts 2 内置的转换器

常用的一些类型转换，在 Struts 2 中已经通过转换器实现，如日期、整数或浮点数等类型。Struts 2 内置的类型转换支持的类型有：

① 常规定义类型，如 int、boolean、double 等；

② 日期类型，可使用当前区域(Locale)的短格式(Short)转换，即 DateFormat.getInstance(DateFormat.SHORT)；

③ 集合(Collection)类型，完成一组字符串对集合类型的转换，将 request.getParameterValues(String arg)返回的字符串数据与 java.util.Collection 转换；

④ 集合(Set)类型，与 List 的转换相似，去掉重复相同的值；

⑤ 数组(Array)类型，将字符串数组的每一个元素转换成特定的类型，并组成一个数组。

除此之外，Struts 2 提供了很好的扩展性，开发者可以非常简单地开发自己的类型转换器，完成字符串和自定义复合类型之间的转换。总之，Struts 2 的类型转换器提供了非常强大的表现层数据处理机制，开发者可以利用 Struts 2 的类型转换机制来完成任意的类型转换。

2. 实现自定义类型转换器

自定义类型转换器的实现，在 Struts 2.0 中的转换器需要实现 ognl.TypeConverter 接口，或者继承 DefaultTypeConverter 实现类，然后重写其 convertValue 方法即可。

public Object convertValue(Map context, Object value, Class toType)方法的参数如下。

- Context：用于获取当前的 ActionContext。
- value：需要转换的值。
- toType：需要转换成的目标类型。

此外，Struts 2 还提供了一个 StrutsTypeConverter 抽象类，其提供了两个不同转换方向的方法：

Object convertToString(Map context,String[] values,Class toClass)和 String convertFromString(Map context,Object o)。在 Struts 2 框架中可以通过抛出 XWorkException 或者 TypeConversionException 异常,发现类型转换的错误。

3. 注册自定义类型转换器

自定义类型转换器实现后,需要注册在 Web 应用中,Struts 2 框架才可以正常使用该类型转换器。

关于类型转换器的注册方式,通常主要有以下几种。

- 局部类型转换器注册:仅仅对某个 Action 的属性起作用。
- 全局类型转换器注册:对所有 Action 的特定类型的属性都会生效。
- 使用 JDK1.6 的注释来注册类型转换器:通过注释方式来生成类型转换器。

(1) 局部类型转换器

资源文件名及内容格式如下。

- 文件名: ActionName-conversion.properties。
- 文件内容:多个 propertyName(属性名)=类型转换器类(含包名),如 date=com.aumy.DateConverter。
- 存放位置:和 ActionName 类相同路径。

(2) 全局类型转换器

资源文件及内容格式如下。

- 文件名: xwork-conversion.properties。
- 内容: 多个"复合类型=对应类型转换器"项组成,如 java.Util.Date=com.aumy.DateConverter。
- 存放位置:WEB-INF/classes/目录下。

7.4 数据验证

在 Struts 2 中最简单的验证数据的方法是使用 validate,在实际开发中外部输入通常都需要进行校验,而表单是应用程序常用的数据提交方式,对其数据进行校验,即可以通过客户端的 JavaScript 技术来完成,也可以使用 Struts 的数据验证框架。

在 Struts 2 中数据验证的方法包括以下两种方法:

① 自定义 Action 类继承 ActionSupport 类,重写 validate()方法;

② 用基于 XML 文件的 validation 框架验证。

ActionSupport 类实现了一个 Validateable 接口,这个接口只有一个 validate 方法。如果 Action 类实现了这个接口,Struts 2 在调用 execute 方法之前首先会调用这个方法。然而,如果发生错误,可以根据错误的水平(level)选择字段级错误,还是动作级错误,并且可使用 addFieldError 或 addActionError 加入相应的错误信息。如果存在 Action 或 Field 错误,Struts 2 会返回"input"(由 Struts 2 自动返回,并不需要开发人员来写),如果返回了"input",Struts 2 就不再调用 execute 方法。如果不存在错误信息,Struts 2 在最后会调用 execute 方法。除了加入错误信息外,还可以使用 addActionMessage 方法加入成功提交后的信息。当提交成功后,可以显示这些信息。

下述 3 个 add 方法都在 ValidationAware 接口中定义,并且在 ActionSupport 类中有一个默认的实现。事实上,在 ActionSupport 类中的实现,实际上是调用了 ValidationAwareSupport 中的相应的方法,也就是这 3 个 add 方法是在 ValidationAwareSupport 类中实现的,代码如下。

```
import com.opensymphony.xwork2.ValidationAwareSupport;
private final ValidationAwareSupport validationAware = new ValidationAwareSupport();
public void addActionError(String anErrorMessage) {
    validationAware.addActionError(anErrorMessage);
}
public void addActionMessage(String aMessage) {
    validationAware.addActionMessage(aMessage);
}
public void addFieldError(String fieldName, String errorMessage) {
    validationAware.addFieldError(fieldName, errorMessage);
}
```

7.4.1 使用 Action 的 validate()方法

在 Web 根目录建立一个主页面(validate.jsp),使用了 Struts 2 的 tag:<s:actionerror>、<s:fielderror>和<s:actionmessage>,分别用来显示动作错误信息、字段错误信息和动作信息。如果信息为空,则不显示,代码如下。

```
<%@ page language="java" %>
<%@ taglib prefix="s" uri="/struts-tags" %>
<html>
  <head>
    <title>数据验证</title>
  </head>
  <body>
    <s:actionerror/>
    <s:actionmessage/>
    <s:form action="validateAction"  theme="simple">
      input:<s:textfield name="msg"/>
    <%--  <s:fielderror key="msg.hello" /> --%>
      <s:fielderror>
            <s:param>msg.hello</s:param>
      </s:fielderror>
         <br/>
      <s:submit/>
    </s:form>
  </body>
</html>
```

然后接着建立实现一个 Action 动作类,Field 错误需要一个 key(一般用来表示是哪一个属性出的错误),而 Action 错误和 Action 消息只要提供一个信息字符串就可以了,代码如下。

```
package com.bcpl.action;
import javax.servlet.http.*;
import com.opensymphony.xwork2.ActionSupport;
import org.apache.struts2.interceptor.*;

public class ValidateAction extends ActionSupport{
    private String msg;
    public String execute() {
        System.out.println(SUCCESS);
        return SUCCESS;
    }
    public void validate(){
        if(!msg.equalsIgnoreCase("hello")){
            System.out.println(INPUT);
            this.addFieldError("msg.hello", "请输入 hello!");
             this.addActionError("处理动作失败!");
        }else{
             this.addActionMessage("提交成功");
        }
    }
    public String getMsg(){
        return msg;
    }
    public void setMsg(String msg){
        this.msg = msg;
    }
}
```

在 struts.xml 中配置 Action，设置返回所调用的信息显示页面。

```
<package name = "demo" extends = "struts-default">
    <action name = "validate" class = "com.bcpl.action.ValidateAction">
        <result name = "success">/error/validate.jsp</result>
        <result name = "input">/error/validate1.jsp</result>
    </action>
</package>
```

假设应用程序的工程应用路径为 myApp，则可通过此 URL 来测试程序：http://localhost:8080/myApp/error/validate.jsp。

除了上述的 validate() 方法之外，还可以使用 ValidationAware 接口的其他方法(由 ValidationAwareSupport 类实现)获得或设置字段错误信息、动作错误信息以及动作消息，如 hasActionErrors 方法判断是否存在动作层的错误，getFieldErrors 获得字段错误信息(一个 Map 对象)，下面是 ValidationAware 接口提供的所有的方法。

```
package com.opensymphony.xwork2;
import java.util.Collection;
import java.util.Map;
public interface ValidationAware{
    void setActionErrors(Collection errorMessages);
    Collection getActionErrors();

    void setActionMessages(Collection messages);
    Collection getActionMessages();
    void setFieldErrors(Map errorMap);
    Map getFieldErrors();
    void addActionError(String anErrorMessage);
    void addActionMessage(String aMessage);
    void addFieldError(String fieldName, String errorMessage);
    boolean hasActionErrors();
    boolean hasActionMessages();
    boolean hasErrors();
    boolean hasFieldErrors();
}
```

7.4.2　使用 Validation 框架验证数据

上述 validate()方法实现虽然简洁，但其有一个致命的缺陷，即模块之间的功能独立、代码之间松散耦合，如果使用上述的 validate()方法实现数据验证，其验证代码和正常的逻辑代码混在一起，并不利于后续的代码维护，也不符合程序的设计逻辑，此外，而且也很难将这些代码用于其他程序的验证。为此，Struts 2 中提供了一个 Validation 框架，它的功能与 Struts 1.x 提供的 Validation 框架类似，也是通过 XML 文件进行配置。

Validation 校验框架数据校验流程为：首先对目标 Action 进行指定——ValidationAction；然后 Validation 框架依据命名规则，找到目标 Action 对应的 ValidationAction-validation.xml 文件，Validation 框架为该 Action 类创建一个验证对象（基于 XML 文件），校验器作用于输入的数据。如果验证失败，错误信息被添加到内部序列中。当所有的校验器都已经执行完毕，如果框架发现有错误信息的产生，则依据配置文件 struts.xml 中的配置，寻找 result 标签中“input”所对应的页面，而并不去调用 Action 类。如果校验通过，则调用 Action 相应的方法，并返回 result 标签中“success”对应的结果。

通过上面的例子，大家可以看到使用该校验框架十分简单方便。校验框架是通过 Validation 拦截器实现，该拦截被注册到默认的拦截器链中。它在 conversionError 拦截器之后，在 validateXxx()之前被调用。这里又出现了一个选择的问题：到底是应该在 Action 中通过 validateXxx()或 validate()实现校验，还是使用 Validation 拦截器。绝大多数情况，我建议大家使用校验框架，只有当框架满足不了要求时再自己编写代码实现。

已有的校验器 Struts 2.0 已经实现了许多常用的校验，下述是在 jar 的 default.xml 中注册的校验器。

```
< validators >
< validator name = "required"
class = "com.opensymphony.xwork2.validator.validators.RequiredFieldValidator" />
< validator name = "requiredstring"
class = "com.opensymphony.xwork2.validator.validators.RequiredStringValidator" />
< validator name = "int"
class = "com.opensymphony.xwork2.validator.validators.IntRangeFieldValidator" />
< validator name = "double"
class = "com.opensymphony.xwork2.validator.validators.DoubleRangeFieldValidator" />
< validator name = "date"
class = "com.opensymphony.xwork2.validator.validators.DateRangeFieldValidator" />
< validator name = "expression"
class = "com.opensymphony.xwork2.validator.validators.ExpressionValidator" />
< validator name = "fieldexpression"
class = "com.opensymphony.xwork2.validator.validators.FieldExpressionValidator" />
< validator name = "email"
class = "com.opensymphony.xwork2.validator.validators.EmailValidator" />
< validator name = "url"
class = "com.opensymphony.xwork2.validator.validators.URLValidator" />
< validator name = "visitor"
class = "com.opensymphony.xwork2.validator.validators.VisitorFieldValidator" />
< validator name = "conversion"
class = "com.opensymphony.xwork2.validator.validators.ConversionErrorFieldValidator" />
< validator name = "stringlength"
class = "com.opensymphony.xwork2.validator.validators.StringLengthFieldValidator" />
< validator name = "regex"
class = "com.opensymphony.xwork2.validator.validators.RegexFieldValidator" />
</ validators >
```

1. HttpServletRequest、HttpServletResponse 对象 Action 类中获取方法

在 Struts 2 中，并没有如 Struts 1.x Action 类的 execute 方法实现一样有 4 个参数(其中包含 request 和 response 两个参数)，它并没有任何参数。因此，就不能简单地从 execute 方法获得 HttpServletResponse 或 HttpServletRequest 对象了。但在 Struts 2 的 Action 类中还有很多方法可以获得这些对象。

(1) 使用 Aware 拦截器

使用 Aware 拦截器需要 Action 类实现相应的拦截器接口。如要获得 HttpServletRequest 对象，需要实现 org.apache.Struts 2.interceptor.ServletRequestAware 接口，就要获得 HttpServletResponse 对象，需要实现 org.apache.Struts 2.interceptor.Servlet.

ResponseAware 接口的具体实现方法如下。

```
import com.opensymphony.xwork2.ActionSupport;
import javax.servlet.http.*;
import org.apache.struts2.interceptor.ServletRequestAware;
import org.apache.struts2.interceptor.ServletResponseAware;
public class MyAction extends ActionSupport implements ServletRequestAware, ServletResponseAware{
```

```
    private javax.servlet.http.HttpServletRequest servletRequest;
    private javax.servlet.http.HttpServletResponse servletResponse;
    public void setServletRequest(HttpServletRequest servletRequest){
        this.servletRequest = servletRequest;
    }
    public void setServletResponse(HttpServletResponse servletResponse){
        this.servletResponse = servletResponse;
    }
    @Override
    public String execute() throws Exception {
    String name = this.servletRequest.getParameter("name");
    System.out.println("name: " + name);
    this.servletResponse.setContentType("text/html;charset = utf-8");
    PrintWriter out = servletResponse.getWriter();
    out.print("HelloWorld!!!" + "你好,北京政法!!!" + "<br>");
    out.write("hello,good!!!");
    out.flush();
    out.close();
    return this.SUCCESS;
    }
}
```

如果一个动作类实现了 ServletRequestAware 和 ServletResponseAware 接口，Struts 2 在调用 execute 方法之前，就会先调用 setServletRequest、setServletResponse 方法，并将 request、response 参数传入这个方法。如果想获得 HttpServletRequest、HttpSession 和 Cookie 等对象，动作类可以分别实现 ServletRequestAware、SessionAware 和 CookiesAware 等接口，这些接口都包含在 org. apache. Struts 2. interceptor 包。

（2）调用 ActionContext 类

此方法相对简单，可以通过 org. apache. Struts 2. ActionContext 类的 get 方法获得相应的对象，方法如下。

```
HttpServletResponse response = (HttpServletResponse) ActionContext.getContext().get(org.apache.struts2.StrutsStatics.HTTP_RESPONSE);
HttpServletRequest request = (HttpServletRequest)ActionContext.getContext().get(org.apache.struts2.StrutsStatics.HTTP_REQUEST);
```

（3）调用 ServletActionContext 类

Struts 2 中获得 HttpServletResponse 及其他对象，还可以通过调用 org. apache. Struts 2 . ServletActionContext 类。直接调用 ServletActionContext 类的 getRequest、getResponse 方法来获得 HttpServletRequest、HttpServletResponse 对象，方法如下。

```
HttpServletRequest request = ServletActionContext.getRequest();
HttpServletResponse response = ServletActionContext.getResponse();
```

2. 服务器端校验(Validation 框架)

上述介绍了 Action 类中获取 HttpServletRequest 和 HttpServletResponse 对象的方法，并就如何通过 validate 方法和 validateXxx 方法进行客户端输入校验进行了介绍，因其代码复用率不高、需要编写大量的代码等缺陷，Struts 2 提供了 Validation 解决框架，进行服务器端的数据校验，本部分以输入信息、邮件地址、年龄等信息来进行服务器端的验证为例，首先创建 Action 类(MyValidateAction. java)。

```
package com.bcpl.action;
import com.opensymphony.xwork2.ActionSupport;
public class NewValidateAction extends ActionSupport{
    private String msg;  //必须输入
    private int age;  //在 13 和 20 之间
    private String email ;  //email 地址是否合法
    public String getEmail() {
    return email;
    }
    public void setEmail(String email) {
    this.email = email;
    }

    public String getMsg(){
        return msg;
    }
    public void setMsg(String msg){
        this.msg = msg;
    }
    public int getAge(){
        return age;
    }
    public void setAge(int age){
        this.age = age;
    }
}
```

上述 MyValidateAction 类封装了 msg、age 和 email 的属性信息，下面通过 Validation 校验框架对封装的属性进行校验。Validation 校验框架校验属性需要配置校验规则的 XML 文件，通常将这个文件放置到和 Action 类的. class 文件相同的目录中，而且配置文件名要使用如下两个规则中的一个来命名。

```
<ActionClassName>-validation.xml
<ActionClassName>-<ActionAliasName>-validation.xml
```

其中，<ActionAliasName>是 struts. xml 中<ation>的 name 属性值。

在配置 XML 文件之前，先就 struts. xml 的信息进行如下配置。

```
<?xml version="1.0" encoding="UTF-8" ?>
<!DOCTYPE struts PUBLIC
    "-//Apache Software Foundation//DTD Struts Configuration 2.0//EN"
    "http://struts.apache.org/dtds/struts-2.0.dtd">
<struts>
    <package name="demo" extends="struts-default" namespace="/wjj">
        <action name="my_validate" class="com.bcpl.action.MyValidateAction">
            <result name="input">/validate_form.jsp</result>
            <result name="success">/validate_form1.jsp</result>
        </action>
    </package>
</struts>
```

然后就本节所使用的校验文件进行命名，本例中使用上述命名规则中的第一种命名规则，所以文件名为 MyValidateAction-validation.xml，文件的内容如下。

```
<?xml version="1.0" encoding="UTF-8"? >
<!DOCTYPE validators PUBLIC "-//OpenSymphony Group//XWork Validator 1.0.2//EN"
  "http://www.opensymphony.com/xwork/xwork-validator-1.0.2.dtd">
  <validators>
        <field name="msg">
              <field-validator type="requiredstring">
                    <message>required</message>
              </field-validator>
        </field>
        <field name="age">
              <field-validator type="int">
                    <param name="min">10</param>
                    <param name="max">60</param>
                    <message>must between 10 and 60</message>
              </field-validator>
        </field>
        <field name="email">
              <field-validator type="requiredstring">
                    <message>required</message>
              </field-validator>
              <field-validator type="email">
                    <message>
                        The email address is not valid.
                    </message>
              </field-validator>
        </field>
</validators>
```

上述配置文件中使用的规则：requiredstring（必须输入）和 int（确定整型范围）以及 email 地址是否合法。

最后编写用户输入信息数据的客户端页面 JSP 页，在工程的 Web 根目录中建立一个 validate_form.jsp 文件，其代码如下。

```
<%@ page language="java" import="java.util.*" %>
<%@ taglib prefix="s" uri="/struts-tags" %>
<html>
  <head>
    <title> validation data</title>
  </head>
  <body>
    <s:form action="my_validate" namespace="/wjj"  validate="true" >
        <s:textfield name="msg" label="name"  />
        <s:textfield name="age" label="age"/>
        <s:textfield name="email" label="email"/>
        <s:submit/>
    </s:form>
  </body>
</html>
```

如果在 struts.xml 的<package>标签中指定了 namespace 属性，如上述所示，则需要在<s:form>中也将 namespace 和 action 分开写，如上述代码所示，而不能将其连在一起，Struts 2 需要分开的 action 和 namespace，如果像下面这样来写，就会发生错误。

```
<s:form action="/wjj/my_validate">
  ...
</s:form>
```

7.5　Struts 2 拦截器

7.5.1　Struts 2 拦截器概述

Struts 2 的拦截器和 Servlet 过滤器的 Filter 类似。在执行 Action 的 execute 方法之前，Struts 2 会首先执行在 struts.xml 中引用的拦截器，在执行完所有引用的拦截器的 intercept 方法后，会执行 Action 的 execute 方法。Struts 2 拦截器类必须从 com.opensymphony.xwork2.interceptor.Interceptor 接口继承，在 Intercepter 接口中有如下 3 个方法需要实现。

```
void init();
String intercept(ActionInvocation invocation) throws Exception;
void destroy();
```

其中，intercept 方法是拦截器的核心方法，所有的拦截器都会调用这个方法。

Struts 2 已经在 struts-default.xml 中预定义了一些自带的拦截器，如 timer、params 等。

如果在<package>标签中继承 struts-default，则当前 package 就会自动拥有 struts-default.xml 中的所有配置，代码如下。

```
<package name="xxx" extends="struts-default" >…</package>
```

在 struts-default.xml 中有一个默认的引用，在默认情况下(也就是<action>中未引用拦截器时)会自动引用一些拦截器,默认的拦截器引用如下所示。

```
<default-interceptor-ref name = "defaultStack"/>
```

默认的拦截器包含如下拦截器:

```
<interceptor-stack name = "defaultStack">
    <interceptor-ref name = "exception"/>
    <interceptor-ref name = "alias"/>
    <interceptor-ref name = "servletConfig"/>
    <interceptor-ref name = "prepare"/>
    <interceptor-ref name = "i18n"/>
    <interceptor-ref name = "chain"/>
    <interceptor-ref name = "debugging"/>
    <interceptor-ref name = "profiling"/>
    <interceptor-ref name = "scopedModelDriven"/>
    <interceptor-ref name = "modelDriven"/>
    <interceptor-ref name = "fileUpload"/>
    <interceptor-ref name = "checkbox"/>
    <interceptor-ref name = "staticParams"/>
    <interceptor-ref name = "params">
        <param name = "excludeParams">dojo\.. * </param>
    </interceptor-ref>
    <interceptor-ref name = "conversionError"/>
    <interceptor-ref name = "validation">
        <param name = "excludeMethods">input,back,cancel,browse</param>
    </interceptor-ref>
    <interceptor-ref name = "workflow">
        <param name = "excludeMethods">input,back,cancel,browse</param>
    </interceptor-ref>
</interceptor-stack>
```

在上述的 defaultStack 中引用的拦截器都可以在<action>中不经过引用就可以使用,但是如果在<action>中引用了任何拦截器后,要使用在 defaultStack 中定义的拦截器,需要在<action>中重新明确引用。

7.5.2 拦截器的应用

1. timer 拦截器

timer 是 Struts 2 中最简单的拦截器,这个拦截器对应的类是 com. opensymphony. xwork2. interceptor. TimerInterceptor。它的功能是记录 execute 方法和其他拦截器(在 timer 后面定义的拦截器)的 intercept 方法执行的时间总和。

在配置文件中的代码配置如下所示。

```
<action name = "hello" class = "packageName.HelloAction" namespace = "test">
    <interceptor-ref name = "logger"/>
    <interceptor-ref name = "timer" />
</action>
```

由于在 timer 后面没有其他的拦截器定义,因此,timer 只能记录 execute 方法的执行时间,在访问 hello 动作时,会在控制台输出类似下面的一条信息:

```
Executed action [/test/hello! execute] took 16 ms.
```

在使用 timer 拦截器时,需要 commons-logging.jar 的支持。将 logger 引用放到 timer 的后面,就可以记录 logger 拦截器的 intercept 方法和 Action 的 execute 方法的执行时间总和,代码如下所示。

```
<action name = "hello" class = "com.ascent.struts.HelloAction">
    <interceptor-ref name = "timer" />
    <interceptor-ref name = "logger"/>
    <result>/hello.jsp</result>
</action>
```

可以使用如下的 Action 类来测试一下 timer 拦截器。

```
import com.opensymphony.xwork2.ActionSupport;

public class HelloAction extends ActionSupport {
        public String execute() throws Exception{
            Thread.sleep(1000); //延迟 1 秒
            return SUCCESS;
        }
}
```

在访问 hello 动作时,会在控制台输出类似下面的一条信息:

```
Executed action [//hello! execute] took 1000 ms.
```

2. params 拦截器

当客户端的一个 form 向服务端提交请求时,如 textfield,代码如下所示。

```
<s:form action = "login"  >
            username  <s:textfield  name = "username"/> <br>
            password  <s:password  name = "password" /> <br>
            <s:submit/>
</s:form>
```

在提交后,Struts 2 将会自动调用 login 动作类中的 setXX 方法,并将文本框中的值通过 setXX 方法的参数传入。实际上,这个操作是由 params 拦截器完成的,params 对应的类是 com.opensymphony.xwork2.interceptor.ParametersInterceptor。

由于 params 已经在 defaultStack 中定义,因此,在未引用拦截器的<action>中是会自动引用 params 的,如下面的配置代码,在访问 login 动作时,Struts 2 是会自动执行相应的 setter 方法的。

```
<action name = "login" class = "com.ascent.struts.LoginAction">
            <result>/success.jsp</result>
            <result name = "input">/login.jsp</result>
</action>
```

但如果在<action>中引用了其他的拦截器，就必须再次引用 params 拦截器，Struts 2 才能调用相应的 setter 方法，配置代码如下所示。

```
<action name = "login" class = "com.ascent.struts.LoginAction">
        <interceptor-ref name = "timer" />
   <interceptor-ref name = "logger"/>
        <interceptor-ref name = "params"/>
        <result>/success.jsp</result>
        <result name = "input">/login.jsp</result>
</action>
```

3. staticParams 拦截器

staticParams 拦截器可以通过配置<params>标签来调用 Action 类的相应的 setter 方法，staticParams 拦截器对应的类是 com . opensymphony. xwork2. interceptor. StaticParameters Interceptor。

staticParams 拦截器配置如下所示。

```
<action name = "login" class = "com.ascent.struts.LoginAction">
      <interceptor-ref name = "timer" />
      <interceptor-ref name = "params"/>

      <param name = "users">zhang</param>
      <interceptor-ref name = "staticParams"/>
</action>
```

如果 login 动作使用上面的配置，在访问 login 动作时，Struts 2 会自动调用 setUsers 方法将“zhang”作为参数值传入 setUsers 方法。

4. 拦截器栈

为了能在多个动作中方便地引用同一个或几个拦截器，可以使用拦截器栈将这些拦截器作为一个整体来引用。拦截器栈要在<package>标签中使用<interceptors>和子标签<interceptor-stack>来定义。可以像使用拦截器一样使用拦截器栈，配置代码如下。

```
<package name = "demo" extends = "struts-default" >
    <interceptors>
        <interceptor-stack name = "mystack">
            <interceptor-ref name = "timer" />
            <interceptor-ref name = "logger" />
            <interceptor-ref name = "params" />
            <interceptor-ref name = "staticParams" />
```

```
        </interceptor-stack>
    </interceptors>

    <action name = "login" class = "com.ascent.struts.LoginAction">
        <param name = "who">zhang3</param>
        <interceptor-ref name = "mystack"/>
    </action>
</package>
```

5. 自定义拦截器

自定义拦截器的实现，必须要实现 com.opensymphony.xwork2.interceptor.Interceptor 接口。

Interceptor 接口有如下 3 个方法。

```
public interface Interceptor extends Serializable {
    void destroy();
    void init();
    String intercept(ActionInvocation invocation) throws Exception;
}
```

其中，init 和 destroy 方法只在拦截器加载和释放(都由 Struts 2 自身处理)时执行一次，而 intercept 方法在每次访问动作时都会被调用。

Struts 2 在调用拦截器时，每个拦截器类只有一个对象实例，而所有引用这个拦截器的动作都共享这一个拦截器类的对象实例，因此，在实现 Interceptor 接口的类中如果使用类变量，要注意同步问题。

invoke 方法和 Servlet 过滤器中调用 FilterChain.doFilter 的方法类似，如果在当前拦截器后面还有其他的拦截器，则 invoke 方法就是调用后面拦截器的 intercept 方法，否则，invoke 会调用 Action 类的 execute 方法(或其他的执行方法)。

下面为实现一个拦截器的父类 ActionInterceptor 的代码。

```
import com.opensymphony.xwork2.ActionInvocation;
import com.opensymphony.xwork2.interceptor.Interceptor;
import javax.servlet.http.*;
import org.apache.struts2.*;

public class ActionInterceptor implements Interceptor{

    public void destroy(){
        System.out.println("destroy");
    }

    public void init() {
        System.out.println("init");
    }

    public String intercept(ActionInvocation invocation) throws Exception{
```

```
        Map session = invocation.getInvocationContext().getSession();
         String isLogin = (String) session.get("isLogin");
         if (null != isLogin && isLogin.equals("true")) {
              System.out.println("拦截器:合法用户登录---");
              return invocation.invoke();
         } else {
              System.out.println("拦截器:用户未登录---");
              return Action.LOGIN;
         }
         return invocation.invoke();
    }
}
```

intercept 方法里可以生成 Action 对象所需要的 dao 层的对象。从上面代码中的 intercept 方法可以看出,在调用 Action 所指定的方法后,来判断返回值。执行 return invockation.invoke(),返回值为 INVOKE,同样,也可以通过继承 AbstractInterceptor 类或继承 MethodFilterInterceptor 类来实现自己的拦截器,实现 doIntercept 方法对 Action 类里的具体方法进行拦截,配置代码如下。

```
    <interceptor name = "methodInterceptor" class = "com.bcpl.action.MethodInterceptor"></
interceptor>
    <interceptor-ref name = "methodInterceptor">
         <param name = "includeMethods">execute,test</param>
         <param name = "excludeMethods">abc</param>
    </interceptor-ref>
```

7.6 Struts 2 文件传输

在 Struts 2 框架之前,通常涉及的上传文件和下载文件功能,一般使用 Apache 下面的 commons 子项目的 FileUpload 组件来进行文件的上传,但其缺陷是代码比较繁琐且不灵活,Struts 2 为文件上传下载提供了更好的实现机制,但 Struts 2 并没有提供自己的请求解析器,它借助其他上传组件 commons-fileupload-1.2.2.jar、commons-io-2.0.1.jar,在此基础上做了进一步封装,以简化文件上传。

7.6.1 创建上传、下载页面

文件上传页面 upload.jsp,代码如下。

```
    <%@ page language = "java" import = "java.util.*" pageEncoding = "GB18030" %>
    <%@ taglib prefix = "s"  uri = "/struts-tags" %>

    <%
    String path = request.getContextPath();
    String basePath = request.getScheme() + "://" + request.getServerName() + ":" + request.
getServerPort() + path + "/";
```

```
%>

<! DOCTYPE HTML PUBLIC "-//W3C//DTD HTML 4.01 Transitional//EN">
<html>
  <head>
    <base href = "<% = basePath%>">

    <! --
    <link rel = "stylesheet" type = "text/css" href = "styles.css">
    -->
  </head>

  <body>
   <! --获取错误信息 -->
    <s:fielderror/>
    <! --为了完成文件上传,设置该表单的 enctype 属性为 multipart/form-data-->
    <s:form action = "upload" method = "post" enctype = "multipart/form-data">
            <s:textfield name = "title" label = "文件标题"/>
            <s:file name = "upload" label = "选择文件"/>
            <s:submit value = "上传"/>
    </s:form>

  </body>
</html>
```

文件下载页面 download.jsp,代码如下。

```
<%@ page language = "java" import = "java.util.*" pageEncoding = "GB18030" %>
<%
String path = request.getContextPath();
String basePath = request.getScheme() + "://" + request.getServerName() + ":" + request.
getServerPort() + path + "/";
%>

<! DOCTYPE HTML PUBLIC "-//W3C//DTD HTML 4.01 Transitional//EN">
<html>
  <head>
    <base href = "<% = basePath%>">
      <title>My JSP 'download.jsp' starting page</title>

  </head>

  <body>
    <h1>文件下载页面</h1>

    <a href = "download.action? filePath = 06.gif">06.gif</a>
  </body>
</html>
```

文件上传成功页面 uploadSucc.jsp，代码如下。

```
<%@ page language="java" contentType="text/html; charset=GBK" %>
<%@ taglib uri="/struts-tags" prefix="s" %>
<html>
    <head>
        <title>上传成功页面</title>
    </head>
    <body>
        文件上传成功了！<br>
        文件名：<s:property value="title"/><br>
        文件为：<img src="upload/<s:property value='uploadFileName'/>"/>
    </body>
</html>
```

7.6.2 创建文件上传、下载 Action 处理类

(1) 创建处理文件上传 Action 类，代码如下。

```
/**
 *用来处理上传文件的动作类
 */
package com.bcpl.action;

import java.io.File;
import java.io.FileInputStream;
import java.io.FileOutputStream;

import org.apache.struts2.ServletActionContext;

import com.opensymphony.xwork2.ActionSupport;
public class UploadAction extends ActionSupport {
    private String title ;
    private File upload ;

    private String uploadFileName ; //规则定义
    private String uploadContentType ;
    private String savePath ; //自定义变量

    @Override
    public String execute() throws Exception {
        //以服务器的文件保存地址和原文件的名，建立上传文件输出流
FileOutputStream fos = new FileOutputStream(this.getSavePath() + "\\" + this.getUploadFileName());

       //以上传文件建立一个文件上传流
       FileInputStream fis = new FileInputStream(this.getUpload());
```

```
        //将上传文件的内容写入服务器
        byte [] buffer = new byte[1024];
        int len = 0;
        while((len = fis.read(buffer))>0){
            fos.write(buffer, 0, len);
        }
        System.out.println("结束上传单个文件----------------------");
        return SUCCESS;

  }

  /**
   * @return the title
   */
public String getTitle() {
        return title;
}
/**
   * @param title the title to set
   */
public void setTitle(String title) {
        this.title = title;
}
/**
  * @return the upload
  */
public File getUpload() {
        return upload;
}
/**
  * @param upload the upload to set
  */
public void setUpload(File upload) {
        this.upload = upload;
}
/**
  * @return the uploadFileName
  */
public String getUploadFileName() {
        return uploadFileName;
}
/**
  * @param uploadFileName the uploadFileName to set
  */
public void setUploadFileName(String uploadFileName) {
        this.uploadFileName = uploadFileName;
```

```
    }
    /**
      * @return the uploadContentType
      */
    public String getUploadContentType() {
         return uploadContentType;
    }
    /**
      * @param uploadContentType the uploadContentType to set
      */
    public void setUploadContentType(String uploadContentType) {
         this.uploadContentType = uploadContentType;
    }
    /**
      * @return the savePath
      */
    public String getSavePath() {
         return ServletActionContext.getRequest().getRealPath(savePath);
    }
    /**
      * @param savePath the savePath to set
      */
    public void setSavePath(String savePath) {
         this.savePath = savePath;
    }
}
```

(2)创建处理文件下载的 Action 类,代码如下。

```
/**
 *用于下载处理的动作类
 */
package com.bcpl.action;
import java.io.InputStream;
import org.apache.struts2.ServletActionContext;
import com.opensymphony.xwork2.ActionSupport;
public class DownloadAction extends ActionSupport {
    //下载文件的名字
    private String filePath;
    //文件存放的根包,自动注册过来
    private String savePath;
    //属性的setter方法

    //下载用的Action应该返回一个InputStream实例
    //该方法对应在一个配置的result里面的inputName,属性值为targetFile
      public  InputStream  getTargetFile ( )  throws  Exception {                  return
ServletActionContext.getServletContext().getResourceAsStream(savePath + "/" + filePath);
```

```
    }

    @Override
    public String execute() throws Exception {
        return super.execute();
    }

    /**
     * @return the filePath
     */
    public String getFilePath() {
        return filePath;
    }

    /**
     * @param filePath the filePath to set
     */
    public void setFilePath(String filePath) {
        this.filePath = filePath;
    }

    /**
     * @return the savePath
     */
    public String getSavePath() {
        return savePath;
    }

    /**
     * @param savePath the savePath to set
     */
    public void setSavePath(String savePath) {
        this.savePath = savePath;
    }

}
```

7.6.3 配置 struts.xml 文件

Struts 2 提供了文件上传拦截器 FileUpload，拦截器实现文件过滤，需要在 Action 中配置或在下面的配置文件中进行配置，如下所示。

```
<?xml version="1.0" encoding="UTF-8" ?>
<!DOCTYPE struts PUBLIC
    "-//Apache Software Foundation//DTD Struts Configuration 2.0//EN"
    "http://struts.apache.org/dtds/struts-2.0.dtd">
```

```
<struts>
    <constant name = "struts.custom.i18n.resources" value = “message"/>
    <package name = "struts2upload" extends = "struts-default">
        <default-action-ref name = "download"></default-action-ref>

        <action name = "upload" class = "com.bcpl.action.UploadAction">
            <! -- 配置上传拦截器 -->
            <interceptor-ref name = "fileUpload">
                <! -- 配置允许上传文件类型 -->
                 <param name = "allowedTypes">image/bmp,image/gif,image/jpg</param>
                <! -- 配置允许上传文件大小  1024 * 1024-->
                <param name = "maximumSize">5048576</param>
            </interceptor-ref>

            <! -- 配置默认拦截器栈,配置拦截器后,默认拦截器栈就不会再自动设置  -->
            <interceptor-ref name = "defaultStack"></interceptor-ref>

            <! -- 设置上传路径 -->
            <param name = "savePath">/upload</param>
            <result>/uploadSucc.jsp</result>
            <! -- 上传过滤出错后自动返回 input -->
            <result name = "input">/upload.jsp</result>
        </action>

        <action name = "download" class = "com.bcpl.action.DownloadAction">
            <! -- 配置下载文件的存放文件路径 -->
            <param name = "savePath">/upload</param>
                        <! -- 配置结果类型为 stream 的结果 -->
            <result name = "success" type = "stream">
                <! -- 配置下载的类型 -->
                <param name = "contentType">image/bmp,image/gif,image/jpg</param>
                <! -- 配置下载文件的位置 -->
                <param name = "inputName">targetFile</param>
                <! -- 配置下载文件的名字 -->
                 <param name = "contentDisposition">filename = "${filePath}"</param>
                <! -- 配置下载文件的缓冲大小 -->
                <param name = "bufferSize">4096</param>
            </result>
        </action>

    </package>
</struts>
```

上述配置中 FileUpload 拦截器有两个参数。

- allowedTypes：指定文件类型，类型间用英文逗号隔开。
- maximumSize：指定上传文件的最大值，单位为字节。

此外，如果自定义拦截器，需要设置默认的拦截器 defaultStack 拦截器。当发生不符合条件的错误时，会自动返回 input 逻辑视图对应的 JSP 页面，所以必须配置 input 逻辑视图对应的 JSP 页面。

7.6.4 错误信息输出

如上述的上传 JSP 代码页面，如果上传文件失败，系统返回到 input 对应的页面，要在 input 对应的页面输出文件过滤失败信息，可以在 input 对应的页面中增加 <s:fielderror/> 来显示错误信息。当发生错误、上传类型不匹配或文件过大等问题时，页面提示信息。

在 Struts 2 的资源文件中默认有两个 key 来定义过大错误信息和类型不匹配的错误信息。

- 默认文件太大的提示信息 key 为：struts. messages. error. file. too. large。
- 类型不允许的提示信息 key 为：struts. messages. error. content. type. not. allowed。

所以在 Struts 2 中，错误信息输出使用国际化信息，需要创建 Struts 的属性文件和国际化资源文件。同时在上述的配置文件中添加国际化的引用：

```
<constant name = "struts.custom.i18n.resources" value = "essage"/>
```

1. 创建 struts. properties 属性文件

代码如下所示。

```
struts.custom.i18n.resources = message
```

2. 创建国际化的资源属性文件

在 src 下创建国际化的资源文件 message_zh_CN. properties 和 message_en_US. properties。

message_zh_CN. properties 文件的代码如下所示。

```
    struts.messages.error.content.type.not.allowed = \u60a8\u4e0a\u4f20\u7684\u6587\u4ef6\u7c7b\u578b\
u53ea\u80fd\u662f\u56fe\u7247\u6587\u4ef6\uff01\u8bf7\u91cd\u65b0\u9009\u62e9\uff01

    struts.messages.error.file.too.large = \u60a8\u8981\u4e0a\u4f20\u7684\u6587\u4ef6\u592a\
u5927\uff0c\u8bf7\u91cd\u65b0\u9009\u62e9\uff01

    struts.messages.error.uploading = \u63d0\u793a\u4fe1\u606f\u662f\u4e0a\u4f20\u8fc7\u7a0b\
u7c7b\u578b\u5141\u8bb8\uff0c\u5927\u5c0f\u6b63\u786e\uff0c\u4f46\u662f\u5728\u4f20\u8f93\
u8fc7\u7a0b\u4e2d\u51fa\u73b0\u9519\u8bef\u7684\u63d0\u793a\u4fe1\u606f\u3002
```

message_en_US. properties 文件的代码如下所示。

```
struts.messages.error.content.type.not.allowed = The type is not allowed!!
struts.messages.error.file.too.large = The file is too large!!
struts.messages.error.uploading = uploading wrong!!
```

7.7 小　结

我们讲解了 Struts 2 中的高级技术，它在大多数的标准下都能良好地运行，同时，我们也可以利用其开发出许多基于 Web 框架的应用程序。尤其是基于它的国际化应用高级技术、服务器端生成的 HTML 和客户端验证的 JavaScript 的核心技术应用及结合的全面深入的分析，有助于我们对其关键技术进行深入了解、应用和开发。

第 8 章 Hibernate 技术

8.1 Hibernate 概述

Hibernate 是一个开放源代码的对象关系映射框架，它对 JDBC 进行了非常轻量级的对象封装，使得 Java 程序员可以随心所欲地使用对象编程思维来操纵数据库。Hibernate 可以应用在任何使用 JDBC 的场合，既可以在 Java 的客户端程序使用，也可以在 Servlet/JSP 的 Web 应用中使用。Hibernate 是一个面向 Java 环境的对象/关系数据库映射工具。对象/关系数据库映射（Object/Relational Mapping，ORM）这个术语表示一种技术，用来把对象模型表示的对象映射到基于 SQL 的关系模型数据结构中去。

目前有好多持久化层中间件：有些是商业性的，如 TopLink；有些是非商业性的，如 JDO、Hibernate、iBatis。Java 开发人员可以方便地通过 Hibernate API 操纵数据库，用来把对象模型表示的对象映射到基于 SQL 的关系模型数据结构中去。Hibernate 不仅仅管理 Java 类到数据库表的映射，还提供数据查询和获取数据的方法，可以大幅度减少开发时人工使用 SQL 和 JDBC 处理数据的时间，Hibernate 应用架构如图 8-1 所示。

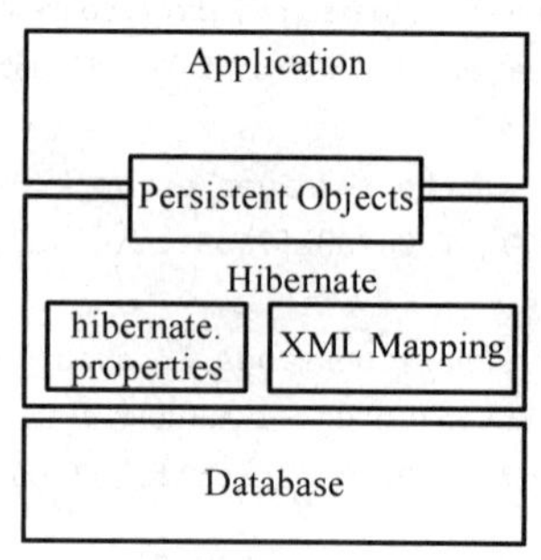

图 8-1 Hibernate 应用架构

对于应用程序来说，所有的底层 JDBC/JTA API 都被抽象了，Hibernate 会替开发者管理所有的细节。Hibernate 特性主要包含以下 7 个方面。

① Persistence for POJOs（Plain Old Java Object）：对 POJO 持久化（简单传统 Java 对象）。

② Flexible and intuitive mapping：灵活与易学的映射。

③ Support for fine-grained object models：支持细粒度的对象模型。

④ Powerful，high performance queries：强大和高效的查询。

⑤ Dual-Layer Caching Architecture（HDLCA）：两层缓存架构。

⑥ Toolset for roundtrip development ：（SQL、Java 代码、XML 映射文件中）进行相互转换的工具。

⑦ Support for detached persistent objects：支持游离，持久对象。

Hibernate 的核心接口一共有 6 个，分别为 Session、SessionFactory、Transaction 、Query、Criteria 和 Configuration，这 6 个核心接口在任何开发中都会用到。通过这些接口，不仅可以对持久化对象进行存取，还能够进行事务控制，如图 8-2 所示，下面对这 6 个核心接口分别加以介绍。

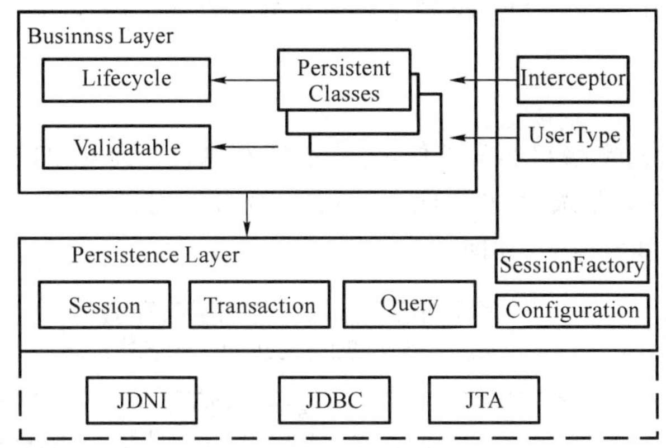

图 8-2　Hibernate 接口框架

(1) SessionFactory

对属于单一数据库编译过的映射文件的一个线程安全的，不可变的缓存快照，它是 Session 的工厂，是 ConnectionProvider 的客户，可能持有一个可选的(第二级)数据缓存，可以在进程级别或集群级别保存且可以在事物中重用的数据。

(2) Session

单线程且生命期短暂的对象，代表应用程序和持久化层之间的一次对话。封装了一个 JDBC 连接，也是 Transaction 的工厂，保存有必需的(第一级)持久化对象的缓存，用于遍历对象图，或者通过标识符查找对象。

(3) Persistent Object and Collection

生命周期短暂的单线程的对象，包含了持久化状态和商业功能。它们可能是普通的 JavaBeans/POJOs，唯一特别的是它们从属于且仅从属于一个 Session。一旦 Session 被关闭，它们都将从 Session 中取消联系，可以在任何程序层自由使用(如直接作为传送到表现层的数据传输对象(DTO))。

(4) Transient Object and Collection

目前没有从属于一个 Session 的持久化类的实例，它们可能是刚刚被程序实例化，还没有来得及被持久化，或者是被一个已经关闭的 Session 实例化。

(5) Transaction

单线程且生命期短暂的对象，应用程序用它来表示一批不可分割的操作，它是底层的 JDBC、JTA 或 CORBA 事务的抽象。一个 Session 在某些情况下可能跨越多个 Transaction 事务。

(6) ConnectionProvider

JDBC 连接的工厂和池，从底层的 Datasource 或者 DriverManager 抽象而来，对应用程序不可见，但可以被开发者扩展/实现。

(7) TransactionFactory

事务实例的工厂，对应用程序不可见，但可以被开发者扩展/实现。单线程且生命期短促的对象，应用程序用它来表示一批工作的原子操作，它是底层的 JDBC、JTA 或 CORBA 事务的抽象。一个 Session 在某些情况下可能跨越多个 Transaction 事务。

(8) Query

Query 接口让你方便地对数据库及持久对象进行查询，它可以有两种表达方式：HQL 语言或本地数据库的 SQL 语句。Query 经常被用来绑定查询参数、限制查询记录数量，并最终执行查询操作。

(9) Criteria

Criteria 接口与 Query 接口非常类似，允许创建并执行面向对象的标准化查询。值得注意的是 Query 接口也是轻量级的，它不能在 Session 之外使用。

Hibernate 的基本执行流程，如图 8-3 所示。

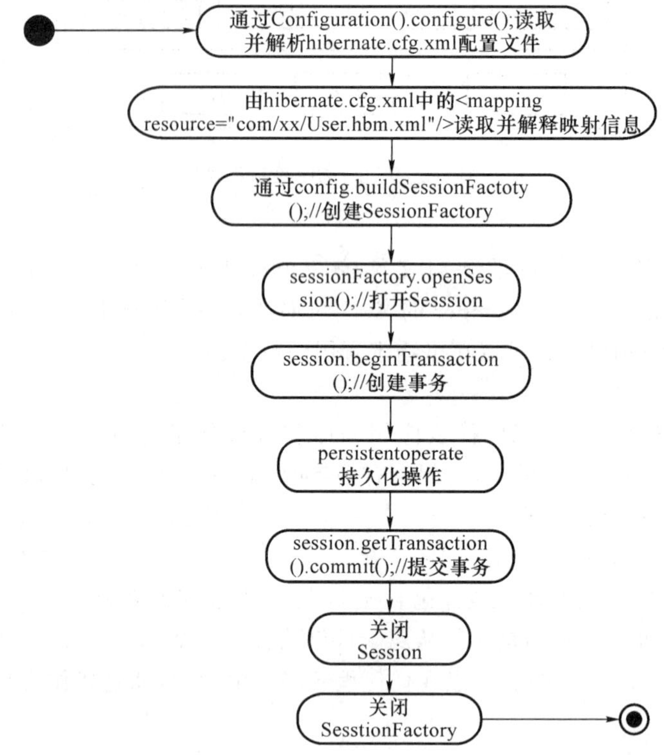

图 8-3　Hibernate 基本执行流程

8.2　Hibernate 对象/关系数据库映射(单表)

8.2.1　持久化层

数据持久化层(Persistent Objects)包括 3 个部分：整体数据库的 hiberenate. cfg. xml 文件、每个表的 POJO/JavaBean 类和每个表的 hbm. xml 文件。

1. hibernate. cfg. xml 文件

首先我们来讨论一个重要的 XML 配置文件：hibernate. cfg. xml。这个文件可以被用于替代以前版本中的 hibernate. properties 文件，如果二者都出现，它会覆盖 properties 文件。

XML 配置文件默认会期望在 classpath 的根目录中找到，代码实例如下：

```
<? xml version = '1.0' encoding = 'UTF - 8'? >
<! DOCTYPE hibernate - configuration PUBLIC
          " - //Hibernate/Hibernate Configuration DTD 3.0//EN"
          "http://hibernate.sourceforge.net/hibernate - configuration - 3.0.dtd">

<! -- Generated by MyEclipse Hibernate Tools. -- >
<hibernate - configuration>
     <session - factory>
         <property name = "connection.username">root</property>
         <property name = "connection.url">jdbc:mysql://localhost:3306/test</property>
         <property name = "dialect">org.hibernate.dialect.MySQLDialect</property>
         <property name = "myeclipse.connection.profile">mysql driver</property>
         <property name = "connection.password">123</property>
         <property name = "connection.driver_class">com.mysql.jdbc.Driver</property>
      </session - factory>
</hibernate - configuration>
```

上述代码配置文件主要是管理数据库的整体信息，如 URL、driver class 和 dialect 等，同时管理数据库中各个表的映射文件（hbm. xml），在 hibernate. cfg. xml 文件基础上配置 Hibernate，如下：

```
SessionFactory sf = new Configuration().configure().buildSessionFactory();
```

或者可以使用一个叫 HibernateSessioFactory 的工具类，它优化改进了 Session Factory 和 Session 的管理，源代码如下。

```
import net.sf.hibernate.HibernateException;
import net.sf.hibernate.Session;
import net.sf.hibernate.cfg.Configuration;

public class HibernateSessionFactory {


     private static String CONFIG_FILE_LOCATION = "/hibernate.cfg.xml";

     /** Holds a single instance of Session */
     private static final ThreadLocal threadLocal = new ThreadLocal();

     /** The single instance of hibernate configuration */
     private static final Configuration cfg = new Configuration();

     /** The single instance of hibernate SessionFactory */
     private static net.sf.hibernate.SessionFactory sessionFactory;


     public static Session currentSession() throws HibernateException {
          Session session = (Session) threadLocal.get();
```

```
        if (session == null) {
            if (sessionFactory == null) {
                try {
                    cfg.configure(CONFIG_FILE_LOCATION);
                    sessionFactory = cfg.buildSessionFactory();
                }
                catch (Exception e) {
                    System.err.println("%%%% Error Creating SessionFactory %%%%");
                    e.printStackTrace();
                }
            }
            session = sessionFactory.openSession();
            threadLocal.set(session);
        }

        return session;
    }


    public static void closeSession() throws HibernateException {
        Session session = (Session) threadLocal.get();
        threadLocal.set(null);

        if (session != null) {
            session.close();
        }
    }
}
```

2. 持久化类

持久化类(Persistent Classes)是应用程序用来解决商业问题的类。持久化类,就如同它的名字暗示的一样,不是短暂存在的,它的实例会被持久性保存于数据库中。如果这些类符合简单的规则,Hibernate 能够工作得最好,这些规则就是简单传统 Java 对象(Plain Old Java Object,POJO)编程模型。

```
package com.bcpl.po;
public abstract classPojouser implements java.io.Serializable {
    private String username;
    private String password;
    publicPojouser() {
    }
    public AbstractProductuser(String username,String password) {
        this.username = username;
        this.password = password;
            }
    public String getUsername() {
        return this.username;
    }
    public void setUsername(String username) {
```

```
        this.username = username;
    }
    public String getPassword() {
        return this.password;
    }
    public void setPassword(String password) {
        this.password = password;
    }
```

上述属性不一定需要声明为 public。Hibernate 可以对 default、protected 或 private 的 get/set 方法对的属性一视同仁地执行持久化。

所有的持久化类都必须具有一个默认的构造方法(可以不是 public 的),这样的话 Hibernate 就可以使用 Constructor. newInstance()来实例化它们。

Hibernate 的关键功能之一:代理(proxies)。它要求持久化类不是 final 的,或者是一个全部方法都是 public 的接口的具体实现。可以对一个 final 的,也没有实现接口的类执行持久化,但是不能对它们使用代理,否则会影响进行性能优化的选择。

3. hbm. xml 文件

hbm. xml 文件是 O/R Mapping 的基础,这个映射文档被设计为易读的并且可以手工修改。映射语言是以 Java 为中心的,意味着映射是按照持久化类的定义来创建的,而非表的定义。

虽然很多 Hibernate 用户选择手工定义 XML 映射文档,也用一些工具来生成映射文档,包括 XDoclet、Middlegen 和 AndroMDA。

关于上述 Pojo 类的 pojouser. hbm. xml 例,配置代码如下。

```
<?xml version="1.0" encoding="utf-8"?>
<!DOCTYPE hibernate-mapping PUBLIC "-//Hibernate/Hibernate Mapping DTD 3.0//EN"
"http://hibernate.sourceforge.net/hibernate-mapping-3.0.dtd">
<hibernate-mapping>
    <class name="com.bcpl.po.Pojouser" table="user" >
        <id name="uid" type="integer">
            <column name="uid" />
            <generator class="native" />
        </id>
        <property name="username" type="string">
            <column name="username" length="32" />
        </property>
        <property name="password" type="string">
            <column name="password" length="32" />
        </property>
    </class>
</hibernate-mapping>
```

关于映射文档的内容和 Hibernate 在运行时用到的文档元素和属性,此外还包括一些额外的可选属性和元素,它们在使用 Schema 导出工具的时候会影响导出的数据库 Schema 结果。

(1) doctype

所有的 XML 映射都需要定义如上所示的 doctype。DTD 可以从上述 URL 中获取,或在

hibernate-x.x.x/src/net/sf/hibernate 目录中，或在 hibernate.jar 文件中找到。Hibernate 总是会在它的 classpath 中首先搜索 DTD 文件。

(2) hibernate-mapping

包括 3 个可选的属性。

schema 属性指明了这个映射所引用的表所在的 schema 名称。假若指定了这个属性，表名会加上所指定的 schema 的名字扩展为全限定名。假若没有指定，表名就不会使用全限定名。

default-cascade 属性指定了未明确注明 cascade 属性的 Java 属性和集合类 Java 会采取什么样的默认级联风格。

auto-import 属性默认让我们在查询语言中可以使用非全限定名的类名。

```
<hibernate-mapping
            schema="schemaName"
            default-cascade="none|save-update"
            auto-import="true|false"
            package="package.name"
 />
```

① schema(可选)：数据库 schema 名称。

② default-cascade(可选，默认为 none)：默认的级联风格。

③ auto-import(可选，默认为 true)：指定是否我们可以在查询语言中使用非全限定的类名(仅限于本映射文件中的类)。

④ package(可选)：指定一个包前缀，如果在映射文档中没有指定全限定名，就使用这个包名。

假若有两个持久化类，它们的非全限定名是一样的(就是在不同的包里面)，应该设置 auto-import="false"。假若说把一个"import"的名字同时对应两个类，Hibernate 会抛出一个异常。

(3) class

可以使用 class 元素来定义一个持久化类。

```
<class
          name="ClassName"
          table="tableName"
          discriminator-value="discriminator_value"
          mutable="true|false"
          schema="owner"
          proxy="ProxyInterface"
          dynamic-update="true|false"
          dynamic-insert="true|false"
          select-before-update="true|false"
          polymorphism="implicit|explicit"
          where="arbitrary sql where condition"
          persister="PersisterClass"
          batch-size="N"
          optimistic-lock="none|version|dirty|all"
          lazy="true|false"
/>
```

① name:持久化类(或者接口)的 Java 全限定名。

② table:对应的数据库表名。

③ discriminator-value(辨别值)(可选,默认和类名一样):一个用于区分不同的子类的值,在多态行为时使用。

④ mutable(可变)(可选,默认值为 true):表明该类的实例可变(不可变)。

⑤ schema(可选):覆盖在根<hibernate-mapping>元素中指定的 schema 名字。

⑥ proxy(可选):指定一个接口,在延迟装载时作为代理使用,可以在这里使用该类自己的名字。

⑦ dynamic-update(动态更新)(可选,默认值为 false):指定用于 update 的 SQL 将会在运行时动态生成,并且只更新那些改变过的字段。

⑧ dynamic-insert(动态插入)(可选,默认值为 false):指定用于 insert 的 SQL 将会在运行时动态生成,并且只包含那些非空值字段。

⑨ select-before-update(可选,默认值为 false):指定 Hibernate 除非确定对象的确被修改了,否则,不会执行 SQL update 操作。如果为 true 时,在执行修改操作时,Hibernate 会在 update 之前执行一次额外的 SQL select 操作,来决定是否应该进行 update。

⑩ polymorphism(多形,多态)(可选,默认值为 implicit(隐式)):界定是隐式还是显式的使用查询多态。

⑪ where(可选):指定一个附加的 SQL where 条件,在抓取这个类的对象时会一直增加这个条件。

⑫ persister(可选):指定一个定制的 ClassPersister。

⑬ batch-size(可选,默认值为 1):指定一个用于根据标识符抓取实例时使用的"batch size"(批次抓取数量)。

⑭ optimistic-lock(乐观锁定)(可选,默认是 version):决定乐观锁定的策略。

⑮ lazy(延迟)(可选):假若设置 lazy="true",就是设置这个类自己的名字作为 proxy 接口的一种等价快捷形式。

若指明的持久化类实际上是一个接口,也可以被完美地接受,其后可以用<subclass>来指定该接口的实际实现类名。可以持久化任何 static(静态的)内部类,但应该使用标准的类名格式。

不可变类,mutable="false"不可以被应用程序更新或者删除,这可以让 Hibernate 做一些性能优化。

可选的 proxy 属性可以允许延迟加载类的持久化实例。Hibernate 开始会返回实现了这个命名接口的 CGLIB 代理。当代理的某个方法被实际调用的时候,真实的持久化对象才会被装载。

implicit(隐式)的多态是指,如果查询中给出的是任何超类、该类实现的接口或者该类的名字,都会返回这个类的实例;如果查询中给出的是子类的名字,则会返回子类的实例。explicit(显式)的多态是指,只有在查询中给出的明确是该类的名字时才会返回这个类的实例;同时只有当在这个<class>的定义中作为<subclass>或者<joined-subclass>出现的子类,才会可能返回。大多数情况下,默认的 polymorphism="implicit"都是合适的。显式的多态在有两个不同的类映射到同一个表的时候很有用。

persister 属性可以让你定制这个类使用的持久化策略。可以指定你自己实现的 net. sf. hibernate. persister. EntityPersister 的子类，甚至可以完全从头开始编写一个 net. sf. hibernate. persister. ClassPersister 接口的实现，可能是用储存过程调用、序列化到文件或者 LDAP 数据库来实现的。

请注意 dynamic-update 和 dynamic-insert 的设置并不会继承到子类，所以在<subclass>或者<joined-subclass>元素中可能需要再次设置，这些设置是否能够提高效率要视情形而定。

使用 select-before-update 通常会降低性能，当它在防止数据库不必要地触发 update 触发器时，这就很有用了。

如果打开了 dynamic-update，可以选择几种锁定的策略：

- version(版本检查)检查 version/timestamp 字段；
- all(全部)检查全部字段；
- dirty(脏检查)只检查修改过的字段；
- none(不检查)不使用乐观锁定。

建议在 Hibernate 中使用 version/timestamp 字段来进行乐观锁定。对性能来说，这是最好的选择，并且这也是唯一能够处理在 Session 外进行操作的策略(就是说，当使用 Session. update()的时候)。记住 version 或 timestamp 属性永远不能使用 null，不管何种 unsaved-value 策略，否则实例会被认为是尚未被持久化的。

(4) id

被映射的类必须声明对应的数据库表主键字段。大多数类有一个 JavaBeans 风格的属性，为每一个实例包含唯一的标识。<id>元素定义了该属性到数据库表主键字段的映射。

```
<id
        name = "propertyName"
        type = "typename"
        column = "column_name"
        unsaved - value = "any|none|null|id_value"
        access = "field|property|ClassName">
        <generator class = "generatorClass"/>
</id>
```

① name(可选)：标识属性的名字。

② type(可选)：标识 Hibernate 类型的名字。

③ column(可选，默认为属性名)：主键字段的名字。

④ unsaved-value(可选，默认为 null)：一个特定的标识属性值，用来标识该实例是刚刚创建的，尚未保存。这可以把这种实例和从以前的 Session 中装载过(可能又作过修改)但未再次持久化的实例区分开来。

⑤ access(可选，默认为 property)：Hibernate 用来访问属性值的策略。

如果 name 属性不存在，会认为这个类没有标识属性。

unsaved-value 属性很重要。如果类的标识属性不是默认为 null 的，应该指定正确的默认值。还有一个另外的<composite-id>声明可以访问旧式的多主键数据，不鼓励使用这种方式。

```
id generator
```

必须声明的<generator>子元素是一个 Java 类的名字，用来为该持久化类的实例生成唯一的标识。如果这个生成器实例需要某些配置值或者初始化参数，用<param>元素来传递。

```
<id name = "id" type = "long" column = "uid" unsaved - value = "0">
        <generator class = "net.sf.hibernate.id.TableHiLoGenerator">
                <param name = "table">uid_table</param>
                <param name = "column">next_hi_value_column</param>
        </generator>
</id>
```

所有的生成器都实现 net. sf. hibernate. id. IdentifierGenerator 接口，这是一个非常简单的接口，某些应用程序可以选择提供它们自己特定的实现。当然，Hibernate 提供了很多内置的实现。下面是一些内置生成器的快捷名字。

① increment

适用于代理主键，由 Hibernate 自动以递增(increment)的方式生成标识符，每次递增为 1，用于为 long、short 或 int 类型生成唯一标识。只有在没有其他进程往同一张表中插入数据时才能使用，在集群下不要使用。

② identity

适用于代理主键，由底层数据库生成标识符，前提是底层数据库支持自动增长字段类型，对 DB2、MySQL、MS SQL Server、Sybase 和 HypersonicSQL 的内置标识字段提供支持，返回的标识符是 long、short 或 int 类型。

③ sequence

适用于代理主键，Hibernate 根据底层数据库的序列生成标识符，前提是底层数据库支持序列。在 DB2、PostgreSQL、Oracle、SAP DB、McKoi 中使用序列(sequence)，而在 Interbase 中使用生成器(generator)，返回的标识符是 long、short 或 int 类型的。

④ hilo

适用于代理主键，Hibernate 根据高低位(hi/low)算法生成标识符。Hibernate 把特定的字段作为"high"。默认情况下选用 hibernate_unique_key 表的 next_hi 字段。使用一个高/低位算法来高效地生成 long、short 或 int 类型的标识符。给定一个表和字段(默认分别是 hibernate_unique_key 和 next_hi)作为高位值的来源。高/低位算法生成的标识符只在一个特定的数据库中是唯一的。在使用 JTA 获得的连接或者用户自行提供的连接中，不要使用这种生成器。

seqhilo(使用序列的高低位)：使用一个高/低位算法来高效地生成 long、short 或 int 类型的标识符，给定一个数据库序列(sequence)的名字。

⑤ uuid. hex

用一个 128 位的 UUID 算法生成字符串类型的标识符，在一个网络中唯一(使用了 IP 地址)。UUID 被编码为一个 32 位 16 进制数字的字符串。

⑥ uuid. string

与 uuid. hex 使用同样的 UUID 算法，UUID 被编码为一个 16 个字符长的任意 ASCII 字符组成的字符串，不能使用在 PostgreSQL 数据库中。

⑦ native

适用于代理主键，根据底层数据库对自动生成标识符的支持能力，来选择 identify、sequence或hilo 中的一个。

⑧ assigned

适用于自然主键，由 Java 应用程序生成标识符。为了能让 Java 应用程序设置 OID，不要把方法声明为 private 类型，应当尽量避免使用自然主键。让应用程序在 save()之前为对象分配一个标识符。

⑨ foreign

使用另外一个相关联的对象的标识符，和<one-to-one>联合一起使用。

⑩ Hi/Lo Algorithm

hilo 和 seqhilo 生成器给出了两种高低算法 Hi/Lo Algorithm 的实现，这是一种很令人满意的标识符生成算法。第一种实现需要一个"特殊"的数据库表来保存下一个可用的"hi"值，第二种实现使用一个 Oracle 风格的序列。

```
<id name = "id" type = "long" column = "cat_id">
        <generator class = "hilo">
                <param name = "table">hi_value</param>
                <param name = "column">next_value</param>
                <param name = "max_lo">100</param>
        </generator>
</id>
<id name = "id" type = "long" column = "cat_id">
        <generator class = "seqhilo">
                <param name = "sequence">hi_value</param>
                <param name = "max_lo">100</param>
        </generator>
</id>
```

如果在为 Hibernate 自行提供 Connection，或者 Hibernate 使用 JTA 获取应用服务器的数据源连接的时候无法使用 hilo。Hibernate 必须能够在一个新的事务中得到一个"hi"值。在 EJB 环境中实现高低位算法的标准方法是使用一个无状态的 session bean。

⑪ UUID Algorithm

UUID 算法(UUID Algorithm)包含 IP 地址、JVM 的启动时间(精确到 1/4 秒)、系统时间和一个计数器值(在 JVM 中唯一)。在 Java 代码中不可能获得 MAC 地址或者内存地址，所以这已经是我们在不使用 JNI 的前提下能做的最好实现了。

标识字段和序列(Identity Columns and Sequences)：对于内部支持标识字段的数据库(DB2、MySQL、Sybase 和 MS SQL)，可以使用 identity 关键字生成。对于内部支持序列的数据库(DB2、Oracle、PostgreSQL、Interbase、McKoi 和 SAP DB)，可以使用 sequence 风格的关键字生成，这两种方式对于插入一个新的对象都需要两次 SQL 查询。

```
<id name = "id" type = "long" column = "uid">
        <generator class = "sequence">
                <param name = "sequence">uid_sequence</param>
        </generator>
</id>
<id name = "id" type = "long" column = "uid" unsaved - value = "0">
        <generator class = "identity"/>
</id>
```

对于跨平台开发，native 策略会从 identity、sequence 和 hilo 中进行选择，取决于底层数据库的支持能力。

⑫ assigned identifiers

如果需要应用程序分配(assigned)一个标识符(identifers)(而非 Hibernate 来生成它们)，可以使用assigned生成器，这种特殊的生成器会使用已经分配给对象的标识符属性的标识符值，用这种特性来分配商业行为的关键字要特别小心。

因为继承关系，使用这种生成器策略的实体不能通过 Session 的 saveOrUpdate()方法保存。作为替代，应该明确告知 Hibernate 是应该被 save 还是 update，分别调用 Session 的 save()或 update()方法。

⑬ composite-id

```
<composite - id
        name = "propertyName"
        class = "ClassName"
        unsaved - value = "any|none"
        access = "field|property|ClassName">
        <key - property name = "propertyName" type = "typename" column = "column_name"/>
        <key - many - to - one name = "propertyName class = "ClassName" column = "column_name"/>
        ...
</composite - id>
```

如果表使用联合(composite)主键，可以把类的多个属性组合成为标识符属性。<composite-id>元素接受<key-property>属性映射和<key-many-to-one>属性映射作为子元素。

```
<composite - id>
        <key - property name = "medicareNumber"/>
        <key - property name = "dependent"/>
</composite - id>
```

持久化类必须重载 equals()和 hashCode()方法，来实现组合的标识符判断等价，也必须实现 Serializable 接口，这种组合关键字的方法意味着一个持久化类是它自己的标识。除了对象自己之外，没有什么方便的引用可用。必须自己初始化持久化类的实例，在使用组合关键字 load()持久化状态之前，必须填充它的联合属性。

- name(可选)：一个组件类型，持有联合标识。
- class(可选，默认为通过反射(reflection)得到的属性类型)：作为联合标识的组件类名。
- unsaved-value(可选，默认为 none)：假如被设置为非 none 的值，就表示新创建，尚未被持久化的实例将持有的值。

⑭ discriminator

在“一棵对象继承树对应一个表”的策略中，<discriminator>元素是必需的，它声明了表的识别器(discriminator)字段。识别器字段包含标志值，用于告知持久化层应该为某个特定的行创建哪一个子类的实例。只能使用受到限制的一些类型：string、character、integer、byte、short、boolean、yes_no和true_false。

```
<discriminator
        column = "discriminator_column"
        type = "discriminator_type"
        force = "true|false"
/>
```

- column(可选，默认为class)：识别器字段的名字。
- type(可选，默认为string)：一个Hibernate字段类型的名字。
- force(强制)(可选，默认为false)：Hibernate指定允许的识别器值。

标识器字段的实际值是根据<class>和<subclass>元素的discriminator-value得来的。

force属性仅仅是在表包含一些未指定应该映射到哪个持久化类的时候才是有用的。这种情况不是经常会遇到的。

⑮ version

<version>元素是可选的，表明表中包含附带版本(version)信息的数据。这在准备使用长事务(long transactions)的时候特别有用。

```
<version
        column = "version_column"
        name = "propertyName"
        type = "typename"
        access = "field|property|ClassName"
        unsaved - value = "null|negative|undefined"
/>
```

- column(可选，默认为属性名)：指定持有版本号的字段名。
- name：持久化类的属性名。
- type(可选，默认是integer)：版本号的类型。
- access(可选，默认是property)：Hibernate用于访问属性值的策略。
- unsaved-value(可选，默认是undefined)：用于标明某个实例时刚刚被实例化的(尚未保存)版本属性值，依靠这个值就可以把这种情况和已经在先前的Session中保存或装载的实例区分开来(undefined指明使用标识属性值进行这种判断)。

版本号必须是以下类型：long、integer、short、timestamp或calendar。

⑯ timestamp

可选的<timestamp>元素指明了表中包含时间戳(timestamp)数据，这用来作为版本的替代。时间戳本质上是一种对乐观锁定的一种不是特别安全的实现。当然，有时候应用程序可能在其他方面使用时间戳。

```
<timestamp
        column = "timestamp_column"
        name = "propertyName"
        access = "field|property|ClassName"
        unsaved - value = "null|undefined"
/>
```

- column(可选,默认为属性名):持有时间戳的字段名。
- name:在持久化类中的 JavaBeans 风格的属性名,其 Java 类型是 Date 或 Timestamp。
- access(可选,默认是 property):Hibernate 用于访问属性值的策略。
- unsaved-value(可选,默认是 null):用于标明某个实例时刚刚被实例化的(尚未保存)版本属性值,依靠这个值就可以把这种情况和已经在先前的 Session 中保存或装载的实例区分开来。

注意:<timestamp>和<version type="timestamp">是等价的。

⑰ property

<property>元素为类声明了一个持久化的、JavaBean 风格的属性。

```
<property
        name = "propertyName"
        column = "column_name"
        type = "typename"
        update = "true|false"
        insert = "true|false"
        formula = "arbitrary SQL expression"
        access = "field|property|ClassName"
/>
```

- name:属性的名字,以小写字母开头。
- column(可选,默认为属性名字):对应的数据库字段名,也可以通过嵌套的<column>元素指定(项目中使用的是这种方式)。
- type(可选):一个 Hibernate 类型的名字。
- update、insert(可选,默认为 true):表明在用于 update 和/或 insert 的 SQL 语句中是否包含这个字段。这二者如果都设置为 false,则表明这是一个"衍生(derived)"的属性,它的值来源于映射到同一个(或多个)字段的某些其他属性,或者通过一个 trigger(触发器),或者其他程序。
- formula(可选):一个 SQL 表达式,定义了这个计算(computed)属性的值,计算属性没有和它对应的数据库字段。
- access(可选,默认值为 property):Hibernate 用来访问属性值的策略。

typename 可以是如下几种。

- Hibernate 基础类型之一(如 integer、string、character、date、timestamp、float、binary、serializable、object 和 blob)。
- 一个 Java 类的名字,这个类属于一种默认基础类型(如 int、float、char、java. lang. String、java. util. Date、java. lang. Integer 和 java. sql. Clob);一个 PersistentEnum 的子类的名字;一个可以序列化的 Java 类的名字;一个自定义类型的类的名字。

如果没有指定类型，Hibernate 会使用反射来得到这个名字的属性，以此来猜测正确的 Hibernate 类型。Hibernate 会对属性读取器(getter 方法)的返回类进行解释，按照规则 2，3，4 的顺序。然而，这并不足够。在某些情况下仍然需要 type 属性(例如，为了区别 Hibernate. DATE 和 Hibernate. TIMESTAMP，或者为了指定一个自定义类型)。

access 属性用来控制 Hibernate 如何在运行时访问属性。在默认情况下，Hibernate 会使用属性的 get/set 方法对。如果指明 access="field"，Hibernate 会忽略 get/set 方法对，直接使用反射来访问成员变量。也可以指定你自己的策略，这就需要你自己实现 net. sf. hibernate. property. PropertyAccessor 接口，再在 access 中设置自定义策略类的名字。

4. 数据存储对象 DAO

当前我们完成了 PO(Persistence Object)持久化层的开发工作。那么，如何使用 PO 呢？这里我们引入 DAO(Data Access Object)数据存取对象的概念，它是 PO 的客户端，负责所有与数据操作有关的逻辑，如数据查询、增加、删除及更新。

Hibernate 的对象状态有 3 种，分别为临时对象(Transient Objects)、持久化对象(Persist Objects)和游离对象(Detached Objects)。

(1) 临时对象：新生成的对象，Session 没有引用指向它，没有放入 Session 缓存中，它在数据库里没有相对应的数据，如：

```
Teacher te = new Teacher();     // te 指向 Transient Object 临时对象
te.setName("zhang");
te.setSex("male");
```

(2) 持久化对象：放入 Session 缓存中，Session 有引用指向该对象，它在数据库里有相对应的数据，与数据库里的数据同步，如：

```
Session session = sf.openSession();
Transaction tx = session.beginTransaction();
session.save(te);              // te 指向的临时对象转变成持久化对象，存入对象的数据到数据库里
```

(3) 游离对象：已经被持久化，但不再处于 Session 缓存中，Session 已没有引用指向该对象，数据库里可能还有相对应的数据，但已不能与数据库里的数据同步。

```
tx.commit();        //存入对象的数据到数据库里
session.close();    // te 指向的持久化对象转变成游离对象
te = null;          // te 指向 null，原来 te 指向的对象已没有任何引用指向它，可以被垃圾回收掉
```

(4) 对象状态转换

通过 new 产生临时对象，save()、saveOrUpdate()将临时对象转换为持久化对象，delete()方法将持久化对象转换为临时对象，close()、evict()、clear() 将持久化对象转为游离对象，其中 evict()从 Session 清除一个对象，clear() 从 Session 清除所有对象。update()、saveOrUpdate()、lock() 将游离对象转换为持久化对象，其中 lock()方法是用来让应用程序把一个未修改的对象重新关联到新 Session 的方法。

get()、load()、find()、iterator()从数据库里获得数据，加载持久化对象。

具体应用代码如下。

```
Session session = sf.openSession();
Transaction tx = session.beginTransaction();
Teacher t = (Teacher) session.get(Teacher.class,new Integer(2));  //从数据库里获得数据生成
对象,t 指向持久化对象
tx.commit();
session.close();
t.setName("wang"); // t 指向 Detached Object 游离对象
Session session2 = sf.openSession();
Transaction tx = session2.beginTransaction();
session2.update(t);  //  t 指向的游离对象转变成持久化对象,用对象的数据改变数据库里相应的记录
tx.commit();
session2.close();
t.setName("chen"); //  t 指向的持久化对象转变成游离对象
```

DAO 应用具体代码如下：

```
package com.bcpl.dao;
import java.util.Collection;
import java.util.List;
import org.hibernate.Query;
import org.hibernate.Session;
import org.hibernate.Transaction;
import com.bcpl.po.HibernateSessionFactory;
import com.bcpl.po.Pojouser;
public class PojouserDAO {

    /**
     * 根据用户 id 查询用户方法
     * @param id
     * @return
     */
    public Productuser findPojouserByID(int id){
        Pojouser pojouser = null;
        Session session = null;
        try{
            session = HibernateSessionFactory.getSession();
            Query query = session.createQuery("from Pojouser as p where p.uid = ?");
            query.setInteger(0,id);
            List list = query.list();
            pojouser = (Pojouser)list.get(0);
        }catch(Exception e){
            e.printStackTrace();
            return null;
        }finally{
            session.close();
        }
        return productuser;
```

```
}
/ * *
 * 查询所有用户方法
 * @return
 * /
public Collection findAllPojousers(){
    Collection collection = null;
    Session session = null;
    try{
        session = HibernateSessionFactory.getSession();
        Query query = session.createQuery("from Pojouser as p where p.delFlag = 0");
        List list = query.list();
        collection = (Collection)list;
    }catch(Exception e){
        e.printStackTrace();
        return null;
    }finally{
        session.close();
    }
    return collection;
}
/ * *
 * 添加用户方法
 * @param pojouser
 * @return flag
 * /
 public int  addPojouser(Pojouser pojouser){
    int flag = 0;
    Session session = null;
    Transaction tr = null;
    try{
        session = HibernateSessionFactory.getSession();
        tr = session.beginTransaction();
        session.save(productuser);
        tr.commit();
        flag = 1;
    }catch(Exception e){
        tr.rollback();
        e.printStackTrace();
        flag = 0;
    }
    if(flag == 1)
        System.out.print("Insert Pojouser Successfule!");
    else
        System.out.print("Insert Pojouser failure!");

    return flag;
```

```
    }
    /**
     * 根据用户 id 删除用户方法
     * @param uid
     * @return
     */
    public int delectPojouser(int uid){
        int flag = 0;
        Session session = null;
        Transaction tr = null;
        try{
            session = HibernateSessionFactory.getSession();
            tr = session.beginTransaction();
            Productuser p = (Pojotuser)session.load(Pojouser.class,new Integer(uid));
            session.delete(p);
            tr.commit();
            flag = 1;
        }catch(Exception e){
            tr.rollback();
            e.printStackTrace();
            flag = 0;
        }
        if(flag == 1)
            System.out.print("delete Pojouser Successfule!");
        else
            System.out.print("delete Pojouser failure!");

        return flag;
    }
    /**
     * 修改用户信息方法
     * @param pojouser
     * @return
     */
    public int updateProductuser(Productuser productuser){
        int flag = 0;
        Session session = null;
        Transaction tr = null;
        try{
            session = HibernateSessionFactory.getSession();
            tr = session.beginTransaction();
            session.update(pojouser);
            tr.commit();
            flag = 1;
        }catch(Exception e){
            tr.rollback();
```

```
                e.printStackTrace();
                flag = 0;
            }
            if(flag == 1)
                System.out.print("update Pojouser Successful!");
            else
                        return flag;
        }
        public static void main(String[] args) {
            PojouserDAO pDAO = new PojouserDAO();

            //根据 id 查询用户测试
            /* Pojouser p = pDAO.findPojouserByID(4);

            //查询所有用户测试
            /* int userSize = pDAO.findAllPojousers().size();

            //添加用户测试
            p2.setUsername("wunan");
            p2.setPassword("wunan");

            //删除用户测试
            //pDAO.delectPojouser(11);

            //更新用户测试
            Pojouser p2 = pDAO.findPojouserByID(10);
            p2.setUsername("hibernatetest");
            p2.setPassword("hibernatetest");
            pDAO.updatePojouser(p2);
        }
    }
```

8.2.2 Session 操作方法

1. Session 保存方法 save()

将临时对象转换为持久化对象，通过把临时对象加入缓存，变成持久化对象，为持久化对象分配唯一的 OID，计划一个 insert 语句。只有当 Session 清理缓存时，才会执行 SQL 的 insert 语句。

在应用程序中不应把持久化对象或游离对象传给 save()。对于持久化对象，操作多余，对于游离对象，会导致表里有两条代表相同业务的记录，不符合业务逻辑。

2. Session 更新方法 update()

将游离对象转换为持久化对象，通过把游离对象重新加入缓存，变成持久化对象。

如果传入的参数是持久化对象，Session 计划一个 update 语句；如果传入的参数是游离对象，游离对象重新加入缓存，变成持久化对象。然后 Session 计划一个 update 语句。只有当 Session 清理缓存时，才会执行 SQL 的 update 语句。

如果在 Session 缓存中已经存在与该游离对象相同 OID 的持久化对象，该游离对象不能加入缓存，Session 会抛异常。此外，当 update()关联一个游离对象时，如果数据库里不存在相应的记录，也会抛异常。

3. Session 删除方法 delete()

将持久化对象转换为临时对象，用于从数据库里删除与对象对应的记录。只有当 Session 清理缓存时，才会执行 SQL 的 delete 语句。

4. Session 加载方法 load() 和 get()

根据 OID 从加载数据库里加载持久化对象。它们的区别是：当数据库里不存在与 OID 相应的记录，load()抛异常，get()返回 null。

对于一个数据库连接，不要创建一个以上的 Session 或 Transaction。此外不要在两个并发的线程中访问同一个 Session，一个 Session 一般只对应一批需要一次性完成的单元操作。

8.3 Hibernate 实体关系映射(多表)

上面讲述了单表映射的实例，现实中我们可能遇到更多的情况是联表操作。表和表之间通过主键/外键建立了联系。

处理联表关系分为两步：首先，在 hbm. xml 配置文件中增加对关系的描述(包括一对多/多对一、一对一以及多对多 3 种)；其次，在 PO 持久化 JavaBean 中增加针对关系的 getter/setter 方法。

Hibernate 实体关系主要有：一对一关系(One-to-One relationship)、一对多关系(One-to-Many relationship)、多对一关系(Many-to-One relationship)、多对多关系(Many-to-Many relationship)。

类之间的关系主要有：关联(Association)和继承(Inheritance)。

关于表之间的关系，在关系数据库里，只存在主外键(primary key & foreign key)参照关系，而且总是由"many"参照"one"，因为只有这样才能消除数据冗余，因此关系数据库实际上只支持多对一或一对多单向关系。

8.3.1 一对一关系

下面以 Customer /Address 进行一对一的关系介绍。

(1) 创建数据库 SAMPLEDB 及表 Customer 和 Address，如下所示。

```
drop database if exists SAMPLEDB;
create database SAMPLEDB;
use SAMPLEDB;

create table CUSTOMER (
    ID bigint not null auto_increment,
    NAME varchar(15),
    primary key (ID)
);

create table ADDRESS(
    ID bigint not null auto_increment,
    STREET varchar(128),
    CITY varchar(128),
    PROVINCE varchar(128),
    ZIPCODE varchar(6),
    primary key (ID)
);
```

(2) 创建数据库表 Customer 映射文件 Customer. hbm. xml，如下所示。

```
<hibernate-mapping>
        <class name = "mypack.Customer" table = "customer" lazy = "false">
                <id name = "id" type = "java.lang.Long">
                        <column name = "ID" />
                        <generator class = "increment" />
                </id>
 <property name = "name" type = "java.lang.String">
                        <column name = "NAME" length = "15" />
                </property>
                <one - to - one name = "address" class = "mypack.Address" cascade = "all" lazy = "false"/>
        </class>
</hibernate - mapping>
```

在上述的配置文件中，Hibernate 标签 class 包含的标签元素及其参数解释如下所示。

```
<class  name = "ClassName"     //持久化类或接口
        table = "tableName"   //对应的数据库表名
        discriminator - value = "discriminator_value"  //一个用于区分不同子类的值，在多态行
                                                  为时使用，默认和表名一样
        mutable = "true|false"   //表明该类的实例可变，默认为 true
        schema = "owner"        //覆盖在根<hibernate - mapping>元素中指定的 schema 名字
        proxy = "ProxyInterface"      //指定一个接口，在延迟装载时作为代理使用
        dynamic - update = "true|false"        //动态更新，默认为 false
        dynamic - insert = "true|false"        //动态插入，默认为 false
        select - before - update = "true|false"      //默认为 false
        polymorphism = "implicit|explicit"        //界定是隐式还是显式的使用查询多态
        where = "arbitrary sql where condition"  //指定一个附加的 where 条件
```

```
        persister = "PersisterClass"
        batch - size = "N"        //批次抓取数量,默认 1
        optimistic - lock = "none|version|dirty|all"    //乐观锁定,默认是 version,决定乐观锁定的策略
        lazy = "true|false">   //延迟加载
      <id…/>
      <property…/>
      <many - to - one…/>
      <one - to - one  …/>
      <map  …/>
</class>
```

(3) 创建数据库表 Address 映射文件 Address. hbm. xml。

```
<hibernate - mapping>
    <class name = "mypack.Address" table = "address">
        <id name = "id" type = "java.lang.Long">
            <column name = "ID" />
            <generator class = "foreign" >
                <param name = "property">customer</param>
            </generator>
        </id>
        <property name = "street" type = "java.lang.String">
            <column name = "STREET" length = "128" />
        </property>
        <property name = "city" type = "java.lang.String">
            <column name = "CITY" length = "128" />
        </property>
        <property name = "province" type = "java.lang.String">
            <column name = "PROVINCE" length = "128" />
        </property>
        <property name = "zipcode" type = "java.lang.String">
            <column name = "ZIPCODE" length = "6" />
        </property>
        <one - to - one name = "customer" class = "mypack.Customer" constrained = "true" />
    </class>
</hibernate - mapping>
```

说明:<one-to-one>元素的 constrained 属性为 true,表明 address 表的 ID 主健同时作为外键参照 Customer 表,在 Address. hbm. xml 文件中,必须为 OID 使用 foreign 标识符生成策略,如果使用了 foreign 标识符生成策略,Hibernate 会保证 Address 对象与关联的 Customer 对象共享同一个 OID。

(4) 创建 Action 处理类 Customer 和 Address,代码如下。

```
public class Customer{
        private String id;
        private String name;
        private Address address;
        getXXX();
        setXXX();
        ...

}
```

```
public class Address{
        private String id;
        private String street;
        private String city;
        private String province;
        private String zipcode;
        private Customer customer;
        getXXX();
        setXXX();
        ...
}
```

(5) 创建 DAO 方法和测试方法。

```
public class OptionDAO {

    public void saveCustomer(Customer customer){}
    public Customer loadCustomer(Long id){}
    public void deleteCustomer(Customer customer){}
    public void deleteAddress(Address address){}
}
```

测试语句,在 main 方法中,代码如下:

```
Customer customer = new Customer();

Address address = new Address();
address.setProvince("province1");
address.setCity("city1");
address.setStreet("street1");
address.setZipcode("100085");
address.setCustomer(customer);
customer.setName("Peter");
customer.setAddress(address);
saveCustomer(customer);
```

8.3.2　一对多、多对一关系

1. 一对多关系

下面表 Customer 和 Orders 之间的一对多关系：Customer 1 <-> * Orders。

(1) 创建数据库 SAMPLEDB 及表 Customer 和 Orders。

```
drop database if exists SAMPLEDB;
create database SAMPLEDB;
use SAMPLEDB;
create table CUSTOMER (
    ID bigint not null auto_increment,
    NAME varchar(15),
    primary key (ID)
);
create table ORDERS (
    ID bigint not null auto_increment,
    ORDER_NUMBER varchar(15),
    CUSTOMER_ID bigint,
    primary key (ID)
);
```

(2) 创建 Action 处理类 Customer 和 Orders，简要代码如下。

```
class   Customer{
    Set orders = new HashSet();
    getXXX();
    setXXX();
      …
  }

  class   Orders{
        Customer customer;
        getXXX();
       setXXX();
      …
  }
```

(3) 创建表 Customer 的映射文件 Customer. hbm. xml。

```
<hibernate-mapping package = "mypack">
    <class name = "mypack.Customer" table = "customer">
        <id name = "id" column = "ID" type = "long">
            <generator class = "identity" />
        </id>
        <property name = "name" column = "NAME" type = "string" />
        <set name = "orders" cascade = "all-delete-orphan" inverse = "true">
            <key column = "CUSTOMER_ID" />
            <one-to-many class = "mypack.Orders" />
        </set>
    </class>
</hibernate-mapping>
```

创建表 Orders 的映射文件 Orders. hbm. xml。

```
<hibernate-mapping package="mypack">
    <class name="mypack.Orders" table="orders">
        <id name="id" column="ID" type="long">
            <generator class="identity" />
        </id>
        <property name="orderNumber" column="ORDER_NUMBER" type="string" />
        <many-to-one name="customer" column="CUSTOMER_ID" class="mypack.Customer"
cascade="save-update" />
    </class>
</hibernate-mapping>
```

如上所述，在 Hibernate 配置文件中可以使用<set>、<list>、<map>、<bag>、<array>和<primitive-array>等元素来定义集合。

关于 Map 集合的参数及解释，如下所示。

```
<map
    name="propertyName"
    table="table_name"
    schema="schema_name"
    lazy="true|false"
    inverse="true|false"
    cascade="all|none|save-update|delete|all-delete-orphan"
    sort="unsorted|natural|comparatorClass"
    order-by="column_name asc|desc"
    where="arbitrary sql where condition"
    outer-join="true|false|auto"
    batch-size="N"
    access="field|property|ClassName"
>

    <key…/>
    <index…/>
    <element…/>
</map>
```

① name：集合属性的名称。

② table(可选，默认为属性的名称)：表示这个集合表的名称(不能在一对多的关联关系中使用)。

③ schema(可选)：表的 schema 的名称，它会覆盖在根元素中定义的 schema。

④ lazy(可选，默认为 false)：允许延迟加载(Lazy Initialization)。

⑤ inverse(可选，默认为 false)：标记这个集合作为双向关联关系中的方向一端。

⑥ cascade(可选，默认为 none)：让操作级联到子实体。

⑦ sort(可选)：指定集合的排序顺序，其可以为自然的(Natural)或者给定一个用来比较的类。

⑧ order-by(可选，仅用于 JDK1.4)：指定表的字段(一个或几个)再加上 asc 或者 desc(可

选),定义 Map、Set 和 Bag 的迭代顺序。

⑨ where(可选):指定任意的 SQL where 条件,该条件将在重新载入或者删除这个集合时使用(当集合中的数据仅仅是所有可用数据的一个子集时,这个条件非常有用)。

⑩ outer-join(可选):指定这个集合,只要可能,应该通过外连接(Outer Join)取得。在每一个 SQL 语句中,只能有一个集合可以被通过外连接抓取。

⑪ batch-size(可选,默认为 1):指定通过延迟加载取得集合实例的批处理块大小(Batch Size)。

⑫ access(可选,默认为属性 property):Hibernate 取得属性值时使用的策略。

(4) 创建 DAO 方法和测试方法。

```
public class OptionDAO {
    public void saveCustomer(Customer customer){}

    public Customer findAllOrders(Long cid){
        session = HibernateSessionFactory.getSession();
        tx = session.beginTransaction();
        Customer c = (Customer) session.load(Customer.class,id);
        Hibernate.initialize(c);
        tx.commit();
        session.close();
        return c;
    }

    public void deleteCustomer(Customer customer){}

    public void deleteOrder(Orders order){}
}
```

测试语句,在 main 方法中,代码如下。

```
Customer customer = new Customer();
customer.setName("Tom");
customer.setOrders(new HashSet());

Orders order1 = new Orders();
order1.setOrderNumber("Tom_Order001");
order1.setCustomer(customer);
customer.getOrders().add(order1);

Orders order2 = new Orders();
order2.setOrderNumber("Tom_Order002");
order2.setCustomer(customer);
customer.getOrders().add(order2);
```

2. 多对一关系

通过<many-to-one>元素,可以定义一种常见的与另一个持久化类的关联,这种关系模型是多对一关联(实际上是一个对象引用)。

```
<many-to-one
        name = "propertyName"
        column = "column_name"
        class = "ClassName"
        cascade = "all|none|save - update|delete"
        outer - join = "true|false|auto"
        update = "true|false"
        insert = "true|false"
        property - ref = "propertyNameFromAssociatedClass"
        access = "field|property|ClassName"
/>
```

① name:属性名。

② column(可选):字段名。

③ class(可选,默认是通过反射得到属性类型):关联的类的名字。

④ cascade(级联)(可选):指明哪些操作会从父对象级联到关联的对象。

⑤ outer-join(外连接)(可选,默认为自动):当设置 hibernate. use_outer_join 的时候,对这个关联允许外连接抓取。

⑥ update、insert(可选,defaults to true):指定对应的字段是否在用于 update 和/或 insert 的 SQL 语句中包含。如果二者都是 false,则这是一个纯粹的"外源性(derived)"关联,它的值是通过映射到同一个(或多个)字段的某些其他属性得到的,或者通过除法器(trigger),或者是其他程序。

⑦ property-ref(可选):指定关联类的一个属性,这个属性将会和本外键相对应,如果没有指定,会使用对方关联类的主键。

⑧ access(可选,默认是 property):Hibernate 用来访问属性的策略。

cascade 属性允许下列值:all、save-update、delete 和 none。设置除了 none 以外的其他值会传播特定的操作到关联的(子)对象中。

outer-join 参数允许下列 3 个不同值。

- auto(默认):使用外连接抓取关联(对象),如果被关联的对象没有代理(proxy)。
- true:一直使用外连接来抓取关联。
- false:永远不使用外连接来抓取关联。

<one-to-many>标记指明了一个一对多的关联,其中 class(必须)为被关联类的名称。

```
<one - to - many class = "ClassName"/>
```

8.3.3 多对多关系

下面介绍表 Authorization 和 Usr 之间的多对多关系:Authorization * <-> * Usr。

(1) 创建数据库 SAMPLEDB 及表 Authorization 和 Usr。

创建 Usr 表,如下所示。

```
CREATE TABLE usr (
 id bigint(20) NOT NULL default '0',
 name varchar(16) default NULL,
 password varchar(16) default NULL,
 phone varchar(16) default NULL,
 deptid bigint(20) default NULL,
 address varchar(64) default NULL,
 title varchar(32) default NULL,
 power varchar(32) default NULL,
 auth varchar(32) default NULL,
 homephone varchar(16) default NULL,
 superauth varchar(8) default NULL,
 groupid bigint(20) default NULL,
 birthdate date default NULL,
 male varchar(8) default NULL,
 email varchar(255) default NULL,
 PRIMARY KEY  (id)
) TYPE = InnoDB;
```

创建 Authorization 表,如下所示。

```
drop database if exists SAMPLEDB;
create database SAMPLEDB;
use SAMPLEDB;

CREATE TABLE authorization (
   id bigint(20) NOT NULL  default '0',
   columnId bigint(20) default NULL,
   auth int(11) default NULL,
   init int(11) default NULL,
   authorize int(11) default NULL,
   PRIMARY KEY  (id)
) TYPE = InnoDB;
```

创建关联表 Userauth,如下所示。

```
CREATE TABLE userauth (
   id bigint(20) NOT NULL default '0',
   userId bigint(20) default NULL,
   authId bigint(20) default NULL,
   PRIMARY KEY  (id)
) TYPE = InnoDB;
```

(2) 创建 Action 处理类 Authorization 和 Usr。

```
class  Authorization{
    private java.lang.Integer authorize;
    ...
    private Set users = new HashSet();
        public Set getUsers(){
            return users;
        }
        public void setUsers(Set users){
            this.users = users;
        }
    }
```

```
class Usr{
    private java.lang.String email;
    ...
    private Set auths = new HashSet();
    public Set getAuths(){
        return auths;
    }
    public void setAuths(Set auths){
        this.auths = auths;
    }
}
```

(3) 创建 Authorization 映射文件 Authorization.hbm.xml。

```
<hibernate-mapping package="mypack">
    <class name="Authorization" table="authorization">
        <id name="id" column="id" type="java.lang.Long">
            <generator class="increment" />
        </id>
        <property name="columnid" column="columnId" type="java.lang.Long" />
        <property name="auth" column="auth" type="java.lang.Integer" />
        <property name="init" column="init" type="java.lang.Integer" />
        <property name="authorize" column="authorize" type="java.lang.Integer" />
        <set name="users" table="userauth" inverse="true" lazy="false" cascade="save-update">
            <key column="authid" />
            <many-to-many class="mypack.Usr" column="userid" />
        </set>
    </class>
</hibernate-mapping>
```

(4) 创建 Usr 映射文件 Usr.hbm.xml。

```
<hibernate-mapping package="mypack">
    <class name="mypack.Usr" table="usr">
        <id name="id" column="id" type="java.lang.Long">
            <generator class="increment" />
        </id>
        <property name="name" column="name" type="java.lang.String" />
        <property name="passwor" column="password" type="java.lang.String" />
        <property name="phone" column="phone" type="java.lang.String" />
        <property name="deptid" column="deptid" type="java.lang.Long" />
        <property name="address" column="address" type="java.lang.String" />
        <property name="title" column="title" type="java.lang.String" />
        <property name="power" column="power" type="java.lang.String" />
        <property name="auth" column="auth" type="java.lang.String" />
        <property name="homephone" column="homephone" type="java.lang.String" />
        <property name="superauth" column="superauth" type="java.lang.String" />
        <property name="groupid" column="groupid" type="java.lang.Long" />
        <property name="birthdate" column="birthdate" type="java.util.Date" />
        <property name="male" column="male" type="java.lang.String" />
        <property name="email" column="email" type="java.lang.String" />
        <set name="auths" table="userauth" lazy="false" cascade="save-update">  //!!!!!!!
            <key column="userid" />
            <many-to-many class="mypack.Authorization" column="authid" />
        </set>
    </class>
</hibernate-mapping>
```

① 对于多对多关联，cascade 属性设置为“save-update”是合理的，但是不允许设置为“all”、“delete”、“all-delete-orphan”。假如删除一个 Authorization 对象时，还级联删除与它关联的所有 Usr 对象，由于这些 Usr 对象有可能还与其他的 Authorization 对象关联，因此当 Hibernate 执行级联删除时，会违反数据库的外键参照完整性。

② 对于多对多关联的两端，必须把其中一端的<set>元素的 inverse 属性设置为 true，inverse 端为真，只能使用<set>和<bag>元素。

(5) 创建 DAO 方法和测试方法。

```
public class HibernateService {
        public List listAll() throws HibernateException{
        List list = null;
        Session   session = null;
        Transaction tx = null;
        session = HibernateSessionFactory.getSession();
        tx = session.beginTransaction();
        list = session.createQuery("from Usr").list();
        tx.commit();
        HibernateSessionFactory.closeSession();
        return list;
    }
```

在上述的 DAO 类中，可以通过 getter/setter 来使用关系，完成联表操作。

测试语句，在 main 方法中，代码如下。

```
public static void main(String[] args) {
        HibernateService hs = new HibernateService();
        List list = null;
        Usr user = null;
        Authorization auth = null;
        try {
            list = hs.listAll();
            Iterator it = list.iterator();
            while(it.hasNext()){
                user = (Usr)it.next();
                System.out.println("******" + user.getName() + "*******");
                Iterator its = user.getAuths().iterator();
                while(its.hasNext()){
                    auth = (Authorization)its.next();
                    System.out.println("#####" + auth.getAuth());
                }
            }
        } catch (HibernateException e) {
            e.printStackTrace();
        }

        }
```

8.4 Hibernate 继承策略

Hibernate 框架继承支持 3 种策略，它们分别为：

① 整个类继承树映射一个表(根类一张表)；

② 每个具体的类映射一个表；

③ 每个类映射一个表。

但在 Hibernate 框架中不支持子类映射，以及连接子类映射在同一个类中也不支持。

1. 整个类继承树映射一个表

在 xxx.hbm.xml 文件中只有一个 class 映射配置支持多态和多态查询，可以用父类的名字查询出所有子类的对象。但子类的非空字段难处理，在一棵继承树上的每个子类都必须声明一个唯一的 discriminator-value。如果没有指定，就会使用 Java 类的全名。

假设有一个 Employee 是抽象类(Abstract Class)，分别有两个子类 HourlyEmployee 和 SalariedEmployee。

(1) 创建处理类 Company 的映射文件 Company.hbm.xml，主要代码如下。

```
<class name = "Company" table = "company">
        <id name = "id" column = "id" type = "integer">
          <generator class = "increment"/>
        </id>

        <property name = "name" column = "name" type = "string"  not - null = "true" />

        <set name = "employeesSet" inverse = "true" cascade = "all">
          <key column = "cid"/>
          <one - to - many class = "Employees" />
        </set>
    </class>
```

(2) 创建处理类 Employees 的映射文件 Employees. hbm. xml，主要代码如下。

```
<class name = "Employees" table = "employees">
          <id name = "id" column = "id" type = "integer">
               <generator class = "increment"/>
          </id>
            <discriminator column = "employee_type" type = "string"  not - null = "true" />
          <property name = "name" column = "name" type = "string"  not - null = "true" />

            <many - to - one name = "company" column = "cid" class = "Company"  not - null = "true" />

          <subclass name = "HourEmployee" discriminator - value = "HE">
               <property name = "rate" column = "rate" type = "double"  not - null = "true" />
          </subclass>
          <subclass name = "SalaryEmployee" discriminator - value = "SE">
               <property name = "salary" column = "salary" type = "double"  not - null = "true" />
      </subclass>
      </class>
```

(3) 创建数据库表 Company 和 Employee，代码如下所示。

```
drop database if exists hibernateinheritancesub;
create database hibernateinheritancesub;
use hibernateinheritancesub;

create table COMPANY (
   ID bigint not null auto_increment,
   NAME varchar(15),
   primary key (ID)
);
create table EMPLOYEE (
   ID bigint not null auto_increment,
   NAME varchar(15),
   EMPLOYEE_TYPE varchar(2),
   RATE double precision,
   SALARY double precision,
   COMPANY_ID bigint,
   primary key (ID)
);
```

在插入数据前，执行 SQL 语句，如下所示。

```
alter table EMPLOYEE   add constraint FK_COMPANY foreign key (COMPANY_ID) references COMPANY (ID);
```

(4) 创建 Action 处理类 Employee、HourlyEmployee 和 SalariedEmployee，代码如下。

```
Employee:
   public abstract class Employees implements Serializable
   {
        private java.lang.Integer id;
        private Company company;

        private java.lang.String employeeType;

        private java.lang.String name;
                ...
   }


HourlyEmployee:

   public class HourEmployee extends Employees {
   private Double rate;
   ...
   }

SalariedEmployee:

   public class SalaryEmployee extends Employees {
   private Double salary;
   ...
   }
```

(5) 在测试 main 方法中，主要代码如下所示。

```
public void saveEmployee(AbstractEmployee employee);
public List findAllHourlyEmployees();
public List findAllSalariedEmployees();
public List findAllEmployees() throws Exception;
public Company loadCompany(long id) throws Exception;   //Hibernate.initialize(company.getEmployees());
```

2. 每个具体的类映射一个表

每个具体的类在 xxx.hbm.xml 文件有一个 class 映射配置，不支持多态查询，不能用父类的名字查询出所有子类的对象，必须分别检索子类的对象，然后把它们合并到一个集合。

(1) 创建上述类 Company 的映射文件 Company.hbm.xml，如下所示。

```
<hibernate-mapping>
    <class name = "com.bcpl.po.Company" table = "company" catalog = "hibernateinheritance">
        <id name = "id" type = "java.lang.Long">
            <column name = "ID" />
            <generator class = "increment" />
        </id>
        <property name = "name" type = "java.lang.String">
            <column name = "NAME" length = "15" />
        </property>
    </class>
</hibernate-mapping>
```

(2) 创建上述类 HourlyEmployee 的映射文件：HourlyEmployee. hbm. xml。

```
<class name = "HourlyEmployee" table = "HOURLY_EMPLOYEE">
        <id name = "id" type = "long" column = "ID">
            <generator class = "increment"/>
        </id>
            <property name = "name" type = "string" column = "NAME" />
            <property name = "rate" column = "RATE" type = "double" />
      <many-to-one  name = "company"  class = "Company" column = "COMPANY_ID"  />
  </class>
```

(3) 创建上述类 SalariedEmployee 的映射文件：SalariedEmployee. hbm. xml。

```
<class name = "SalariedEmployee" table = "SALARIED_EMPLOYEE">
        <id name = "id" type = "long" column = "ID">
            <generator class = "increment"/>
        </id>
            <property name = "name" type = "string" column = "NAME" />
          <property name = "salary" column = "SALARY" type = "double" />
        <many-to-one  name = "company"  class = "Company" column = "COMPANY_ID"  />
  </class>
```

(4) 创建的类 Company、SalariedEmployee 和 HourlyEmployee 同上所述。

(5) 创建数据库。

```
drop database if exists hibernateinheritance;
create database HIBERNATEINHERITANCE;
use HIBERNATEINHERITANCE;

create table COMPANY(
   ID bigint not null auto_increment,
   NAME varchar(15),
   primary key (ID)
);
create table HOURLY_EMPLOYEE(
   ID bigint not null auto_increment,
   NAME varchar(15) ,
   RATE double precision,
   COMPANY_ID bigint,
   primary key (ID)
);
create table SALARIED_EMPLOYEE(
   ID bigint not null auto_increment,
   NAME varchar(15),
   SALARY double precision,
   COMPANY_ID bigint,
   primary key (ID)
);
```

说明：在插入数据前，先执行下面的 SQL 语句，给表 Hourly_Employee 和 Salaried_

Employee 添加外键。

```
alter table HOURLY_EMPLOYEE  add constraint FK1_COMPANY foreign key (COMPANY_ID) references COMPANY(ID);
alter table SALARIED_EMPLOYEE add constraint FK2_COMPANY foreign key (COMPANY_ID) references COMPANY(ID);
```

3. 每个类映射一个表

每个类映射一个表，在 xxx.hbm.xml 文件中只有一个 class 映射配置，用外键参照关系表示继承关系，支持多态查询，可以用父类的名字查询出所有子类的对象。

（1）配置类 Employee 和表 Employee 直接的映射，配置代码如下。

```
<class name = "Employee" table = "employee">
        <id name = "id" column = "ID" type = "long">
            <generator class = "increment"/>
        </id>

      <property name = "name" type = "string" column = "NAME" />

      <many - to - one  name = "company"  column = "COMPANY_ID" class = "Company"/>

      <joined - subclass name = "HourlyEmployee" table = "HOURLY_EMPLOYEE" >
          <key column = "EMPLOYEE_ID" />
          <property name = "rate" column = "RATE" type = "double" />
      </joined - subclass>

      <joined - subclass name = "SalariedEmployee"  table = "SALARIED_EMPLOYEE" >
          <key column = "EMPLOYEE_ID" />
          <property name = "salary" column = "SALARY" type = "double" />
      </joined - subclass>

  </class>
```

（2）创建类 Employee、HourlyEmployee 和 SalariedEmployee 以及表 Employee、HourlyEmployee 和 SalariedEmployee，同时，创建映射文件 Employee.hbm.xml，都同前面所述一样。

（3）创建数据库 hibernateinheritancejoin 及表 Company、Employee、Hourly_Employee 和 Salaried_Employee。

```
drop database if exists hibernateinheritancejoin;
create database hibernateinheritancejoin;
use hibernateinheritancejoin;

create table COMPANY (
   ID bigint not null auto_increment,
   NAME varchar(15),
   primary key (ID)
);
create table EMPLOYEE (
   ID bigint not null auto_increment,
   NAME varchar(15),
```

```
    COMPANY_ID bigint,
    primary key (ID)
);

create table HOURLY_EMPLOYEE (
    EMPLOYEE_ID bigint not null,
    RATE double precision,
    primary key (EMPLOYEE_ID)
);

create table SALARIED_EMPLOYEE (
    EMPLOYEE_ID bigint not null,
    SALARY double precision,
    primary key (EMPLOYEE_ID)
);
```

说明：同前述一样，需要给表 Employee 添加外键限制与表 Company 关联，给表 Hourly_Employee、Salaried_Employee 添加外键限制，与表 Employee 关联。具体 SQL 语句如下。

```
alter table EMPLOYEE  add constraint FK_COMPANY foreign key (COMPANY_ID) references COMPANY(ID);
alter table HOURLY_EMPLOYEE add constraint FK_EMPLOYEE1 foreign key (EMPLOYEE_ID) references EMPLOYEE(ID);
alter table SALARIED_EMPLOYEE add constraint FK_EMPLOYEE2 foreign key (EMPLOYEE_ID) references EMPLOYEE(ID);
```

注意：每个类一张表，用外键参照关系表示它们之间的继承关系。如果不需要支持多态查询和多态关联，可以采用每个具体类对应一个表的映射方式；如果需要支持多态查询和多态关联，而且子类包含的属性不多，可以采用根类对应一个表的映射方式；如果需要支持多态查询和多态关联，而且子类包含的属性多，可以采用每个类对应一个表的映射方式；如果继承关系中包含接口，可以把它当作抽象类来处理。

8.5　Hibernate 应用开发

本书 Hibernate 的开发使用的是 MyEclipse 与 MySQL 数据库的集成环境，下面就 Hibernate 在集成开发工具 MyEclipse 中进行开发的步骤进行简要介绍。

开发步骤共分 5 步，分别为创建 Hibernate 的配置文件、创建表单(在数据库中)、创建持久化类、创建“对象-关系”映射文件和实现通过 Hibernate API 编写访问数据库的代码。

(1) 创建 Hibernate 的配置文件 Hibernate. cfg. xml ，具体如下。

```
<?xml version='1.0' encoding='UTF-8'?>
<!DOCTYPE hibernate-configuration PUBLIC
          "-//Hibernate/Hibernate Configuration DTD 2.0//EN"
          "http://hibernate.sourceforge.net/hibernate-configuration-2.0.dtd">
<!-- Generated by MyEclipse Hibernate Tools. -->
<hibernate-configuration>
```

```
    <session - factory>
        <! -- mapping files -- >
        <property name = "myeclipse.connection.profile">mysql</property>
        <property name = "connection.url">jdbc:mysql://localhost:3306/bookstoresql</property>
        <property name = "connection.username">root</property>
        <property name = "connection.password">123</property>
        <property name = "connection.driver_class">com.mysql.jdbc.Driver</property>
        <property name = "dialect">net.sf.hibernate.dialect.MySQLDialect</property>
      </session - factory>
</hibernate - configuration>
```

（2）创建数据库表

创建数据库表 book，其 SQL 语句如下。

```
create table book( id bigint not null auto_increment,title varchar(50) not null,price  double,
description varchar(200),primary key(id));
```

（3）创建持久化类

```
public abstract class AbstractBook implements Serializable
public class Book extends AbstractBook
```

用 MyEclipse 自动从数据库表单生成两个类(一个抽象，一个具体)，如果自己手工建，可以只生成一个具体类。

（4）创建“对象-关系”映射文件 Book.hbm.xml

```
<hibernate - mapping package = "com.bcpl.hibernate.po">
      <class name = "Book" table = "book">
            <id name = "id" column = "id" type = "long">
                  <generator class = "increment"/>
            </id>

            <property name = "isbnNumber" column = "isbnNumber" type = "string"  not - null = "true" />
            <property name = "title" column = "title" type = "string"  not - null = "true" />
            <property name = "price" column = "price" type = "double" />
            <property name = "description" column = "description" type = "string" />
      </class>

</hibernate - mapping>
```

在一般 Java 项目里，配置文件存放的路径为：自己的项目名称/hibernate.cfg.xml、自己的项目名称/hibernate.properties 或自己的项目名称/包名/xxx.hbm.xml。

注意：项目名称必须设置在 classpath 中，在 Web 项目里，配置文件存放的路径为：自己的 Web 应用名/WEB-INF/classes/hibernate.cfg.xml、Web 应用名/WEB-INF/classes/hibernate.properties 或自己的 Web 应用名/WEB-INF/classes/包名/xxx.hbm.xml。

（5）创建 DAO

通过用 MyEclipse 工具生成的 HibernateSessionFactory 来设定配置，并获得 Session，代码如下。

```
public void  addBook(Book b){
           Transaction tx = null;
          try{
              Session sess = HibernateSessionFactory.currentSession(); //HibernateSessionFactory.getSession();
             tx = sess.beginTransaction();
             sess.save(b);
             tx.commit();
          }catch(Exception e){
                    e.printStackTrace();
                if(tx! = null){
                      try{
                           tx.rollback();
                      }catch (Exception ee){   ee.printStackTrace();}
                }
         }finally{
              try{
                 HibernateSessionFactory.closeSession();
              }catch (Exception ee){   ee.printStackTrace();}
         }
     }
```

或者自己生成 SessionFactory 对象，并获得 Session。

第一种用法，装载 hibernate. properties 文件。

① 生成 SessionFactory 对象

```
//第一种用法，装载 hibernate.properties 文件
  SessionFactory sfactory = null;
  try{
     Configuration config = new Configuration();
     config.addClass(packageName.Book.class);
     //config.addClass("Book.hbm.xml");

     sfactory = config.buildSessionFactory();
}catch(Exception e)
     {  e.printStackTrace();
     }
```

② 创建 hibernate. properties 文件，内容如下。

```
hibernate.dialect = net.sf.hibernate.dialect.MySQLDialect
hibernate.connection.driver_class = com.mysql.jdbc.Driver
hibernate.connection.url = jdbc:mysql://localhost:3306/test
hibernate.connection.username = root
hibernate.connection.password = 123
hibernate.show_sql = true
```

第二种用法，装载 hibernate. cfg. xml 文件。

① 生成 SessionFactory 对象

```
SessionFactory sfactory = null;
        try{
        Configuration config = new Configuration();
        config.configure();
        sfactory = config.buildSessionFactory();

        //sfactory2 = new Configuration().configure().buildSessionFactory();
    }catch(Exception e){
     e.printStackTrace();
      }
```

Hibernate 会自动在 classpath 中寻找相应的配置文件(hibernate.properties 或 hibernate.cfg.xml)。

② 获得 Session

```
Session se = sfactory.openSession();
Transaction tx = null;
try{
           tx = se.beginTransaction();
           //do some work
           se.delete("from Book as test  ");    // !!! delete all books

           tx.commit();
      }catch(Exception e){
              if(tx! = null)
                    tx.rollback();
               throw e;
          }finally{
               se.close();
          }
```

8.6 小　　结

本章只是对 Hibernate 的基本应用、框架、实体关系、关系映射及继承策略进行了简述，同时，也对 Hibernate 的基本应用开发步骤进行了实例介绍，但 Hibernate 中的实体、值、组件 Component 集合及映射，限于篇幅，并没有进行介绍。此外，管理 Hibernate 的缓存、缓存的作用、缓存的级别和缓存的范围以及 Hibenate 缓存所选用的插件、配置缓存的步骤和缓存策略等部分，都是实践开发需要掌握的内容，读者可以根据自己的开发需要，进行拓展性的阅读和补充。

第 9 章　Spring 技术

9.1　Spring 概述

Spring Framework(简称 Spring)是于 2003 年兴起的一个轻量级的 J2EE 开发的开源框架,由 Rod Johnson 在其著作 Expert One-On-One J2EE Development and Design 中阐述的部分理念和原型衍生而来,它是为了解决企业应用开发的复杂性而创建的。Spring 使用基本的 JavaBean 来完成以前只可能由 EJB 完成的事情。然而,Spring 的用途不仅限于服务器端的开发,更严格地讲它是针对 Bean 的生命周期进行管理的轻量级容器(Lightweight Container),可以单独利用 Spring 构筑应用程序,也可以和 Struts、Webwork 和 Tapestry 等众多 Web 应用程序框架组合使用,并且可以与 Swing 等桌面应用程序 API 组合。所以 Spring 并不仅仅只能应用在 J2EE 中,也可以应用在桌面应用及小应用程序中。针对 Spring 开发的组件不需要任何外部库。

Spring 框架的优势主要体现在:Spring 能有效地组织中间层对象;能消除在许多工程中常见的对 Singleton 的过多使用,同时消除各种各样自定义格式的属性文件的需要,使配置信息一元化;可以帮助我们真正意义上实现针对接口编程,在 Spring 应用中的大多数业务对象没有依赖于 Spring,其构建的应用程序易于单元测试;它支持 JDBC 和 O/R Mapping 产品(Hibernate),采用的是 MVC Web 框架,提供一种清晰、无侵略性的 MVC 实现方式;具有 JNDI 抽象层,便于改变实现细节,可以方便地在远程服务和本地服务间切换,简化了访问数据库时的例外处理;能使用 AOP 提供声明性事务管理。

Spring 框架是一个分层架构,由 7 个定义好的模块组成。Spring 模块构建在核心容器之上,核心容器定义了创建、配置和管理 Bean 的方式,如图 9-1 所示。

Spring 框架的每个模块(或组件)都可以单独存在,或者与其他一个或多个模块联合实现,框架中每个模块的功能如下。

核心容器:核心容器提供 Spring 框架的基本功能。核心容器的主要组件是 BeanFactory,它是工厂模式的实现。BeanFactory 使用控制反转(IoC)模式将应用程序的配置和依赖性规范与实际的应用程序代码分开。

Spring 上下文:Spring 上下文是一个配置文件,向 Spring 框架提供上下文信息。Spring 上下文包括企业服务,如 JNDI、EJB、电子邮件、国际化、校验和调度功能。

Spring AOP:通过配置管理特性,Spring AOP 模块直接将面向方面的编程功能集成到了 Spring 框架中。所以,可以很容易地使 Spring 框架管理的任何对象支持 AOP。Spring AOP 模块为基于 Spring 的应用程序中的对象提供了事务管理服务。通过使用 Spring AOP,不用依赖 EJB 组件，就可以将声明性事务管理集成到应用程序中。

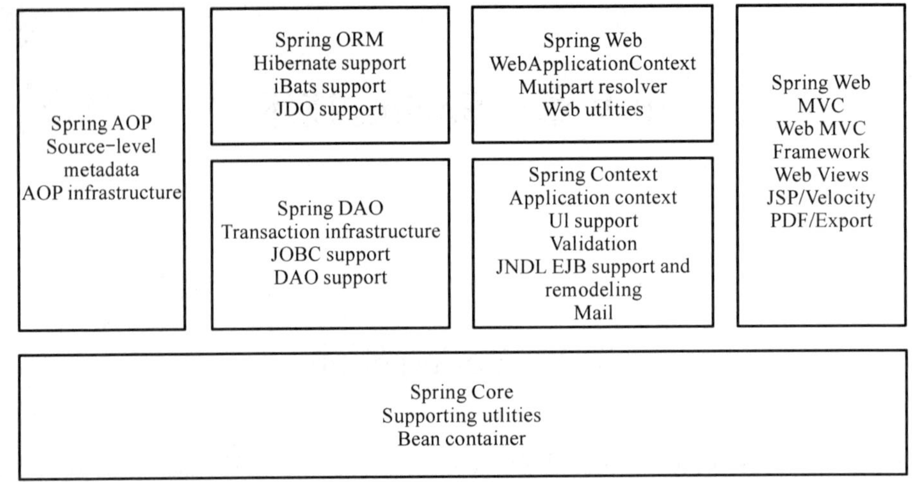

图 9-1　Spring 框架

Spring DAO：JDBC DAO (Data Access Object)抽象层提供了有意义的异常层次结构，可用该结构来管理异常处理和不同数据库供应商抛出的错误消息。异常层次结构简化了错误处理，并且极大地降低了需要编写的异常代码数量。Spring DAO 的面向 JDBC 的异常遵从通用的 DAO 异常层次结构。

Spring ORM：Spring 框架插入了若干个 Object/Relation Mapping 框架，从而提供了 ORM 的对象关系映射工具，其中包括 JDO、Hibernate 和 iBatis SQL Map，所有这些都遵从 Spring 的通用事务和 DAO 异常层次结构。

Spring Web 模块：Web 上下文模块建立在应用程序上下文模块之上，为基于 Web 的应用程序提供了上下文，所以，Spring 框架支持与 Jakarta Struts 的集成。Web 模块还简化了处理多部分请求以及将请求参数绑定到域对象的工作。

Spring MVC 框架：MVC 框架是一个全功能的构建 Web 应用程序的 MVC 实现。通过策略接口，MVC 框架变成为高度可配置的，MVC 容纳了大量视图技术，其中包括 JSP、Velocity、Tiles、iText 和 POI。

Spring 框架的功能可以用在任何 J2EE 服务器中，大多数功能也适用于不受管理的环境。Spring 支持不绑定到特定 J2EE 服务的可重用业务和数据访问对象。显然，这样的对象可以在不同 J2EE 环境(Web)、独立应用程序和测试环境之间重用。

Spring 框架具有以下特点：

① 设计良好的分层结构；

② 以 IoC 为核心，提倡面向接口编程；

③ 良好的架构设计；

④ 可以代替 EJB；

⑤ 实现了 MVC；

⑥ 可以和其他框架良好的结合，如 Hibernate、Struts 等。

Spring 框架的核心思想是控制反转 IoC(Inversion of Control) 和 AOP(Aspect Oriented

Programming)，也就是面向方面编程的技术，AOP 是基于 IoC 基础上，是对 OOP 的有益补充。

Spring 本身是一个轻量级容器，和 EJB 容器不同，Spring 的组件就是普通的 Java Bean (POJO-Plain Old Java Object)，这使得单元测试可以不再依赖容器，编写更加容易。Spring 负责管理所有的 Java Bean 组件，同样支持声明式的事务管理。我们只需要编写好 Java Bean 组件，然后将它们"装配"起来就可以了，组件的初始化和管理均由 Spring 完成，只需在配置文件中声明即可。这种方式最大的优点是各组件的耦合极为松散，并且无须我们自己实现 Singleton 模式。

9.2　IoC(控制反转)模式

IoC(Inversion of Control)，即控制反转，就是由容器控制程序之间的关系，而非传统实现中由程序代码直接操控。所谓"控制反转"即控制权由应用代码中转到了外部容器，控制权的转移，是所谓反转。IoC 也称为"依赖注入(Dependency Injection)"，所谓依赖注入，即组件之间的依赖关系由容器在运行期决定，由容器动态地将某种依赖关系注入组件之中。

分离关注(Separation of Concerns，SOC)是 IoC 模式和 AOP 产生最原始动力，通过功能分解可得到关注点，这些关注可以是组件(Components)、方面(Aspects)或服务(Services)。

一种新的思维编程方式：Interface Driven Design 接口驱动。接口驱动有很多好处，可以提供不同灵活的子类实现，增加代码稳定性和健壮性等，但是接口一定是需要实现的，也就是如下语句迟早要执行：

```
AInterface a = new AInterfaceImp();
```

AInterfaceImp 是接口 AInterface 的一个子类，IoC 模式可以延缓接口的实现，根据需要实现，例如，接口如同空的模型套，在必要时，需要向模型套注射石膏，这样才能成为一个模型实体，因此，我们将人为控制接口的实现称为"注射"。其实 IoC 模式也是解决调用者和被调用者之间的一种关系，上述 AInterface 实现语句表明当前是在调用被调用者 AInterfaceImp，由于被调用者名称写入了调用者的代码中，这产生了一个接口实现的原罪：彼此联系，调用者和被调用者有紧密联系，在 UML 中是用依赖 Dependency 表示。但是这种依赖在分离关注的思维下是不可忍耐的，必须切割，实现调用者和被调用者解耦，新的 IoC 模式 Dependency Injection 模式由此产生了，Dependency Injection 模式是依赖注入的意思，也就是将依赖先剥离，然后在适当时候再注入。

简单应用过程步骤如下。

(1) 创建接口

```
interface
    public interface IHelloWorld {
        public void sayHello();
    }
```

(2) 创建接口的实现类

```
public class HelloWorld2 implements IHelloWorld{
    private String msg;
    public String getMsg() {
        return msg;
    }
    public void setMsg(String msg) {  //通过 setter 方法进行注入
        this.msg = msg;
    }
    public void sayHello(){
        System.out.println("Hello," + msg);
    }
}
```

(3) 编写 Spring 配置文件 ApplicationContext.xml

```
<beans>
    <bean  id = "Hello" class = "HelloWorld2">
        <property name = "msg">
            <value>Hello World! </value>
        </property>
    </bean>
</beans>
```

(4) 实现测试类

```
public class TestHelloWorld {
    public static void main(String[] args) {
     ApplicationContext ac = new FileSystemXmlApplicationContext("applicationContext.xml");
     IHelloWorld hw = (IHelloWorld)ac.getBean("Hello");
     hw.sayHello();
    }
    }
```

从上述应用中可以看出,我们的组件并不需要实现框架指定的接口,就可以轻松地将组件从 Spring 中脱离,甚至不需要任何修改。另外组件间的依赖关系减少,极大地改善了代码的可重用性。Spring 的依赖注入机制,可以在运行期为组件配置所需资源,而无须在编写组件代码时就加以指定,从而在相当程度上降低了组件之间的耦合。

上述 IoC 控制模式依赖注入有 3 种类型,分别为接口注入、set 注入和构造注入。

Spring 框架支持 set 注入和构造注入,下面我们把上述的 Helloworld 应用转为构造注入,其过程如下。

① 创建类,实现上述代码的接口,代码如下。

```
public classBHelloWorld implements IHelloWorld{
    public String msg;
    public BHelloWorld(String msg){
        this.msg = msg;
    }
    public void sayHello(){
        System.out.print(msg);
    }
  }
```

② 在 Spring 配置文件:ApplicationContext. xml,其内容如下。

```
<bean  id = "BHello" class = "BHelloWorld">
        <constructor - arg index = "0">
                <value>C Hello World! </value>
        </constructor - arg>
    </bean>
```

constructor-arg 用来表示用构造方式注入参数,index="0"表示是构造方法中的第一个参数。

③ 编写测试类,代码如下。

```
public class TestHelloWorld {
        public static void main(String[] args) { //ClassPathXmlApplicationContext
          ApplicationContext ac = new FileSystemXmlApplicationContext("applicationContext.xml");
          IHelloWorld hw = (IHelloWorld)ac.getBean("BHello");
          hw.sayHello();
        }
        }
```

9.3　Spring 核心容器

Spring 采用一种动态的、灵活的方式来设计框架,其中大量使用了反射技术,在 Spring 中要解决的一个问题就是如何管理 Bean。因为 IoC 的思想要求 Bean 之间不能够直接调用,而应该采用一种被动的方式进行协作,所以 Bean 的管理是 Spring 中的核心部分。

Spring 中两个最基本最重要的包是 org. springframework. context 和 org. springframework. beans. factory,它们是 Spring 的 IoC 应用的基础。在这两个包中最重要的是 BeanFactory 和 ApplicationContext(接口)。Spring 通过上述 org. springframework. beans 包中 BeanWrapper 类来封装动态调用的细节问题,BeanFactory 来管理各种 Bean,使用 ApplicationContext 类框架来管理 Bean,ApplicationContext 在 BeanFactory 之上增加了其他功能,如国际化、获取资源事件传递等。

9.3.1　BeanFactory

如上所述,Spring IoC 设计的核心是 org. springframework. beans 包,它的设计目标是与 JavaBean 组件一起使用。这个包通常不是由用户直接使用,而是由服务器将其用作其他多数功能的底层中介。其中 BeanFactory 接口,它是工厂设计模式的实现,允许通过名称创建和检索对象。BeanFactory 也可以管理对象之间的关系。

BeanFactory 支持两个对象模型。

单态模型:它提供了具有特定名称的对象的共享实例,可以在查询时对其进行检索。Singleton 是默认的也是最常用的对象模型,对于无状态服务对象很理想。

原型模型:它确保每次检索都会创建单独的对象。在每个用户都需要自己的对象时,原型模型最适合。

Bean 工厂的概念是 Spring 作为 IoC 容器的基础,IoC 将处理事情的责任从应用程序代码转移到框架。Spring 框架使用 JavaBean 属性和配置数据来指出必须设置的依赖关系。

BeanFactory 实际上是实例化,配置和管理众多 Bean 的容器。这些 Bean 通常会彼此合作,因而它们之间会产生依赖。BeanFactory 使用的配置数据可以反映这些依赖关系(一些依赖可能不像配置数据一样可见,而是在运行期作为 Bean 之间程序交互的函数)。

一个 BeanFactory 可以用接口 org. springframework. beans. factory. BeanFactory 表示,这个接口有多个实现。最常使用的简单的 BeanFactory 实现是 org. springframework. beans. factory. xml. XmlBeanFactory(ApplicationContext 是 BeanFactory 的子类,所以通常使用的是 ApplicationContext 的 XML 形式)。

虽然大多数情况下,几乎所有被 BeanFactory 管理的用户代码都不需要知道 BeanFactory,但是 BeanFactory 还是以某种方式实例化。可以使用下面的代码实例化 BeanFactory。

```
InputStream is = new FileInputStream("beans.xml");
XmlBeanFactory factory = new XmlBeanFactory(is);
```

或者

```
ClassPathResource res = new ClassPathResource("beans.xml");
XmlBeanFactory factory = new XmlBeanFactory(res);
```

或者

```
ClassPathXmlApplicationContext appContext = new ClassPathXmlApplicationContext(
        new String[] {"applicationContext.xml","applicationContext-part2.xml"});
// of course,an ApplicationContext is just a BeanFactory
BeanFactory factory = (BeanFactory) appContext;
```

在很多情况下,用户代码不需要实例化 BeanFactory,因为 Spring 框架代码会做这件事。例如,Web 层提供支持代码,在 J2EE Web 应用启动过程中自动载入一个 Spring ApplicationContext,这个声明过程在这里描述。

下面就 BeanFactory 的配置进行介绍。

一个最基本的 BeanFactory 配置由一个或多个它所管理的 Bean 定义组成。在一个 XmlBeanFactory 中,根节点 Beans 中包含一个或多个 Bean 元素。

```
<?xml version="1.0" encoding="UTF-8"?>
<!DOCTYPE beans PUBLIC "-//SPRING//DTD BEAN//EN" "http://www.springframework.org/dtd/spring-beans.dtd">

<beans>

  <bean id="…" class="…">
    ...
  </bean>
  <bean id="…" class="…">
    ...
  </bean>

  ...
</beans>
```

9.3.2 BeanWrapper

BeanWrapper 是 Bean 的包装器。它的主要工作，就是对任何一个 Bean 进行属性（包括内嵌属性）的设置和方法的调用。在 BeanWrapper 的默认实现类 BeanWrapperImpl 中，虽然代码繁琐，但其完成的工作非常集中。

如下面的 Bean 类应用，代码如下。

```
public class A {

    public Object getObject(){

        Class c1 = Class.forName("com.mypro.spring.bean.IHelloworld");
        Method method = c1.getMethod("setMessage",new Class[]{String.class});
        Object obj = (Object) c1.newInstance();
        method.invoke(obj,new Object[]{"Hello World"});

        return obj;

    }

}
```

通过使用 BeanWrapper 类来封装动态调用的细节问题，代码如下。

```
public class B {

    public Object getObject()
    {
     Class c1 = Class.forName("com.mypro.spring.bean.IHelloworld");
     BeanWrapper wrapper = new BeanWrapperImpl();
     wrapper.setProperty("message","Hello World");
     return wrapper.getWrappedInstance();
    }
}
```

Spring 通过使用 BeanWrapper 类，封装了细节问题，使用统一的方式来管理 Bean 的属性。有了对单个 Bean 的包装，还需要对多个的 Bean 进行管理。在 Spring 中，把 Bean 纳入到一个核心库中进行管理。Bean 的创建有两种方法：一种是一个 Bean 创建多个实例，另一种是一个 Bean 只创建一个实例。如果从设计模式角度分析考虑，前者可以采用 Prototype，后者可以采用 Singleton。

注意：反射可以非常灵活地根据类的名称创建一个对象，所以 Spring 只使用了 Prototype 和 Singleton 两个基本的模式。上述的 org.springframework.beans.factory 包中的 BeanFactory 定义了统一的 getBean 方法，使用户能够维护统一的接口，而不需要关心当前的 Bean 是来自 Prototype 产生的独立的 Bean，还是 Singleton 产生的共享的 Bean。

9.3.3 ApplicationContext

Beans 包提供了以编程的方式管理和操控 Bean 的基本功能，而 Context 包增加了 ApplicationContext，它以一种更加面向框架的方式增强了 BeanFactory 的功能。多数用户以一种完全的声明式方式来使用 ApplicationContext，甚至不用去手工创建它，可以依赖类似 ContextLoader 的支持类，在 J2EE 的 Web 应用的启动进程中，用它来启动 ApplicationContext。此外，还可以以编程的方式创建一个 ApplicationContext。

Context 包的基础是位于 org. springframework. context 包中的 ApplicationContext 接口。它是由 BeanFactory 接口集成而来，提供 BeanFactory 所有的功能。为了以一种更像面向框架的方式工作，Context 包使用分层和有继承关系的上下文类，包括：

① MessageSource，提供对 i18n 消息的访问；

② 资源访问，如 URL 和文件；

③ 事件传递给实现了 ApplicationListener 接口的 Bean；

④ 载入多个(有继承关系)上下文类，使得每一个上下文类都专注于一个特定的层次，如应用的 Web 层。

由于 ApplicationContext 包括了 BeanFactory 所有的功能，所以通常建议先于 BeanFactory 使用，除了有限的一些场合，例如，在一个 Applet 中，内存的消耗是关键的，每 kb 字节都很重要。ApplicationContext 在 BeanFactory 的基本功能上增建的功能如下所述。

(1) 使用 MessageSource

ApplicationContext 接口继承 MessageSource 接口，所以提供了 messaging 功能(i18n 或者国际化)。同 NestingMessageSource 一起使用，就能够处理分级的信息，这些是 Spring 提供的处理信息的基本接口，以下是其定义的方法。

- String getMessage (String code，Object[] args，String default，Locale loc)：这个方法是从 MessageSource 取得信息的基本方法。如果对于指定的 Locale 没有找到信息，则使用默认的信息。传入的参数 args 被用来代替信息中的占位符，这个是通过 Java 标准类库的 MessageFormat 实现的。
- String getMessage (String code，Object[] args，Locale loc)：与上一个方法实质一样，其区别是没有默认值可以指定，如果信息找不到，就会抛出一个 NoSuchMessage Exception。
- String getMessage(MessageSourceResolvable resolvable，Locale locale)：上面两个方法使用的所有属性都是封装到一个称为 MessageSourceResolvable 的类中，可以通过这个方法直接使用它。

当 ApplicationContext 被加载的时候，它会自动查找在 Context 中定义的 MessageSource Bean，这个 Bean 必须称为 MessageSource。如果找到了这样的一个 Bean，所有对上述方法的调用将会被委托给找到的 MessageSource。如果没有找到 MessageSource，ApplicationContext 将会尝试查它的父类是否包含这个名字的 Bean。如果有，它将会把找到的 Bean 作为 MessageSource。如果它最终没有找到任何的信息源，一个空的 StaticMessageSource 将会被实例化，使它能够接受上述方法的调用。

Spring 目前提供了两个 MessageSource 的实现，它们是 ResourceBundleMessageSource 和 StaticMessageSource。两个都实现了 NestingMessageSource，以便能够嵌套地解析信息。StaticMessageSource 很少被使用，但是它提供以编程的方式向 Source 增加信息。Resource

BundleMessageSource 用得更多一些，其代码如下。

```
<beans>
    <bean id = "messageSource"
            class = "org.springframework.context.support.ResourceBundleMessageSource">
        <property name = "basenames">
            <list>
                <value>format</value>
                <value>exceptions</value>
                <value>windows</value>
            </list>
        </property>
    </bean>
</beans>
```

上述配置表明在 classpath 有 3 个 resource bundle，分别为 format、exceptions 和 windows。使用 JDK 通过 ResourceBundle 解析信息的标准方式，任何解析信息的请求都会被处理。

(2) 事件传递

ApplicationContext 中的事件处理是通过 ApplicationEvent 类和 ApplicationListener 接口来提供的。如果上下文中部署了一个实现了 ApplicationListener 接口的 Bean，每次一个 ApplicationEvent 发布到 ApplicationContext 时，那个 Bean 就会被通知。实质上，这是标准的 Observer 设计模式，Spring 提供了 3 个标准事件，如表 9-1 所示。

表 9-1　内置事件

事　件	解　释
ContextRefreshedEvent	当 ApplicationContext 已经初始化或刷新后发送的事件。这里初始化意味着：所有的 Bean 被装载、Singleton 被预实例化以及 ApplicationContext 已准备好
ContextClosedEvent	当使用 ApplicationContext 的 close()方法结束上下文的时候发送的事件。这里结束意味着 Singleton 被销毁
RequestHandledEvent	一个与 Web 相关的事件，告诉所有的 Bean 一个 HTTP 请求已经被响应了(这个事件将会在一个请求结束后被发送)。注意，这个事件只能应用于使用了 Spring 的 DispatcherServlet 的 Web 应用

同样也可以实现自定义的事件。通过调用 ApplicationContext 的 publishEvent()方法，并且指定一个参数，这个参数可以是自定义的事件类的一个实例。如下 ApplicationContext 代码所示。

```
<bean id = "email" class = "example.EmailBean">
    <property name = "blackList">
        <list>
                <value>zhang@163.com</value>
                <value>wang@126.com</value>
                <value>john@bcpl.cn</value>
        </list>
```

```
    </property>
</bean>

<bean id = "blackListListener" class = "example.BlackListNotifier">
    <property name = "notificationAddress">
        <value>test@bcpl.cn</value>
    </property>
</bean>
```

实际的 Bean 如下所示。

```
public class EmailBean implements ApplicationContextAware {

    /** the blacklist */
    private List blackList;

    public void setBlackList(List blackList) {
        this.blackList = blackList;
    }

    public void setApplicationContext(ApplicationContext ctx) {
        this.ctx = ctx;
    }

    public void sendEmail(String address,String text) {
        if (blackList.contains(address)) {
            BlackListEvent evt = new BlackListEvent(address,text);
            ctx.publishEvent(evt);
            return;
        }
    }
}

public class BlackListNotifier implement ApplicationListener {

    /** notification address */
    private String notificationAddress;

    public void setNotificationAddress(String notificationAddress) {
        this.notificationAddress = notificationAddress;
    }

    public void onApplicationEvent(ApplicationEvent evt) {
        if (evt instanceof BlackListEvent) {
            // notify appropriate person
        }
    }
}
```

(3) 在 Spring 中使用资源

很多应用程序都需要访问资源,Spring 以一种协议无关的方式访问资源。ApplicationContext 接口包含一个方法 getResource(String)完成工作。

Resource 类定义了几个方法,这几个方法被所有的 Resource 实现所共享,如表 9-2 所示。

表 9-2 资源功能

方法	解释
getInputStream()	用 InputStream 打开资源,并返回这个 InputStream
exists()	检查资源是否存在,如果不存在返回 false
isOpen()	如果这个资源不能打开多个流将会返回 true,因为除了基于文件的资源,一些资源不能被同时多次读取,它们就会返回 false
getDescription()	返回资源的描述,通常是全限定文件名或者实际的 URL

Spring 提供了几个 Resource 的实现,它们都需要一个 String 表示的资源的实际位置。依据这个 String,Spring 将会自动选择正确的 Resource 实现。当向 ApplicationContext 请求一个资源时,Spring 首先检查指定的资源位置,寻找任何前缀。根据不同的 ApplicationContext 的实现,不同的 Resource 实现可被使用。Resource 最好是使用 ResourceEditor 来配置,如 XmlBeanFactory。

使用 ApplicationContext 类框架来管理 Bean 的代码如下所示。

```
public classTest{
    public Object getObject(String classname){
        ApplicationContext context = new FileSystemXmlApplicationContext("applicationContext.xml");
        return context.getBean(classname);
    }
}
```

9.3.4 Web Context 应用

Spring 提供了可配置 ApplicationContext 的加载机制有两种,分别为 ContextLoaderListener 和 ContextLoaderServlet。

如果对 web.xml 进行声明,其代码如下。

```
<listener>
        <listener-class>
            org.springframework.web.context.ContextLoaderListener
        </listener-class>
</listener>
```

或者进行如下声明。

```
<servlet>
        <servlet-name>context</servlet-name>
<servlet-class>
org.springframework.web.context.ContextLoaderServlet
</servlet-class>
</servlet>
```

其加载过程：Tomcat 等自动会加载在 /WEB-INF/applicationContext. xml 路径下的 applicationContext. xml 文件。

如果需要自定义，可以通过如下方式。

```
<context-param>
        <param-name>contextConfigLocation</param-name>
        <param-value>/WEB-INF/config/applicationContext.xml</param-value>
</context-param>
```

然后可以通过一个类的方法 WebApplicationContextUtils. getWebApplicationContext() 来获得 ApplicationContext 对象。

```
ServletContext sc = null;
sc.getAttribute("contextConfigLocation");
context1.getWebApplicationContext(sc);
```

9.4 Bean 应用

9.4.1 Bean 定义及应用

一个 XmlBeanFactory 中的 Bean 定义包括以下两个方面的内容。

- classname：这通常是 Bean 的真正的实现类。但是如果一个 Bean 使用一个静态工厂方法所创建而不是被普通的构造函数创建，那么这实际上就是工厂类的 classname。
- Bean 行为配置元素：它声明这个 Bean 在容器的行为方式(如 Prototype 或 Singleton、自动装配模式、依赖检查模式、初始化和析构方法)。

构造函数的参数和新创建 Bean 需要的属性：举一个例子，一个管理连接池的 Bean 使用的连接数目(即可以指定为一个属性，也可以作为一个构造函数参数)，或者池的大小限制。

和这个 Bean 工作相关的其他 Bean：比如它的合作者(同样可以作为属性或者构造函数的参数)。这个也被称为依赖。Bean 的属性元素如下：

```
<bean  id="beanId"
       name="beanName"
       class="beanClass"
       parent="parentBean"
       abstract="true|false"
       singleton="true|false"
       lazy-init="true|false"
       autowire="byName|byType|constructor|autodetect|no"
       dependency-check="none|all|object|simple"
       depends-on="dependsOnBean"
       init-method="methodName"
       destory-method="methodName"
       fatory-method="beanClass"/>
```

上面列出的概念直接转化为组成 Bean 定义的一组元素，这些元素说明如表 9-3 所示。

表 9-3　Bean 属性元素

属性名	作　用
id	Bean 的唯一标识符
name	用来为 id 创建一个或多个别名，多个别名之间用逗号或空格分开
class	用来定义类的名称(包名＋类名)
parent	子类 Bean 定义时所引用的父类 Bean，此时 class 属性失效，子类 Bean 会继承父类 Bean 的所有属性(子类和父类属于同一个 Java 类)
abstract	用来定义 Bean 是否为抽象 Bean，如为 true 表示不会被实例化，一般用于父类，供子类继承
singleton	定义是否为单例模式，如为 true，表示在 BeanFactory 作用范围内，只维护这个 Bean 的一个实例；如果为 false，Bean 是 Prototype 状态，BeanFactory 为每次 Bean 请求创建一个新的 Bean 实例
lazy-init	如为 true，表示来延迟初始化 Bean；如果对所有的 Bean 都延迟初始化，修改属性为： ＜beans default-lazy-init＝"true"＞
autowire	byName 模式：通过 Bean 的属性名字进行自动装配 byType 模式：通过在配置文件中查找一个属性类型一样的 Bean 来进行自动装配 constructor 模式：是指根据构造函数的参数自动装配 autodetect 模式：通过对 Bean 检查类的内部来选择是 constructor 还是 byType 模式 no 模式：不使用自动装配
dependency-check	simple 模式：对基本类型、字符串、集合进行依赖检查 object 模式：对依赖的对象进行依赖检查 all 模式：对全部属性进行依赖检查 none 模式：不进行依赖检查
depends-on	Bean 在初始化时依赖的对象(依赖的对象在这个 Bean 初始化之前创建)
init-method	用来定义 Bean 的初始化方法
destory-method	用来定义 Bean 的销毁方法，在 BeanFactory 关闭时调用
fatory-method	定义创建该 Bean 对象的工厂方法，此时，class 属性失效

注意：Bean 初始化可以通过表中指定 init-method 属性来完成，Bean 的定义通过 org. springframework. beans. factory. config. BeanDefinition 以及它的各种子接口来实现。然而，绝大多数的用户代码不需要与 BeanDefination 直接接触，具体详细介绍如下。

1. Bean 类

class 属性通常是强制性的，有两种用法。在绝大多数情况下，BeanFactory 直接调用 Bean 的构造函数来“new”一个 Bean(相当于调用 new 的 Java 代码)，class 属性指定了需要创建的 Bean 的类。在比较少的情况下，BeanFactory 调用某个类的静态的工厂方法来创建 Bean，class 属性指定了实际包含静态工厂方法的那个类(至于静态工厂方法返回的 Bean 的类型是同一个类还是完全不同的另一个类，这并不重要)。

(1) 通过构造函数创建 Bean

当使用构造函数创建 Bean 时,所有普通的类都可以被 Spring 使用并且和 Spring 兼容。这就是说,被创建的类不需要实现任何特定的接口或者按照特定的样式进行编写,仅仅指定 Bean 的类就足够了。然而,根据 Bean 使用的 IoC 类型,可能需要一个默认的(空的)构造函数。

另外,BeanFactory 并不局限于管理真正的 JavaBean,它也能管理任何你想让它管理的类。虽然很多使用 Spring 的人喜欢在 BeanFactory 中用真正的 JavaBean (仅包含一个默认的(无参数的)构造函数,在属性后面定义相对应的 setter 和 getter 方法),但是在 BeanFactory 中也可以使用特殊的非 Bean 样式的类。如果你需要使用一个遗留下来的完全没有遵守 JavaBean 规范的连接池, Spring 同样能够管理它。

使用 XmlBeanFactory 可以像下面这样定义 Bean class。

```
<bean id = "exampleBean"
        class = "examples.ExampleBean"/>
<bean name = "anotherExample"
        class = "examples.ExampleBeanTwo"/>
```

至于为构造函数提供(可选的)参数,以及对象实例创建后设置实例属性,将会在后面叙述。

(2) 通过静态工厂方法创建 Bean

当定义一个使用静态工厂方法创建的 Bean,同时使用 class 属性指定包含静态工厂方法的类时,需要 factory-method 属性来指定工厂方法名。Spring 调用这个方法(包含一组可选的参数)并返回一个有效的对象,之后这个对象就完全和构造方法创建的对象一样。用户可以使用这样的 Bean 定义在遗留代码中调用静态工厂。

下面是一个 Bean 定义的例子,声明这个 Bean 要通过 factory-method 指定的方法创建。注意这个 Bean 定义并没有指定返回对象的类型,只指定包含工厂方法的类。在这个例子中,createInstance 必须是 static 方法。

```
<bean id = "exampleBean"
        class = "examples.ExampleBean2"
        factory - method = "createInstance"/>
```

至于为工厂方法提供(可选的)参数,以及对象实例被工厂方法创建后设置实例属性,将会在后面叙述。

(3) 通过实例工厂方法创建 Bean

使用一个实例工厂方法(非静态的)创建 Bean 和使用静态工厂方法非常类似,调用一个已存在的 Bean(这个 Bean 应该是工厂类型)的工厂方法来创建新的 Bean。

使用这种机制,class 属性必须为空,而且 factory-bean 属性必须指定一个 Bean 的名字,这个 Bean 一定要在当前的 Bean 工厂或者父 Bean 工厂中,并包含工厂方法,而工厂方法本身仍然要通过 factory-method 属性设置。下面是一个例子。

```
<! -- The factory bean,which contains a method called
        createInstance -- >
<bean id = "myFactoryBean"
      class = "…">
   ...
</bean>
<! -- The bean to be created via the factory bean -- >
<bean id = "exampleBean"
      factory - bean = "myFactoryBean"
      factory - method = "createInstance"/>
```

虽然我们要在后面讨论设置 Bean 的属性，但是这个方法意味着工厂 Bean 本身能够被容器通过依赖注射来管理和配置。

2. Bean 的标志符

每一个 Bean 都有一个或多个 id(也称为标志符)，这些 id 在管理 Bean 的 BeanFactory 或 ApplicationContext 中必须是唯一的。一个 Bean 差不多总是只有一个 id，但是如果一个 Bean 有超过一个的 id，也可以认为是别名。id 和 name 都可以用来指定 id，这两者中至少有一个，区别是 id 的命名必须符合 xml id 中合法字符，name 则没有限制，而且可以使用 name 指定多个 id(用逗号或分号分隔)。

在一个 XmlBeanFactory 中(包括 ApplicationContext 的形式)，可以用 id 或者 name 属性来指定 Bean 的 id(s)，并且在这两个或其中一个属性中至少指定一个 id。id 属性允许指定一个 id，并且它在 XML DTD(定义文档)中作为一个真正的 XML 元素的 ID 属性被标记，所以 XML 解析器能够在其他元素指回向它的时候做一些额外的校验。正因如此，用 id 属性指定 Bean 的 id 是一个比较好的方式。然而，XML 规范严格限定了在 XML ID 中合法的字符。通常这并不是真正的限制，但是如果有必要使用这些字符(在 ID 中的非法字符)，或者想给 Bean 增加其他的别名，那么可以通过 name 属性指定一个或多个 id(用逗号或者分号分隔)。

3. Bean 的属性

在定义 Bean 的属性时除了直接指定 Bean 的属性外还可以参考配置文件中定义的其他 Bean。

```
<bean  id = "DateHello" class = "DateHelloWorld">
         <property name = "msg">
              <value>Hello World! </value>
         </property>
         <property name = "date">
              <bean id = "date" class = "java.util.Date"/>
         </property>
  </bean>
```

```
<bean  id = "DateHello" class = "DateHelloWorld" depends - on = "date">
         <property name = "msg">
              <value>Hello World! </value>
         </property>
         <property name = "date">
              <ref bean = "date"/>
         </property>
  </bean>
  <bean id = "date" class = "java.util.Date"/>
```

4. null 值的处理

把属性设为 null 值有两种方法。

方法一：

```
<bean  id = "DateHello" class = "DateHelloWorld">
     <property name = "msg">
          <value>null</value>
     </property>
</bean>
```

方法二：

```
<bean  id = "DateHello" class = "DateHelloWorld">
          <property name = "msg">
               <null/>
          </property>
     </bean>
```

5. Singleton 的使用

Beans 被定义为两种部署模式中的一种：Singleton 或 Non-singleton（后一种也称为 Prototype）。如果一个 Bean 是 Singleton 形态的，那么就只有一个共享的实例存在，所有和这个 Bean 定义的 id 符合的 Bean 请求都会返回这个唯一的、特定的实例。

如果 Bean 以 Non-singleton(Prototype)模式部署，对这个 Bean 的每次请求都会创建一个新的 Bean 实例，这对于如每个 user 需要一个独立的 user 对象这样的情况是非常理想的。

Beans 默认被部署为 Singleton 模式，除非指定。要记住把部署模式变为 Non-singletion 后，每一次对这个 Bean 的请求都会导致一个新创建的 Bean，而这可能并不是真正想要的，所以仅仅在绝对需要的时候才把模式改为 Prototype。

在下面这个例子中，两个 Bean 一个被定义为 Singleton，而另一个被定义为 Non-singleton。客户端每次向 BeanFactory 请求都会创建新的 exampleBean，而 AnotherExample 仅仅被创建一次，在每次对它请求时都会返回这个实例的引用。

```
<bean id = "exampleBean"
       class = "examples.ExampleBean" singleton = "false"/>
<bean name = "yetAnotherExample"
       class = "examples.ExampleBeanTwo" singleton = "true"/>
```

注意：当部署一个 Bean 为 Prototype 模式时，这个 Bean 的生命周期就会有稍许改变。通过定义，Spring 无法管理一个 Non-singleton/Prototype Bean 的整个生命周期，因为当它创建之后，它被交给客户端而且容器不再跟踪它。

6. depends-on 的使用

Bean 的 depends-on 可以用来在初始化这个 Bean 前，强制执行一个或多个 Bean 的初始化，如下所示。

```
<bean  id = "DateHello" class = "DateHelloWorld" depends - on = "date">
```

9.4.2　Bean 的生命周期

Bean 的生命周期包含定义 JavaBean、初始化 JavaBean 和销毁 JavaBean 3 个阶段，如图 9-2 所示。

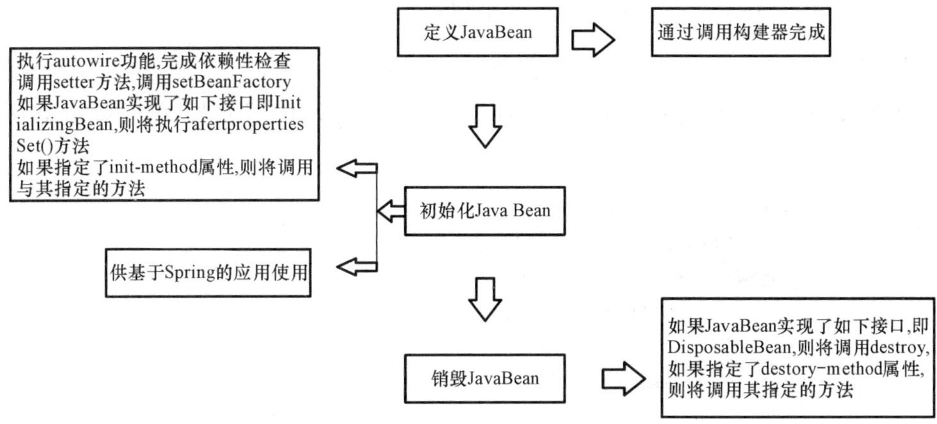

图 9-2 Bean 生命周期

其生命周期中 Bean 的应用如下所示。

1. Bean 的定义

以 beans 标签开始配置 Bean,beans 标签中包含一个或多个 Bean,具体如下所示。

```
<beans>
        //定义一个 Bean,id 是唯一标志,class 是 Bean 的来源
            <bean  id = "DateHello" class = "DateHelloWorld">
           //配置 Bean 的开始
           <property name = "msg">
               <null/>
           </property>
           //定义 Bean 的结束
        </bean>
            //配置 Bean 的结束
    </beans>
```

2. Bean 的初始化

Bean 初始化的方式有两种:在配置文件中指定 init-method 属性来完成和实现org.springframework. beans. factory. InitialingBean 接口并实现其中的 afterPropertiesSet()方法。

(1) 在配置文件中指定 init-method 属性来完成。

```
public class HelloWorld {
        private String msg;
        public void init(){
                msg = "Hello World";
        }
        public String getMsg() {
            return msg;
        }
        public void setMsg(String msg) {
            this.msg = msg;
        }
        public void sayHello(){
            System.out.println(msg);
        }
    }
```

然后在 ApplicationContext. xml 文件中添加：

```
<bean  id = "DateHello" class = "HelloWorld" init - method = "init">
```

（2）实现 org. springframework. beans. factory. InitialingBean 接口并实现其中的 afterPropertiesSet() 方法。

```
public class HelloWorld implements InitializingBean{
        private String msg;
        public void afterPropertiesSet(){
                msg = "Hello World";
        }
        public String getMsg() {
            return msg;
        }
        public void setMsg(String msg) {
            this. msg = msg;
        }
        public void sayHello(){
            System. out. println(msg);
        }
    }
```

然后在 ApplicationContext. xml 文件中添加：

```
<bean  id = "DateHello" class = "HelloWorld">
</bean>
```

3. Bean 的使用

Bean 的调用有 3 种方式，分别如下所示。

（1）使用 BeanWrapper

```
HelloWorld helloWorld = new HelloWorld ();
BeanWrapper bw = new BeanWrapperImpl(helloWorld);
bw. setPropertyValue("msg","HelloWorld");
System. out. println(bw. getPropertyValue("msg"));
```

（2）使用 BeanFactory

```
//InputStream is = new FileInputStream("applicationContext. xml");//是错误的写法
ClassPathResource   is   =   new   ClassPathResource("applicationContext. xml");
XmlBeanFactory factory = new XmlBeanFactory (is);
HelloWorld helloWorld = (HelloWorld)factory. getBean ("HelloWorld");
helloWorld. sayHello();
```

（3）使用 ApplicationContext

```
ApplicationContext ac = new FileSystemXmlApplicationContext(". /src/applicationContext. xml");
IHelloWorld hw = (IHelloWorld)ac. getBean("CHello");
hw. sayHello();
```

4. Bean 的销毁

Bean 的销毁有两种方式，分别如下所示。

（1）在配置文件中指定 destory-method 属性来完成。

```
public class HelloWorld {
    private String msg;
    public void init(){
            msg = "Hello World";
    }
    public void cleanup(){
            msg = "";
    }
    public String getMsg() {
        return msg;
    }
    public void setMsg(String msg) {
        this.msg = msg;
    }
    public void sayHello(){
        System.out.println(msg);
    }
}
```

然后在 ApplicationContext.xml 文件中进行引用，代码如下：

```
<bean  id = "DateHello" class = "HelloWorld" init - method = "init" destory - method = "cleanup" >
```

（2）实现 org.springframework.beans.factory.DisposableBean 接口，实现其中的 destory()方法。

```
public class HelloWorld implements InitializingBean,DisposableBean{
    private String msg;
    public void afterPropertiesSet(){
            msg = "Hello World";
    }
    public void destory(){
            msg = "";
    }
    public String getMsg() {
        return msg;
    }
    public void setMsg(String msg) {
        this.msg = msg;
    }
    public void sayHello(){
        System.out.println(msg);
    }
}
```

然后在 ApplicationContext.xml 文件中进行引用，具体如下所示。

```
<bean  id = "DateHello" class = "HelloWorld">
  </bean>
```

9.4.3 Bean 的依赖方式

Bean 标签中的子标签 ref 指定依赖的方式有两种，分别为使用 local 属性指定和使用 bean 属性指定。

1. 使用 local 属性指定

local 属性的值必须与被参考引用的 Bean 的 id 一致，如果在同一个 XML 文件里没有匹配的元素，XML 解析将产生一个错误。local 属性应用如下所示。

```
<bean  id="DateHello" class="DateHelloWorld">
     <property name="msg">
          <value>Hello World! </value>
     </property>
     <property name="date">
          <ref local="date"/>
     </property>
</bean>
<bean id="date" class="java.util.Date"/>
```

2. 使用 bean 属性指定

用 ref 元素中的 bean 属性指定被参考引用的 Bean 是 Spring 中最常见的形式，bean 属性的值可以与被引用的 Bean 的 id 相同也可以与 name 相同。Bean 属性应用如下所示。

```
<bean  id="DateHello" class="DateHelloWorld">
     <property name="msg">
          <value>Hello World! </value>
     </property>
     <property name="date">
          <ref bean="date"/>
     </property>
</bean>
<bean id="date" class="java.util.Date"/>
```

3. Bean 的自动装配

Bean 的自动装配有 5 种模式，分别为 byName、byType、constructor、autodetect 和 no。可以使用 bean 元素的 autowire 属性来指定 Bean 的装配模式，显示的指定依赖如 property 和 constructor-arg 元素总会覆盖自动装配。对于大型应用不鼓励使用自动装配。

4. Bean 依赖检查模式

Bean 默认的是不检查依赖关系，可以使用 bean 元素的 dependency-check 属性来指定 Bean 的依赖检查，共有 4 种模式，分别如下所示。

(1) 使用 simple 模式

使用 simple 模式是指对基本类型、字符串和集合进行依赖检查，如下所示。

```
<bean  id="DateHello" class="DateHelloWorld" autowire="autodetect" dependency-check="simple">
</bean>
<bean id="date" class="java.util.Date"/>
```

上述 simple 模式只会对 msg 进行检查。

(2) 使用 object 模式

使用 object 模式是指对对象进行依赖检查,如下所示。

```
<bean  id = "DateHello" class = "DateHelloWorld" autowire = "autodetect" dependency - check = "object">
</bean>
<bean id = "date" class = "java.util.Date"/>
```

上述 object 模式只会对 date 进行检查。

(3) 使用 all 模式

使用 all 模式是指对所有属性进行依赖检查, 如下所示。

```
<bean  id = "DateHello" class = "DateHelloWorld" autowire = "autodetect" dependency - check = "all">
</bean>
<bean id = "date" class = "java.util.Date"/>
```

上述 all 模式只会对 msg 和 date 进行检查。

(4) 使用 none 模式

使用 none 模式是指对所有属性不进行依赖检查,如下所示。

```
<bean  id = "DateHello" class = "DateHelloWorld" autowire = "autodetect" dependency - check = "none">
</bean>
<bean id = "date" class = "java.util.Date"/>
```

上述 none 模式不会对 msg 和 date 进行检查。

一般情况下依赖检查和自动装配结合使用,当 bean 属性都有默认值或不需要对 Bean 的属性是否被设置到 Bean 上检查时,依赖检查的作用就不大了。

9.4.4 集合注入的方式

对于集合 List、Set 、Map 以及 Properties 的元素,集合注入的方式有不同的配置方式,分别如下所示。

1. List 集合

(1) 创建 Action 类

```
public class HelloWorld{
            //定义一个 List 变量 msg
            List msg = null;
            public void setMsg(List msg){
                  this.msg = msg;
            }
      }
```

(2) 配置对应的 XML 文件

```
<bean  id = "Hello" class = "HelloWorld">
          <property name = "msg">
               <list>
                       <value>Hello World! </value>
                       <value>Hello World2! </value>
               </list>
          </property>
</bean>
```

2. Set 集合

（1）创建 Action 类

```
public class HelloWorld{
        //定义一个 Set 变量 msg
        Set msg = null;
        public void setMsg(Set msg){
            this.msg = msg;
        }
    }
```

（2）配置对应的 XML 文件

```
<bean  id = "Hello" class = "HelloWorld">
        <property name = "msg">
            <set>
                <value>Hello World! </value>
                <value>Hello World2! </value>
            </set>
        </property>
    </bean>
```

3. Map 集合

（1）创建 Action 类

```
public class HelloWorld{
        //定义一个 Map 变量 msg
        Map msg = null;
        public void setMsg(Map msg){
            this.msg = msg;
        }
    }
```

（2）配置对应的 XML 文件

```
<bean  id = "Hello" class = "HelloWorld">
        <property name = "msg">
            <map>
                <entry key = "h1">
                    <value>Hello World! </value>
                </entry>
                <entry key = "h2">
                    <value>Hello World2! </value>
                </entry>
            </map>
        </property>
    </bean>
```

4. Properties

(1) 创建 Action 类

```
public class HelloWorld{
    //定义一个 Properties 变量 msg
    Properties msg;
    public void setMsg(Properties msg){
        this.msg = msg;
    }
}
```

(2) 配置对应的 XML 文件

```
<bean  id = "Hello" class = "HelloWorld">
    <property name = "msg">
        <props>
            <prop key = "h1">Hello World! </prop>
            <prop key = "h2">Hello World2! </prop>
        </props>
    </property>
</bean>
```

通常 Java 中所有的对象都必须创建,或者说,使用对象之前必须创建,但是现在可以从 IoC 容器中直接获得一个对象然后直接使用,无须事先创建它们。这种变革,就如同我们无须考虑对象销毁一样,因为 Java 的垃圾回收机制帮助我们实现了对象销毁。现在又无须考虑对象创建,对象的创建和销毁都无须考虑了,这给编程带来的影响是巨大的。

使用 IoC 容器,我们只需从 IoC 容器中抓取一个类的对象然后直接使用它们。当然,在使用之前,我们需要做一个简单的配置,把将来需要使用的类全部告诉 IoC 容器,写在 IoC 容器配置文件 xxx. xml 中。如果项目中有非常多的类,调用关系很复杂,而且调用关系随时都可能变化,那么,使用无须照顾调用关系的 IoC 容器无疑是减轻开发负担的首选。

9.5　Spring Bean 应用开发

(1) 创建 JavaBean 类 HelloWorld. java。

```
Package  com.bcpl.spring;
public class HelloWorld {

      private String hello;

      public void setHello(String hello) {
        this.hello = hello;
      }

       public String getHello() {
        return hello;
      }
}


HelloWorld  h = new HelloWorld();
h.setHello("hello,Beijing");
System.out.println(h.getHello());
```

(2) 编写配置文件 bean. xml,该配置文件将应用程序的业务逻辑组合在一起,具体如下所示。

```
<beans>
    <bean id = "helloBean" class = "com.bcpl.spring.HelloWorld"  >
        <property name = "hello"><value>Hello,zhang </value></property>
    </bean>
</beans>
```

(3) 创建应用程序。

```
HelloWorld h = (HelloWorld) AppUtil.getBean("helloBean");
System.out.println(h.getHello());
```

或者创建更为复杂的应用,如下所示。

① 创建接口和对应的继承实现类,如下所示。

```
interface  People{
    public String getName();
    public void setName(String name);

    public int  getAge();
    public void setAge(int age);
}

class Teacher implements People{
  //...
}

class Engineer implements People{
  //...
}

public class PeopleUse{

    private People people;


    public void setPeople( People people) {
      this.people = people;
    }

    public People getPeople() {
      return people;
    }

    public String getPeopleAllInfo(){
            return people.getName() + people.getAge();
    }

}
```

② 编写配置文件 bean. xml,该配置文件将应用程序的业务逻辑组合在一起,具体如下所示。

```
<? xml version = "1.0" encoding = "UTF - 8"? >
<! DOCTYPE beans PUBLIC " - //SPRING/DTD BEAN/EN" "http://www.springframework.org/dtd/spring - beans.dtd">
<beans>

    <bean id = "helloBean" class = "com.bcpl.spring.HelloWorld"  >
        <property name = "hello">
            <value>
                Hello!
            </value>
        </property>
    </bean>

    <bean id = "t" class = "com.bcpl.spring.Teacher" >
        <property name = "name"> <value > Lisan</value></property>
        <property name = "age"> <value >27</value></property>
    </bean>

    <bean id = "e" class = "com.bcpl.spring.Engineer"  />

    <bean id = "li" class = "com.bcpl.spring.PeopleUse"  >
        <property name = "people">
            <ref  bean = "t"/>
        </property>
    </bean>
</beans>
```

bean.xml 中定义了 JavaBean 的别名与来源类别，<property>标签中设定了希望注入至 JavaBean 的字符串值，bean.xml 必须在 classpath 可以存取到的目录中，可以是现行的工作目录，在 Web 程序中可以是在 classes 目录下，如果使用的是单机程序的方式，将之置于现行的工作目录中。

③ 创建简单的测试程序 BeanTest.java。

```
package com.bcpl.spring;

import  org.springframework.beans.factory.BeanFactory;
import  org.springframework.core.io.Resource;
import  org.springframework.core.io.ClassPathResource;
import  org.springframework.beans.factory.xml.XmlBeanFactory;

public class BeanTest {
    public static void main(String[] args)  throws Exception{
      Resource resource = new ClassPathResource("bean.xml");//bean.properties
      BeanFactory factory = new XmlBeanFactory(resource);

          testHello(factory);
    }

    public static void testHello(BeanFactory factory){
          HelloWorld hello = (HelloWorld) factory.getBean("helloBean");
          System.out.println(hello.getHello());
    }
```

```
        public static void testLiPeopele(BeanFactory factory){
                PeopleUse l = (PeopleUse) factory.getBean("li");
                System.out.println(l.getTeaacherAllInfo());
        }

}
```

上述从比较低层次的角度来使用 Spring 的 IoC 容器功能，藉由 BeanFactory 来读取 XML 文件并完成依赖的关联注入，其依赖指的是 HelloBean 相依于 String 对象，通过 setter 所保留的接口，我们使用 setter injection 来完成这个依赖注入。

BeanFactory 是整个 Spring 的重点所在，整个 Spring 的核心都围绕着它，上述使用的是 XmlBeanFactory，负责读取 XML 文件，当然我们也可以使用 Properties 文件。

在 BeanFactory 读取 Bean 的组态设定并完成关系维护之后，我们可以藉由 getBean()方法并指定 Bean 的别名来取得实例，从而调用它的方法。

9.6 小　结

Spring 技术重点介绍了 IoC、核心容器、Bean 的定义及应用，Spring 框架也是采用 MVC 模式，其应用组件有着清晰的分离，分为表示层、持久层、域模块层（业务层），Spring 充当业务层，主要有轻量级的 IoC 容器和 AOP 支持，Spring 区别于其他框架，它是一个非侵入式的分层框架，建立于依赖注入之上。另外本章 Spring 技术中并没有介绍 AOP 代理服务、自动代理创建、AdvisorAutoProxyCreator 的应用以及 Spring 框架中层与层之间的衔接、中间件（使用第三方的 Web 框架）等内容，由于篇幅原因，并没有涉及，部分 Spring 高级技术可以从后续章节进行了解或拓展性学习。

第 10 章　Spring 高级技术与集成

10.1　Spring 持久层

在直接使用 JDBC 的程序里必须自行取得 Connection 与 Statement 对象、执行 SQL、捕捉异常、关闭相关资源，当使用 Spring 的 JdbcTemplate 时，两行代码即可完成，Spring 的 JdbcTemplate 封装了传统的 JDBC 程序执行流程，并做了异常处理与资源管理等动作，需要的只是给它一个 DataSource，而这只要在 Bean 的配置文件中完成依赖注入。

```
<bean id = "myDataSource" class =
"org.springframework.jdbc.datasource.DriverManager DataSource" destroy-method = "close">
        <property name = "driverClassName" value = "com.mysql.jdbc.Driver" />
        <property name = "url" value = "jdbc:mysql://localhost:3306/test" />
        <property name = "username" value = "root" />
        <property name = "password" value = "123" />
    </bean>

    <bean id = "myJdbcTemplate" class = "org.springframework.jdbc.core.JdbcTemplate">
        <property name = "dataSource">
            <ref bean = "myDataSource"/>
        </property>
    </bean>

    <bean id = "useJdbcBean" class = "com.bcpl.spring.UseJDBCTemplate">
        <property name = "jdbcTemplate">
            <ref bean = "myJdbcTemplate" />
        </property>
    </bean>
```

然后创建应用测试类，如下所示。

```
import org.springframework.jdbc.core.JdbcTemplate;

public class UseJDBCTemplate {
    private JdbcTemplate jdbcTemplate;

    public void setJdbcTemplate(JdbcTemplate jdbcTemplate) {
```

```
                this.jdbcTemplate = jdbcTemplate;
        }

    public JdbcTemplate getJdbcTemplate() {
                return this.jdbcTemplate;
        }

        public void  test() {
              List rows = jdbcTemplate.queryForList("SELECT * FROM Student");
              Iterator it = rows.iterator();
              while(it.hasNext()) {
                  Map userMap = (Map) it.next();
                  System.out.print(userMap.get("name") + "\t");
                  System.out.print(userMap.get("sex") + "\n");
              }
        }
}
```

JdbcTemplate 只是将我们使用 JDBC 的流程封装起来而已，包括了捕捉异常、SQL 的执行、查寻结果的转换与传回等。Spring 大量使用 Template Method 模式来封装固定流程的动作，XXXTemplate 等类都是基于这种方式实现的。除了大量使用 Template Method 来封装一些低层操作细节，Spring 也大量使用 callback 方式来呼叫相关类的方法以提供传统 JDBC 相关类的功能，使得传统 JDBC 的使用者也能清楚了解 Spring 所提供的相关封装类方法的使用。

Spring 的 JDBC 封装等功能基本上可以独立于 Spring 来使用，JdbcTemplate 所需要的就是一个 DataSource 对象，在不使用 Spring IoC 容器时，可以单独使用 spring-dao.jar 中的东西。

除了 JdbcTemplate 之外，Spring 还提供了其他的 Template 类，如对 Hibernate、JDO 和 iBatis 等的 Template 实现，另外，在事务处理方面，Spring 提供了编程式与声明式的事务处理功能，大大简化了持久层程序的复杂度，并提供了更好的维护。

10.1.1 数据源的注入

就 Spring 持久层的封装，还通过 XML 实现 DataSource 数据源的注入，其共有 3 种方式，如下所示。

1. 使用 Spring 自带的 DriverManagerDataSource

```
<beans>
    <bean id = "dataSource" class = "org.springframework.jdbc.datasource.DriverManagerDataSource">
              <property name = "driverClassName">
          <value>com.mysql.jdbc.Driver</value>
          </property>
              <property name = "url">
              <value>jdbc:mysql://localhost:3306/testdb</value>
          </property>
          <property name = "username">
              <value>root</value>
          </property>
          <property name = "password">
              <value>123</value>
          </property>
    </bean>
```

```
<bean id = "dataSource" class = "org.springframework.jdbc.datasource.DriverManagerDataSource">
        <property name = "username">
            <value> $ {jdbc.username}</value>
        </property>
        <property name = "password">
            <value> $ {jdbc.password}</value>
        </property>
        <property name = "driverClassName">
            <value> $ {jdbc.driverClassName}</value>
        </property>
        <property name = "url">
            <value> $ {jdbc.url}</value>
        </property>
    </bean>
```

在 src 下创建 jdbc.properties 属性文件。

```
jdbc.driverClassName = com.mysql.jdbc.Driver
jdbc.url = jdbc:mysql://localhost:3306/test
jdbc.username = root
jdbc.password = 123
```

```
<bean id = "propertyConfigurer" class =
"org.springframework.beans.factory.config.Property PlaceholderConfigurer">
    <property name = "location">
        <value>classpath:jdbc.properties</value>
    </property>
</bean>
```

```
<bean id = "transactionManager" class = "org.springframework.jdbc.datasource.DataSource
TransactionManager">
    <property name = "dataSource">
        <ref bean = "dataSource"/>
    </property>
    </bean>

    <bean id = "helloDAO" class = "HelloDAO">
    <property name = "dataSource">
            <ref bean = "dataSource"/>
    </property>
    </bean>

<! -- 声明式事务处理 -->
<bean id = "helloDAOProxy" class = "org.springframework.transaction.interceptor.TransactionProxyFactoryBean">
    <property name = "transactionManager">
        <ref bean = "transactionManager"/>
    </property>
    <property name = "target">
        <ref bean = "helloDAO"/>
    </property>
    <property name = "transactionAttributes">
        <props>
            <! -- 对 create 方法进行事务管理,PROPAGATION_REQUIRED 表示如果没有事务就新建
            一个事务 -->
            <prop key = "create * ">PROPAGATION_REQUIRED</prop>
        </props>
    </property>
</bean>
</beans>
```

2. 使用 DBCP 连接池的方式

DBCP 连接池方式需要加载 commons-pool-1. 2. jar 和 commons-collections. jar 包,其应用代码如下所示。

```
<beans>
    <bean id = "dataSource" class = "org.apache.commons.dbcp.BasicDataSource" destroy - method = "close">
        <property name = "driverClassName">
            <value>com.mysql.jdbc.Driver</value>
        </property>
        <property name = "url">
            <value>jdbc:mysql://localhost:3306/testdb</value>
        </property>
        <property name = "username">
            <value>root</value>
        </property>
        <property name = "password">
            <value>123</value>
        </property>
    </bean>

    <bean id = "transactionManager" class =
    "org.springframework.jdbc.datasource.DataSource TransactionManager">
        <property name = "dataSource">
        <ref bean = "dataSource"/>
        </property>
        </bean>

    <bean id = "helloDAO" class = "HelloDAO">
        <property name = "dataSource">
            <ref bean = "dataSource"/>
        </property>
        </bean>

    <! -- 声明式事务处理 -- >
    <bean id = "helloDAOProxy" class =
    "org.springframework.transaction.interceptor.TransactionProxyFactoryBean">
        <property name = "transactionManager">
        <ref bean = "transactionManager"/>
        </property>
        <property name = "target">
        <ref bean = "helloDAO"/>
        </property>
        <property name = "transactionAttributes">
        <props>
            <! -- 对 create 方法进行事务管理,PROPAGATION_REQUIRED 表示如果没有事务就新
            建一个事务 -- >
            <prop key = "create * ">PROPAGATION_REQUIRED</prop>
        </props>
        </property>
    </bean>
    </beans>
```

3. JNDI 方式

通过 Tomcat 提供的 JNDI 方式，使用页面的方式访问，具体如下所示。

```
<Resource
      name = "jdbc/oracle"
      type = "javax.sql.DataSource"
      driverClassName = "oracle.jdbc.driver.OracleDriver"
      password = "123"
      maxIdle = "2"
      maxWait = " - 1"
      username = "ascent"
      url = "jdbc:oracle:thin:@localhost:1521:orcl"
      maxActive = "4"/>
</Context>
```

```
<beans>
        <bean id = "dataSource" class = "org.springframework.jndi.JndiObjectFactoryBean">
        <property name = "jndiName">
            <value>java:comp/env/jdbc/oracle</value>
        </property>
        </bean>
        <bean id = "transactionManager" class =
        "org.springframework.jdbc.datasource.Data SourceTransactionManager">
        <property name = "dataSource">
        <ref bean = "dataSource"/>
        </property>
        </bean>
        <bean id = "helloDAO" class = "HelloDAO">
        <property name = "dataSource">
        <ref bean = "dataSource"/>
        </property>
        </bean>
        <! -- 声明式事务处理 -- >
        <bean id = "helloDAOProxy" class =
        "org.springframework.transaction.interceptor.TransactionProxyFactoryBean">
        <property name = "transactionManager">
        <ref bean = "transactionManager"/>
        </property>
        <property name = "target">
        <ref bean = "helloDAO"/>
        </property>
        <property name = "transactionAttributes">
        <props>
            <! -- 对 create 方法进行事务管理，PROPAGATION_REQUIRED 表示如果没有事务就新
            建一个事务 -- >
            <prop key = "create * ">PROPAGATION_REQUIRED</prop>
        </props>
        </property>
        </bean>
    </beans>
```

10.1.2 Spring 定时器

在 Spring 框架中定时器有两种实现方式，它们分别是 Java 的 java. util. Timer 类和 OpenSymphony 的 Quartz。

创建定时任务共分两步，第一步是创建定时任务，第二步是运行定时任务。而运行定时任务有两种方式，一是通过程序直接启动，二是通过 Web 监听定时任务。

(1) 通过程序直接启动定时任务

下面通过 JSP 页面来启动下面的 run 方法，首先创建一个任务类，具体代码如下所示。

```
public class MyTask extends TimerTask {
            public void run() {
                System.out.print("I am runing");
            }
        }
                public class Main {
            public static void main(String[] args){
                Timer timer = new Timer();
                timer.schedule(new MyTask(),10000,1000);
            }
        }
```

然后在 JSP 页面中嵌入，嵌入代码形式如下所示。

```
<%
            Timer timer = new Timer();
            timer.schedule(new MyTask(),10000,1000);
%>
```

(2) 通过 Web 监听定时任务

首先创建任务类 MyTask extends 和定时器类 BindLoader，代码如下所示。

```
public class MyTask extends TimerTask {
        public void run() {
            System.out.print("I am runing");
        }
    }
public class BindLoader implements ServletContextListener {
        Timer timer = new Timer();
        public void contextInitialized(ServletContextEvent arg0) {
            timer.schedule(new MyTask(),10000,1000);
        }

            public void contextDestroyed(ServletContextEvent arg0) {
            timer.cancel();
        }

    }
```

在 web.xml 中，添加监听，配置如下。

```
<listener>
        <listener-class>
            BindLoader
        </listener-class>
</listener>
```

下面就实际项目开发中，Spring 定时任务开发，可以通过调用配置好的 ApplicationContext.xml 文件，进行定时任务的启动，具体步骤如下。

(1) 创建任务类 MyTask extends。

代码同上，也如下所示。

```
public class MyTask extends TimerTask {
        public void run() {
            System.out.print("I am runing");
        }
    }
```

(2) 在 ApplicationContext.xml 文件中进行配置，具体如下所示。

```
<beans>
    <!-- 注册定时实体 -->
    <bean id="mianTask" class="MyTask"/>
    <!-- 注册定时信息 -->
    <bean id="stTask" class="org.springframework.scheduling.timer.ScheduledTimerTask">
    <!-- 指定推迟时间 -->
    <property name="delay">
        <value>2000</value>
    </property>
    <!-- 指定重复时间 -->
    <property name="period">
        <value>86400000</value>
    </property>
    <!-- 执行具体任务 -->
    <property name="timerTask">
        <ref local="mianTask"/>
    </property>
    </bean>
    <!-- 配置调度器，这个任务我们只能规定每隔 24 小时运行一次，无法精确到某时启动 -->
    <bean id="timerFactory" class="org.springframework.scheduling.timer.TimerFactoryBean">
        <!-- 指定定时器列表 -->
    <property name="scheduledTimerTasks">
        <list>
            <ref local="stTask"/>
        </list>
    </property>
    </bean>
  </beans>
```

(3) 配置 web.xml 文件,代码如下所示。

```
<context-param>
    <param-name>contextConfigLocation</param-name>
    <param-value>classpath:applicationContext.xml</param-value>
</context-param>
<listener>
    <listener-class>
        org.springframework.web.context.ContextLoaderListener
    </listener-class>
</listener>
```

10.2 Spring AOP

Spring AOP 是 Aspect Oriented Programming 的缩写,指的是面向方面编程。AOP 实际是 GoF 设计模式的延续,设计模式追求的是调用者和被调用者之间的解耦,AOP 即为实现此目标,AOP 设计的目标并不是为了取代 OOP,它们两者承担的角色不同,将职责各自分配给 Object 与 Aspect,会使得程序中各个组件的角色更为清楚。

关于 AOP 和 OOP,它们之间的区别有以下几个方面:

① 在 AOP 里,每个关注点的实现并不知道是否有其他关注点关注它,这是 AOP 和 OOP 的主要区别;

② 在 AOP 里组合的流向是从横切关注点到主关注点,在 OOP 中组合流向是从主关注点到横切关注点;

③ AOP 和 OOP 所关注的对象不同,AOP 是 OOP 有益的补充,它们之间并不是相互对立的。

10.2.1 AOP 概念和通知

1. AOP 相关概念

(1) 连接点(JiontPiont)

程序执行过程中明确的点,如方法的调用或特定的异常被抛出。连接点是指 Aspect 的切入点,如某个方法被调用、某个成员被存取或某个异常被抛出。

(2) 切入点(PiontCut)

它指定一个通知将被引发的一系列连接点的集合。PiontCut 就是 JionPiont 点的集合,它是程序中需要注入的 Advice 的集合,指明 Advice 在什么条件下才被触发。AOP 框架允许开发者指定切入点。切入点是 AOP 的关键,使 AOP 区别于其他使用拦截的技术,如提供声明式事务管理的 Around 通知可以被应用到跨越多个对象的一组方法上,因此切入点构成了 AOP 的结构要素。

(3) Advisor

它是 PiontCut 和 Advice 的配置器,它包含切入点(PiontCut)和通知(Advice),是把通知注入到 PiontCut 位置的代码。

2. 通知类型

(1) 通知(Advice)

某个连接点采用的处理逻辑,也就是向连接点注入的代码。Advice 是在 JointPoint 上所要调用的处理建议(在 JointPoint 上所采取的动作,许多 AOP 框架通常以 interceptor 来实现 Advice)。

通知的类型有 Around 通知、Before 通知、Throws 通知和 After returning 通知。

- Around 通知:包围一个连接点的通知,如方法调用。Aroud 通知在方法调用前后完成自定义的行为。它们负责选择继续执行连接点或通过返回它们自己的返回值或抛出异常来短路执行。
- Before 通知:在一个连接点之前执行的通知,但这个通知不能阻止连接点前的执行(除非它抛出一个异常)。
- Throws 通知:在方法抛出异常时执行的通知。Spring 提供强制类型的 Throws 通知,因此可以书写代码捕获感兴趣的异常(和它的子类),不需要从 Throwable 或 Exception 强制类型转换。
- After returning 通知:在连接点正常完成后执行的通知,例如,一个方法正常返回,没有抛出异常。

Around 通知是最通用的通知类型,大部分基于拦截的 AOP 框架,只提供 Around 通知。

Spring 提供所有类型的通知,使用最为合适的通知类型来实现需要的行为。如果只是需要用一个方法的返回值来更新缓存,最好实现一个 After returning 通知而不是 Around 通知,虽然 Around 通知也能完成同样的事情。使用最合适的通知类型使编程模型变得简单,并能减少潜在错误,例如,不需要调用在 Around 通知中所需使用的 MethodInvocation 的 proceed()方法,因此调用失败。

上述这些通知类型会分别在以下的时刻被调用:

① Around 通知在 JointPoint 前后被调用;

② Before 通知在 JointPoint 前被调用;

③ Throws 通知在 JointPoint 抛出异常时被调用;

④ After returning 通知在 JointPoint 执行完毕后被调用。

在 JointPoint 上所采取的动作,许多 AOP 框架通常以 Interceptor 来实现 Advice。Interceptor 的一个例子是 Servlet 中的 Filter 机制,在 Filter 机制下,当请求来临时,会被 Filter 先拦截并进行处理,之后传给下一个 Filter,最后才是真正处理请求的 Servlet,实现 AOP 时所使用的 Interceptor 策略与 Filter 类似,所不同的是 Filter 被绑定于 Servlet API。在 Spring 中,在真正执行某个方法前,会先插入 Interceptor,每个 Interceptor 会执行自己的处理,然后将执行流程传给下一个 Interceptor,如果没有下一个 Interceptor,就执行真正呼叫的方法。下面就不同的通知,进行具体应用代码。

(2) Interception Around 通知

```
public class LogUserManager implements MethodInterceptor {
        Logger logger = Logger.getLogger(this.getClass().getName());
        public Object invoke(MethodInvocation arg0) throws Throwable {
        logger.log(Level.INFO,arg0.getArguments()[0] + "开始处理数据:");
        Object result = arg0.proceed();
        logger.log(Level.INFO,arg0.getArguments()[0] + "处理数据结束!");
        return result;
    }
}
```

(3) Before 通知

```
public class BeforeUserManager implements MethodBeforeAdvice {
    Logger logger = Logger.getLogger(this.getClass().getName());
    public void before(Method arg0,Object[] arg1,Object arg2) throws Throwable {
        logger.log(Level.INFO,arg1[0] + "开始数据操作:");
    }

}
```

(4) After Return 通知

```
public class AfterUserManager implements AfterReturningAdvice {


    Logger logger = Logger.getLogger(this.getClass().getName());
    public void afterReturning(Object arg0,Method arg1,Object[] arg2,Object arg3) throws Throwable {
        logger.log(Level.INFO,arg2[0] + "结束数据操作!");
    }

}
```

(5) Throw 通知

```
public class ThrowsUserManager implements ThrowsAdvice {
    Logger logger = Logger.getLogger(this.getClass().getName());
    public void afterThrowing(Method arg1,Object[] arg2,Object arg3,Throwable subclass ) throws Throwable{
        System.out.println(arg2[0] + "处理数据异常");
    }
}
```

(6) Introduction 通知

Introduction 通知在 JiontPoint 调用完毕后执行，实现 Introduction 通知要实现 Introduction-Advisor 接口和 IntroductionInterceptor 接口。

下面通过一个 Spring AOP 实现的例子。通过 Before 通知的实现，表示 Advice 的代码在被调用的 public 方法开始前被执行。以下是这个 Before 通知实现的具体代码。

```
package com.bcpl.springaop.test;
import java.lang.reflect.Method;
import org.springframework.aop.MethodBeforeAdvice;
public class TestBeforeAdvice implements MethodBeforeAdvice {
public void before(Method m,Object[] args,Object target)
 throws Throwable {
  System.out.println("Hello world! (by "
     + this.getClass().getName()
     + ")");
  }
 }
```

接口 MethodBeforeAdvice 只有一个方法 Before 需要实现，它定义了 Advice 的实现。Before 方法共用 3 个参数。

- 参数 Method m 是 Advice 开始后执行的方法。
- Object[] args 是传给被调用的 public 方法的参数数组。
- Object target 是执行方法 m 对象的引用。

下面的 BeanImpl 类中，每个 public 方法调用前，都会执行 Advice。

```
package com.bcpl.springaop.test;
public class BeanImpl implements Bean {
 public void theMethod() {
  System.out.println(this.getClass().getName()
     + "." + new Exception().getStackTrace()[0].getMethodName()
     + "()"
     + " says HELLO!");
 }
}
```

类 BeanImpl 实现了下面的接口 Bean。

```
package com.bcpl.springaop.test;
public interface Bean {
 public void theMethod();
}
```

在 Spring 中面向接口与面向实现编程相比，面向接口更为实用。PointCut 和 Advice 通过配置文件来实现，下面是通过 main 方法的主实现，如下所示。

```
package com.bcpl.springaop.test;
import org.springframework.context.ApplicationContext;
import org.springframework.context.support.FileSystemXmlApplicationContext;
public class Main {
 public static void main(String[] args) {
  //Read the configuration file
  ApplicationContext ctx
    = new FileSystemXmlApplicationContext("springconfig.xml");
  //Instantiate an object
  Bean x = (Bean) ctx.getBean("bean");
  //Execute the public method of the bean (the test)
  x.theMethod();
 }
}
```

上述读入和处理配置文件后，会得到一个创建工厂的对象 ctx。任何一个 Spring 管理的对象都必须通过这个工厂来创建。

然后用配置文件便可把程序的每一部分组装起来。

```
<?xml version="1.0" encoding="UTF-8"?>
<!DOCTYPE beans PUBLIC "-//SPRING//DTD BEAN//EN" "http://www.springframework.org/dtd/spring-beans.dtd">
<beans>
  <!--CONFIG-->
  <bean id="bean" class="org.springframework.aop.framework.ProxyFactoryBean">
```

```
  <property name = "proxyInterfaces">
     <value>com.bcpl.springaop.test.Bean</value>
    </property>
    <property name = "target">
     <ref local = "beanTarget"/>
    </property>
    <property name = "interceptorNames">
     <list>
       <value>theAdvisor</value>
     </list>
    </property>
  </bean>

  <!-- CLASS -->
  <bean id = "beanTarget" class = "com.bcpl.springaop.test.BeanImpl"/>

<!-- ADVISOR -->
  <!-- Note: An advisor assembles pointcut and advice -->
<bean id = "theAdvisor" class = "org.springframework.aop.support.RegexpMethodPointcutAdvisor">
   <property name = "advice">
     <ref local = "theBeforeAdvice"/>
   </property>
   <property name = "pattern">
     <value>com\.bcpl\.springaop\.test\.Bean\.theMethod</value>
   </property>
</bean>
<!-- ADVICE -->

<bean id = "theBeforeAdvice" class = "com.bcpl.springaop.test.TestBeforeAdvice"/>
</beans>
```

4 个 Bean 定义的次序并不关键。一个 Advice、一个包含了正则表达式 PointCut 的 advisor、一个主程序类和一个配置好的接口，通过工厂 ctx，这个接口返回自己本身实现的一个引用。

BeanImpl 和 TestBeforeAdvice 都是直接配置。用一个唯一的 ID 创建一个 Bean 元素，并指定了一个实现类。

advisor 通过 Spring framework 提供的一个 RegexMethodPointcutAdvisor 类来实现。我们用 advisor 的一个属性来指定它所需的 advice-bean。第二个属性则用正则表达式定义了 PointCut，确保良好的性能和易读性。

最后配置的是 Bean，它可以通过一个工厂来创建。Bean 的定义看起来比实际上要复杂。Bean 是 ProxyFactoryBean 的一个实现，它是 Spring framework 的一部分，这个 Bean 的行为通过以下的 3 个属性来定义：

① 属性 proxyInterface 定义了接口类；

② 属性 target 指向本地配置的一个 Bean，这个 Bean 返回一个接口的实现；

③ 属性 interceptorNames 是唯一允许定义一个值列表的属性，这个列表包含所有需要在 beanTarget 上执行的 advisor。

注意:advisor 列表的次序是非常重要的。

10.2.2 Spring 切入点

Spring 的切入点模型能够使切入点独立于通知类型被重用,同样的切入点有可能接受不同的通知。

在 Spring 框架中 org.springframework.aop.Pointcut 接口,用来指定通知到特定的类和方法目标,其定义如下所示。

```
public interface Pointcut {

    ClassFilter getClassFilter();
    MethodMatcher getMethodMatcher();
}
```

将 PointCut 接口分成两个部分有利于重用类和方法的匹配部分,并且组合细粒度的操作(如和另一个方法匹配器执行一个“并”的操作)。

ClassFilter 接口被用来将切入点限制到一个给定的目标类的集合。如果 matches()永远返回 true,所有的目标类都将被匹配。

```
public interface ClassFilter {

    boolean matches(Class clazz);
}
```

MethodMatcher 接口定义如下所示。

```
public interface MethodMatcher {
    boolean matches(Method m,Class targetClass);
    boolean isRuntime();
    boolean matches(Method m,Class targetClass,Object[] args);
}
```

matches(Method,Class)方法被用来测试这个切入点是否匹配目标类的给定方法,这个测试可以在 AOP 代理创建的时候执行,避免在所有方法调用时都需要进行测试。如果两个参数的匹配方法对某个方法返回 true,并且 MethodMatcher 的 isRuntime()也返回 true,那么 3 个参数的匹配方法将在每次方法调用的时候被调用,这使切入点能够在目标通知被执行之前立即查看传递给方法调用的参数。

大部分 MethodMatcher 都是静态的,意味着 isRuntime()方法返回 false,这种情况下 3 个参数的匹配方法永远不会被调用。

注意:Spring 支持的切入点的运算有并和交,并表示只要任何一个切入点匹配的方法,交表示两个切入点都要匹配的方法。

切入点可以用 org.springframework.aop.support.Pointcuts 类的静态方法来组合,或者使用同一个包中的 ComposablePointcut 类。

Spring 中的切入点的实现有 3 种,它们分别是静态切入点、动态切入点和自定义切入点。

- 静态切入点:静态切入点只限于给定的方法和目标类,不考虑方法的参数。
- 动态切入点:动态切入点不仅限于给定的方法和目标类,还可以指定方法的参数。动态切入点有很大的性能损耗,一般很少使用。

- 自定义切入点可以任意组合。

10.2.3 AOP 基本应用

在实践项目开发中,应用程序通常包含两种类型的代码:一是核心业务代码,一是和业务关系不大的代码如日志、事务处理等。AOP 的思想就是使这两种代码分离,从而降低了两种代码的耦合性,达到了易于重用和维护的目的。

如创建 HelloSpeaker.java,实现如下所示。

```
import java.util.logging.*;
public class HelloSpeaker {
    private Logger logger = Logger.getLogger(this.getClass().getName());

    public void hello(String name) {

            logger.log(Level.INFO,"hello method starts…");    //日志记录
            System.out.println("Hello," + name);
      logger.log(Level.INFO,"hello method ends…");      //日志记录
    }
}
```

HelloSpeaker 在执行 hello()方法时,我们希望能记录该方法已经执行以及结束,最简单的做法就是如上在执行的前后加上记录动作,然而 Logger 介入了 HelloSpeaker 中,记录这个动作并不属于 HelloSpeaker,这使得 HelloSpeaker 的职责加重。

如采用先定义一个接口:

```
public interface IHello {
    public void hello(String name);}
```

接着实现该接口,代码如下所示。

```
public class HelloSpeaker implements IHello {
    public void hello(String name) {
         System.out.println("Hello," + name);
    }
}



public class  Greeting implements IHello{
    public void hello(String name){
      System.out.println("Greeting," + name);
    }
}
```

然后实现一个代理对象 HelloProxy,具体代码如下所示。

```
import java.util.logging.*;
public class HelloProxy implements IHello {
    private Logger logger = Logger.getLogger(this.getClass().getName());
    private IHello helloObject;

    public HelloProxy(){}

    public HelloProxy(IHello helloObject) {
        this.helloObject = helloObject;//把被代理对象传入
    }

    public void  setHelloObject(IHello helloObject){
        this.helloObject = helloObject;
    }

    public  IHello getHelloObject(){
        this.helloObject = helloObject;
    }

    public void hello(String name) {
        logger.log(Level.INFO,"hello method starts…"); //日志记录
        helloObject.hello(name);//调用被代理对象的方法
        logger.log(Level.INFO,"hello method ends…");//日志记录
    }
}
```

然后在 main 方法中执行,代码如下所示。

```
IHello helloProxy = new HelloProxy(new HelloSpeaker());  //生成代理对象,并给它传入一个被代理的对象
helloProxy.hello("world");

//IHello  h = factory.getBean("hello");  // IoC
//h.hello("world");


IHello helloProxy = new HelloProxy(new Greeting());      //生成代理对象,并给它传入一个被代理的对象
helloProxy.hello("world");
```

代理对象 HelloProxy 将代理真正的 HelloSpeaker 来执行 hello(),并在其前后加上记录的动作,这使得我们的 HelloSpeaker 在写时不必介入记录动作,HelloSpeaker 可以专心于它的职责。

这是静态代理的基本范例,然而,代理对象的一个接口只服务于一种类的对象,而且如果要代理的方法很多,我们要为每个方法进行代理,静态代理在程序规模稍大时就必定无法胜任。

Java 在 JDK1.3 之后加入协助开发动态代理功能的类,我们不必为特定对象与方法写特定的代理,使用动态代理,可以使得一个 handler 服务于各个对象,首先,一个 handler 必须实现 java.lang.reflect.InvocationHandler。

```
import java.util.logging.*;
import java.lang.reflect.*;

public class LogHandler implements InvocationHandler {  //
    private Logger logger = Logger.getLogger(this.getClass().getName());
    private Object delegate;   //被代理的对象

    public Object bind(Object delegate) {  //自定义的一个方法,用来绑定被代理对象
        this.delegate = delegate;
        return Proxy.newProxyInstance(
                                delegate.getClass().getClassLoader(),
                                delegate.getClass().getInterfaces(),
                                this); //通过被代理的对象生成它的代理对象,并同 handler 绑定在一起
    }

    public Object invoke(Object proxy,Method method,Object[] args) throws Throwable {
        Object result = null;
        try{
            logger.log(Level.INFO,"method starts…" + method);  //日志记录
            result = method.invoke(delegate,args);     //调用被代理对象的方法
            logger.log(Level.INFO,"method ends…" + method);     //日志记录
        } catch (Exception e){
            logger.log(Level.INFO,e.toString());
        }
        return result;
    }
}
```

InvocationHandler 的 invoke()方法会传入被代理对象的方法名称与参数,实际上要执行的方法交由 method.invoke(),并在其前后加上记录动作,method.invoke()返回的对象是实际方法执行过后的回传结果。

动态代理必须有接口:

```
public interface IHello {
    public void hello(String name);
}
```

实现该接口:

```
public class HelloSpeaker implements IHello {
    public void hello(String name) {
        System.out.println("Hello," + name);
    }
}
```

继续执行,代码如下所示。

```
LogHandler logHandler    = new LogHandler();
IHello helloProxy = (IHello) logHandler.bind(new HelloSpeaker());   //传入被代理对象,传回代理对象
helloProxy.hello("John");
```

用 IoC 方式实现，代码如下所示。

```
IHello helloProxy = (IHello) factory.getBean("hello");
  helloProxy.hello("John");
```

```
public interface ITest {
      public int add(int x,int y);
}


public class Test implements ITest {
      public int add(int x,int y) {
            return x + y;
      }
}


LogHandler logHandler    = new LogHandler();
ITest tProxy = (ITest) logHandler.bind(new Test());
System.out.println(tProxy.add(2,4));
```

IoC 方式实现，代码如下所示。

```
IHello helloProxy = (IHello) factory.getBean("add");
System.out.println(tProxy.add(2,4));
```

LogHandler 不再服务于特定对象与接口，而 HelloSpeaker 也不用插入任何有关于记录的动作，它不用意识到记录 log 动作的存在。上面的例子中，HelloSpeaker 本来必须插入记录 log 动作，这使得 HelloSpeaker 的职责加重，并混淆其原来的角色，为此，使用代理将记录 log 的动作提取出来，以分清记录 log 的动作和 HelloSpeaker 的职责与角色。

在这里，记录这个动作是我们所关注的 AOP 中的 Aspect 所指的，就是像记录 log 这类的动作，将这些动作(或特定职责)视为关注的中心，将其设计为通用、不介入特定对象和职责清楚的组件，这既是所谓的面向 Aspect，每个对象仅代表一个实际的个体，将对象视为关注的中心，将其设计为通用、职责清楚的元件。前一个例子中的记录(log)动作插入至 HelloSpeaker 对象的 hello()中，记录 log 部分是不属于 HelloSpeaker 职责的，它被硬生生切入 HelloSpeaker 中。所以就整个方法的执行流程来说，如果执行流程是纵向的，则记录这个动作硬生生地横切入其中，这个横切入的部分我们就称之为 Aspect，它是横切关注点(Crosscutting Concern，一个 Concern 可以是权限检查和事务等)的模块化，将那些散落在对象中各处的程序聚集起来，所以 Aspect 用中文表达为横切面或切面。AOP 关注于 Aspect，将这些 Aspect 当作中心进行设计，使其从职责被混淆的对象中分离出来，除了使原对象的职责更清楚之外，被分离出来的 Aspect 也可以设计得通用化，可用于不同的场合。

10.3 创建 AOP 代理

Spring 中 AOP 代理方式有两种，分别为动态代理（代理接口）和 CGLIB 代理，动态代理是指代理的是接口，Spring 默认的是动态代理。

用 ProxyFactoryBean 创建 AOP 代理，使用类 org. springfamework. aop. framewor k. ProxyFactoryBean 是创建 AOP 代理的基本方式。下面就基本创建方式进行介绍。

首先在 ApplicationContext. xml 文件中进行配置，一个 Bean 里只能用一种代理方式，采用自动代理或手动代理，手动代理分为指定所有的方法和指定具体的方法两种，具体配置如下所示。

```
<? xml version = "1.0" encoding = "UTF - 8"? >
<! DOCTYPE beans PUBLIC " - //SPRING//DTD BEAN//EN" "http://www.springframework.org/dtd/spring - beans.dtd">
<beans>
    <bean id = "log" class = "lesson2. LogUserManager" />
    <bean id = "logBefore" class = "lesson2. BeforeUserManager" />
    <bean id = "logAfter" class = "lesson2. AfterUserManager" />
    <bean id = "logThrows" class = "lesson2. ThrowsUserManager" />
    <bean id = "userManager" class = "lesson2. UserManager" />
    <bean id = "bookManager" class = "lesson2. BookManager" />



    <! -- 一个 Beans 里只能用一种代理方式 -->
    <! --===========================================================================-->
    <! -- 1. 自动代理 -->
    <bean id = "autoProxy" class =
    "org. springframework. aop. framework. autoproxy. DefaultAdvisor AutoProxyCreator"/>
    <bean id = "logBeforAdvisor" class =
    "org. springframework. aop. support. RegexpMethod PointcutAdvisor">
        <property name = "advice">
            <ref bean = "log"/>
        </property>
        <property name = "patterns">
            <value>. * save. * </value>
        </property>
    </bean>

    <! --===========================================================================-->
    <! -- 2.1. 手动代理 ，使用 ProxyFactoryBean 代理目标类中的所有方法 -->
    <beans>
        <bean id = "log" class = "logAround"/>
        <bean id = "logBefore" class = "logBefore"/>
```

```
    <bean id = "logAfter" class = "logAfter"/>
    <bean id = "logThrow" class = "logThrow"/>
    <bean id = "timebook" class = "TimeBook"/>
    <! - 设定代理类 -- >
    <bean id = "logProxy" class = "org.springframework.aop.framework.ProxyFactoryBean">
        <! -- 代理的是接口 -- >
        <property name = "proxyInterfaces">
                <value>TimeBookInterface</value>
        </property>
              <! -- 要代理的目标类 -- >
        <property name = "target">
                <ref bean = "timebook"/>
        </property>
        <! -- 程序中的 Advice -- >
        <property name = "interceptorNames">
                <list>
                    <value>logBefore</value>
                    <value>logAfter</value>
                    <value>logThrow</value>
                </list>
        </property>
    </bean>
</beans>

<! --=====================================================================-- >
<! -- 2.2. 手动代理,代理目标类的指定方法 -- >
<bean id = "logAdvisor" class = "org.springframework.aop.support.RegexpMethodPointcutAdvisor">
   <property name = "advice">
        <! -- 在执行其他几种通知时,只要把 ref bean = "其他通知了的 Bean 的 id" -- >
     <ref bean = "logThrows"/>
   </property>

   <! -- 指定要代理的方法 -- >
   <property name = "patterns">
     <value>. * update. * </value>
   </property>
</bean>

<! -- 设定代理类 -- >
<bean id = "logProxy" class = "org.springframework.aop.framework.ProxyFactoryBean" >
   <! -- 这里的设置可以省略,如果指定下列代码,则在测试程序里 TestUserManger 就只能通过
   接口来产生 Bean 了 -- >
```

```
        <property name = "proxyInterfaces">
            <value>lesson2.IUserManager</value>
        </property>
        <! -- 要代理的目标类 -->
        <property name = "target">
            <ref bean = "userManager" />
        </property>
        <! -- 在程序中的通知(Advice) -->
        <property name = "interceptorNames">
            <list>
                <! -- 在没有设定代理方法时用:<value>log</value> -->
                <value>logAdvisor</value>
            </list>
        </property>
    </bean>

</beans>
```

10.4　Spring 事务处理

Spring 框架提供了一致的事务管理抽象，事务处理是由多个步骤组成，这些步骤之间有一定的逻辑关系，作为一个整体的操作过程，所有的步骤必须同时成功或失败，如提交，当所有的操作步骤都被完整执行后，称为该事务被提交。回滚表示由于某个操作失败，导致所有的步骤都没被提交则事务必须回滚，回到事务执行前的状态。

如前面章节所述事务具有 4 个特性，分别为：原子性(Atomicity)、一致性(Consistency)、隔离性(Isolation) 和持久性(Durablity)。

Spring 中的事务处理是基于动态 AOP 机制的实现，其中的事务通知由元数据(目前基于 XML 或注解)驱动。代理对象与事务元数据结合产生了一个 AOP 代理，它使用一个 PlatformTransactionManager 实现配合 TransactionInterceptor，在方法调用前后实施事务。Spring 中的事务处理分为编程式事务处理和声明式事务处理两种方式。事务的属性值如表 10-1 所示。

表 10-1　事务的属性值

属性名	含　义
PROPAGATION_REQUIRED	表示如果没有事务就新建一个事务
PROPAGATION_SUPPORTS	表示如果没有事务就以非事务方式执行
PROPAGATION_MANDATORY	表示如果没有事务就抛出一个异常
PROPAGATION_REQUIRES _NEW	表示新建一个事务，如果当前事务存在就把当前事务挂起
PROPAGATION_NOT_SUPPORTS	表示以非事务方式执行，如果当前事务存在就把当前事务挂起
PROPAGATION_NEVER	表示以非事务方式执行，如果当前事务存在就抛出一个异常

10.4.1 编程式事务处理

在 Spring 中,Spring 提供两种方式的编程式事务管理,分别为使用 TransactionTemplate 和直接使用一个 PlatformTransactionManager 实现。

1. TransactionTemplate 实现事务

通过 TransactionTemplate 类能够以编程的方式实现事务控制。创建 TransactionTemplate 类的实例需要提供一个 PlatformTransactionManager 类的实例。

下面就声明式事务处理 TransactionTemplate 类方式进行介绍,实例代码如下所示。

首先创建 HelloDAO. java 文件,具体代码如下所示。

```
public class HelloADO {

        private DataSource dataSource;
        private PlatformTransactionManager platformTransactionManager;
        public void setDataSource(DataSource dataSource) {
            this.dataSource = dataSource;
        }

        public void setPlatformTransactionManager(PlatformTransactionManager platformTransactionManager) {
                this.platformTransactionManager = platformTransactionManager;
        }

            public void create(String msg){
            DefaultTransactionDefinition def = new DefaultTransactionDefinition();
            TransactionStatus status = platformTransactionManager.getTransaction(def);

              try{
            JdbcTemplate jt = new JdbcTemplate(dataSource);
            jt.update("update login set name = 'login' where password = 'password'");
            jt.update("delete from login where name = 'noTransaction'");
            /** 在下面的语句中产生了错误,所以在脱离事务处理的情况下,其后的语句都不被执行;
             * 如果是在事务处理的机制下,只要有一处错误,这个事务就不会被提交! */
            jt.update("insert into login(nadme,password) values('aa','bbbbbb')");
            jt.update("update login set name = 'loverly' where password = 'qq'");

               }catch (Exception e){
            platformTransactionManager.rollback(status);
               }finally {
            platformTransactionManager.commit(status);
               }
            }

        }
```

然后配置 ApplicationContext. xml 文件,具体配置如下所示。

```
<beans>
        <bean id="dataSource" class="org.springframework.jdbc.datasource.DriverManagerDataSource">
        <property name="driverClassName">
            <value>com.mysql.jdbc.Driver</value>
        </property>
        <property name="url">
            <value>jdbc:mysql://localhost:3306/test</value>
        </property>
        <property name="username">
            <value>root</value>
        </property>
        <property name="password">
            <value>123</value>
        </property>
        </bean>
        <bean id="platformTransactionManager" class=
        "org.springframework.jdbc.datasource.Data SourceTransactionManager">
        <property name="dataSource">
            <ref bean="dataSource"/>
        </property>
        </bean>
        <bean id="helloDAO" class="HelloDAO">
        <property name="dataSource">
            <ref bean="dataSource"/>
        </property>
        <property name="platformTransactionManager">
            <ref bean="platformTransactionManager"/>
        </property>
        </bean>
    </beans>
```

2. PlatformTransactionManager 实现事务

通常直接使用 org. springframework. transaction. PlatformTransactionManager 的实现来管理事务，只需通过 Bean 引用简单地传入一个 PlatformTransactionManager 实现，然后使用 TransactionDefinition 和 TransactionStatus 对象，就可以启动一个事务，提交或回滚。具体应用如下所示。

```
DefaultTransactionDefinition tef = new DefaultTransactionDefinition();
tef.setPropagationBehavior(TransactionDefinition.PROPAGATION_REQUIRED);

TransactionStatus status = txManager.getTransaction(def);
try {
    //执行业务逻辑
}
catch (MyException ex) {
    txManager.rollback(status);
    throw ex;
}
txManager.commit(status);
```

10.4.2　声明式事务处理

声明式事务处理在理念上和非侵入性的轻量级容器的观念是一致，即最少影响应用代码，如果在应用中存在大量事务操作，那么声明式事务管理通常便为首选，它将事务管理与业务逻辑分离，而且在 Spring 中配置也不难。

声明式事务处理有两种方式，一种是通过类的方式，一种是通过接口的方式，下面就它们分别进行介绍。

1. 通过类的方式

首先创建类 HelloADO. java 文件，代码如下所示。

```
public class HelloDAO {
        private DataSource dataSource ;
        private JdbcTemplate jdbcTemplate;
        public void setDataSource(DataSource dataSource) {
            this.dataSource = dataSource;
            jdbcTemplate = new JdbcTemplate(dataSource);
        }
        public void createStudent(String name){
            jdbcTemplate.update("insert into st(name,password) values('zhang','123')");
            jdbcTemplate.update("insert into st(name,password) values('zhang',")");
        }
        public void createTeacher(String name){
            jdbcTemplate.update("insert into st(name,password) values('test','123')");
            jdbcTemplate.update("insert into st(name,password) values('test',")");
        }
    }
```

然后配置 ApplicationContext. xml 文件，具体配置如下所示。

```
<beans>
        <bean id = "ds" class = "org.springframework.jdbc.datasource.DriverManagerDataSource">
            <property name = "driverClassName">
                <value>com.mysql.jdbc.Driver</value>
            </property>
            <property name = "url">
                <value>jdbc:mysql://localhost:3306/tw</value>
            </property>
            <property name = "username">
                <value>root</value>
            </property>
            <property name = "password">
                <value>123</value>
            </property>
        </bean>
        <bean id = "tm" class = "org.springframework.jdbc.datasource.DataSourceTransactionManager">
            <property name = "dataSource">
                <ref bean = "ds"/>
            </property>
        </bean>
```

```
<bean id = "helloDAO" class = "HelloDAO">
    <property name = "dataSource">
        <ref bean = "ds"/>
    </property>
</bean>
<! -- 声明式事务处理 -->
<bean id =
" helloDAOProxy" class = "org. springframework. transaction. interceptor. Transaction
ProxyFactoryBean">
    <property name = "transactionManager">
    <ref bean = "tm"/>
    </property>
    <property name = "target">
        <ref bean = "helloDAO"/>
    </property>
    <property name = "transactionAttributes">
        <props>
        <! -- 对 create 方法进行事务管理,PROPAGATION_REQUIRED 表示如果没有事
        务就新建一个事务 -->
            <prop key = "create * ">PROPAGATION_REQUIRED</prop>
    <prop key = "delete * ">PROPAGATION_REQUIRED</prop>
    <prop key = "find * ">PROPAGATION_REQUIRED,readOnly</prop>
        </props>
    </property>
</bean>
</beans>
```

2. 通过接口的方式

首先接口 IJdbcTemplate. java 文件,代码如下所示。

```
public interface IJdbcTemplate{
        public void setJdbcTemplate(JdbcTemplate jdbcTemplate);
        public JdbcTemplate getJdbcTemplate();
        public void insertStudent(Student student) ;
        public void insertStudents();
}
```

然后创建类 HelloADO. java 文件,代码如下所示。

```
public class HelloDAO implements IJdbcTemplate{


        private JdbcTemplate jdbcTemplate;
        public void setDataSource(DataSource dataSource) {
            jdbcTemplate = new JdbcTemplate(dataSource);
        }
       public JdbcTemplate getJdbcTemplate() {
            return this. jdbcTemplate;
        }
      public void insertStudent(Student student) {
            jdbcTemplate. update("INSERT INTO student(name,sex) VALUES ('" +
                student. getName() + " ' ,'" +       student. getSex() +    " ') ");
                System. out. println("insertStudent ok" );
        }
```

```
    public void insertStudents() {
            jdbcTemplate.update("INSERT INTO student(name,sex) VALUES ('lis','male') ");
            jdbcTemplate.update("INSER INTO student(name,sex) VALUES ('john','male') ");
            System.out.println("insertStudents ok" );
        }
}
```

然后配置 ApplicationContext.xml 文件，具体配置如下所示。

```
<beans>
    <bean id = "myDataSource" class =
    " org.springframework.jdbc.datasource.DriverManager DataSource" destroy - method = "
    close">
        <property name = "driverClassName" value = "com.mysql.jdbc.Driver" />
        <property name = "url" value = "jdbc:mysql://localhost:3306/test" />
        <property name = "username" value = "root" />
        <property name = "password" value = "123" />
    </bean>

    <bean id = "transactionTest" class = "com.bcpl.spring.HelloDAO">
        <property name = "dataSource">
            <ref bean = "myDataSource"/>
        </property>
    </bean>

    <bean id = "transManager" class = "org.springframework.jdbc.datasource.DataSourceTransactionManager">
        <property name = "dataSource">
            <ref bean = "myDataSource"/>
        </property>
    </bean>

    <bean id = "transactionInterceptor"class =
    "org.springframework.transaction.interceptor.TransactionInterceptor">
        <property name = "transactionManager">
            <ref bean = "transManager"/>
        </property>
        <property name = "transactionAttributeSource">
            <value>
                com.bcpl.spring.IJdbcTemplate. * = PROPAGATION_REQUIRED
            </value>
        </property>
    </bean>

    <bean id = "userDAOProxy" class = "org.springframework.aop.framework.ProxyFactoryBean">
        <property name = "proxyInterfaces">
            <value>com.bcpl.spring.IJdbcTemplate</value>
        </property>
        <property name = "target">
            <ref bean = "transactionTest"/>
        </property>
        <property name = "interceptorNames">
            <list>
                <value>transactionInterceptor</value>
            </list>
        </property>
    </bean>
</beans>
```

10.5　Spring 和 Struts 集成应用

在讲述 Spring 和 Struts 集成之前，首先要就集成框架 SSH 进行简要介绍，SSH 是 Struts-Spring-Hibernate 的缩写，是集成 SSH 框架的系统，从职责上分为 4 层：表示层、业务逻辑层、数据持久层和域模块层。以帮助开发人员在短期内搭建结构清晰、可复用性好和维护方便的 Web 应用程序。其中使用 Struts 作为系统的整体基础架构，负责 MVC 的分离，在 Struts 框架的模型部分控制业务跳转，利用 Hibernate 框架对持久层提供支持，Spring 做管理，管理 Struts 和 Hibernate。具体做法是：用面向对象的分析方法根据需求提出一些模型，将这些模型实现为基本的 Java 对象，然后编写基本的 DAO(Data Access Objects)接口，并给出 Hibernate 的 DAO 实现，采用 Hibernate 架构实现的 DAO 类来实现 Java 类与数据库之间的转换和访问，最后由 Spring 做管理，管理 Struts 和 Hibernate。SSH 集成框架的架构如图 10-1 所示。

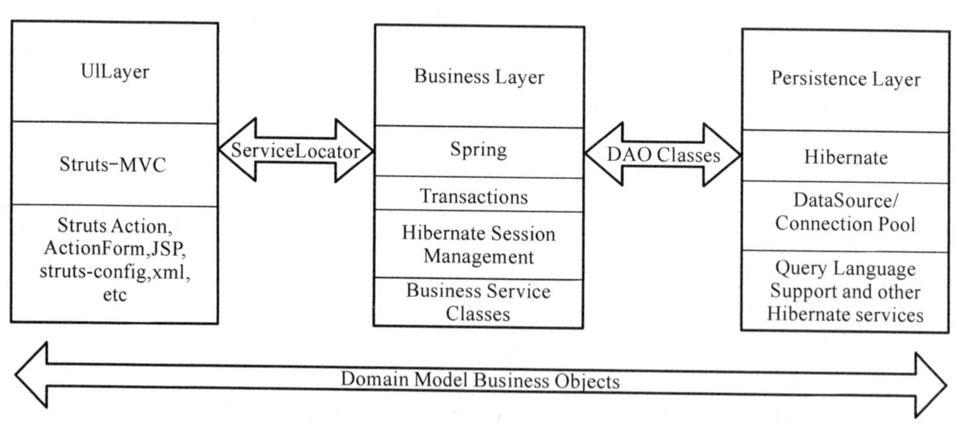

图 10-1　SSH 集成框架的系统架构

SSH 集成系统的基本业务流程如下所示。

① 在表示层中，首先通过 JSP 页面实现交互界面，负责接收请求(Request)和传送响应(Response)，然后 Struts 根据配置文件(struts-config. xml)将 ActionServlet 接收到的 Request 委派给相应的 Action 处理。

② 在业务层中，管理服务组件的 Spring IoC 容器负责向 Action 提供业务模型(Model)组件和该组件的协作对象数据处理(DAO)组件完成业务逻辑，并提供事务处理和缓冲池等容器组件以提升系统性能和保证数据的完整性。

③ 在持久层中，则依赖于 Hibernate 的对象化映射和数据库交互，处理 DAO 组件请求的数据，并返回处理结果。

采用上述开发模型，不仅实现了视图、控制器与模型的彻底分离，而且还实现了业务逻辑层与持久层的分离。这样无论前端如何变化，模型层只需很少的改动，并且数据库的变化也不会对前端有所影响，大大提高了系统的可复用性。而且由于不同层之间耦合度小，有利于团队成员并行工作，大大提高了开发效率。

在 SSH 集成框架中，Spring 和 Struts 的集成整合有 3 种方式，如下所示。

- 通过 Spring 的 ActionSupport 类。

- 通过 Spring 的 DelegatingRequestProcessor 类。
- 通过 Spring 的 DelegatingActionProxy 类。

下面就它们的集成整合方式，分别进行介绍。

1. 通过 Spring 的 ActionSupport 类

创建的方式是 Action 类不再继承 Struts 的 Action 而是继承 Spring 提供的 ActionSupport，然后在 Action 中获得 Spring 的 ApplicationContext。

其缺点是 Action 和 Spring 耦合在一起，而且 Action 不在 Spring 控制之内，也不能处理多个动作在一个 Action 中的情况。

创建工程的步骤，如下所示。

(1) 创建工程加入 Spring。

(2) 在工程中加入 Struts。

(3) 修改 Struts 配置文件 struts-config.xml，注册 ContextLoaderPlugIn 插件。具体配置如下：

```
<plug-in className="org.springframework.web.struts.ContextLoaderPlugIn">
          <set-property property=
          "contextConfigLocation" value="/WEB-INF/applicationContext.xml"/>
          </plug-in>
```

(4) 创建 Action 类，Action 的创建通过 Spring 的 ActionSupport 类而不是 Struts 的 Action 类进行扩展，创建一个新的 Action 类，具体代码如下所示。

```
public class LoginAction extends ActionSupport {


public ActionForward execute(ActionMapping mapping,ActionForm form,
HttpServletRequest request,HttpServletResponse response) {
LoginForm loginForm = (LoginForm) form;
//使用 getWebApplicationContext() 方法获得一个 ApplicationContext。为了获得业务服务，我使用
它在配置文件中查找一个 Spring Bean
ApplicationContext ac = this.getWebApplicationContext();//获得 ApplicationContext
              //查找一个 Spring Bean
LoginInterface li = (LoginInterface)ac.getBean("loginInterface");//获得 Bean
boolean you = li.checkUser(loginForm.getName(),loginForm.getPassword());
if(you){
      request.setAttribute("msg","welcome");
      return mapping.findForward("show");
          }else{
      request.setAttribute("msg","failed");
      return mapping.findForward("show");
            }
          }

      }
```

（5）配置 ApplicationContext. xml 文件，配置如下所示。

```
<beans>
        <bean id = "loginInterface" class = "spring. LoginImp"/>
</beans>
```

2. 通过 Spring 的 DelegatingRequestProcessor 类

创建思路是通过 Spring 的 DelegatingRequestProcessor 代替 Struts 的 RequstProcessor，把 Struts 的 Action 置于 Spring 的控制之下。其缺点是开发人员可以自己定义 RequestProcessor，这样就需要手工整合 Struts 和 Spring。

创建工程的具体步骤如下所示。

（1）创建工程，在工程中加入 Spring。

（2）在工程中加入 Struts。

（3）修改 Struts 配置文件 struts-config. xml，注册 ContextLoaderPlugIn 插件。具体配置如下所示。

```
<struts - config>
                <form - beans >
                <form - bean name = "loginForm" type = "com. bcpl. struts. form. LoginForm" />
                </form - beans>

                <action - mappings >
                <action
                      attribute = "loginForm"
                      input = "/login. jsp"
                      name = "loginForm"
                      path = "/login"
                      scope = "request"
                      type = "com. bcpl. struts. action. LogAction">
                      <forward name = "show" path = "/show. jsp" />
                </action>
                </action - mappings>

                <controller processorClass = "org. springframework. web. struts. DelegatingRequestProcessor">
                        </controller>
                <message - resources parameter = "com. bcpl. struts. ApplicationResources" />
                <plug - in className = "org. springframework. web. struts. ContextLoaderPlugIn">
                <set - property property = "contextConfigLocation" value =
                "/WEB - INF/application Context. xml"/>
                </plug - in>

        </struts - config>
```

(4) 创建 Action 类,具体代码如下所示。

```
public class LogAction extends Action {
                private LoginInterface logInterface ;

                public ActionForward execute( ActionMapping mapping,ActionForm form,
                    HttpServletRequest request,HttpServletResponse response) {
                    LoginForm loginForm = (LoginForm) form;
                    boolean yu = logInterface.checkUser(loginForm.getName(),loginForm.getPassword());
                    if(yu){
                         request.setAttribute("msg","welcome");
                         return mapping.findForward("show");
                    } else {
                         request.setAttribute("msg","failed");
                         return mapping.findForward("show");
                    }
                }
                public void setLogInterface(LoginInterface logInterface) {
                    this.logInterface = logInterface;
                }

           }
```

(5) 配置 ApplicationContext.xml 文件,配置如下所示。

```
<beans>
     <bean id = "loginInterface" class = "spring.LoginImp"/>
     <! -- 要和 Struts 的路径对应 -- >
     <bean name = "/login" class = "com.bcpl.struts.action.LogAction">
        <property name = "logInterface">
            <ref bean = "loginInterface"/>
        </property>
     </bean>
  </beans>
```

3. 通过 Spring 的 DelegatingActionProxy 类

创建思路是通过 Spring 的 DelegatingActionProxy 代替 Struts 的 Action,把 Struts 的 Action 置于 Spring 的控制之下,这种方式比较灵活,并且它可以利用 Spring AOP 特性的优点。创建应用的具体步骤如下所示。

(1) 创建工程,在工程中加入 Spring。

(2) 在工程中加入 Struts。

(3) 修改 Struts 配置文件 struts-config.xml,注册 ContextLoaderPlugIn 插件,具体代码如下所示。

```
<struts-config>
        <data-sources />
            <form-beans >
            <form-bean name="loginForm" type="com.bcpl.struts.form.LoginForm" />
            </form-beans>
            <action-mappings >
            <action
                attribute="loginForm"
                input="/form/login.jsp"
                name="loginForm"
                path="/login"
                scope="request"
                type="org.springframework.web.struts.DelegatingActionProxy">
                <forward name="show" path="/show.jsp" />
            </action>
            </action-mappings>
            <message-resources parameter="com.bcpl.struts.ApplicationResources" />
            <plug-in className="org.springframework.web.struts.ContextLoaderPlugIn">
            <set-property property=
            "contextConfigLocation" value="/WEB-INF/application Context.xml"/>
            </plug-in>
</struts-config>
```

(4) 创建 Action 类,具体如下所示。

```
public class LogAction extends Action {
            private LoginInterface logInterface;

            public ActionForward execute( ActionMapping mapping,ActionForm form,
                HttpServletRequest request,HttpServletResponse response) {
                LoginForm loginForm = (LoginForm) form;
                boolean you = logInterface.checkUser(loginForm.getName(),loginForm.getPassword());
                if(you){
                    request.setAttribute("msg","welcome");
                    return mapping.findForward("show");
                } else{
                    request.setAttribute("msg","failed");
                    return mapping.findForward("show");
                }
            }
            public void setLogInterface(LoginInterface logInterface) {
                this.logInterface = logInterface;
            }

    }
```

（5）配置 ApplicationContext. xml 文件，配置如下所示。

```
<beans>
  <bean id = "loginInterface" class = "spring.LoginImp"/>
      <! -- 要和 Struts 的路径对应 -->
      <bean name = "/login" class = "com.bcpl.struts.action.LogAction">
          <property name = "logInterface">
              <ref bean = "loginInterface"/>
            </property>
      </bean>
  </beans>
```

10.6　Struts-Spring-Hibernate 的集成应用

在 MyEclipse 中创建 Web 工程，工程名为 Struts 2Hello，在 lib 目录里放入下载的 Struts 2 需要的 jar 包，添加的包有 commons-fileupload-1. 3. 1. jar、commons-io-2. 2. jar、commons-logging-api-1. 1. 3. jar、freemarker-2. 3. 19. jar、javassist-3. 11. 0. GA. jar、ognl-3. 0. 6. jar、Struts 2-core-2. 3. 16. 3. jar 和 xwork-core-2. 3. 16. 3. jar。工程的目录结构及引用 jar 包如前述章节的图 5-3 所示。

1. 创建 Struts 工程

具体开发步骤如下所示。

（1）创建 Web Project 工程 ，加入 Struts 2. 3，其他包如上所述。

（2）编辑 web. xml 文件，具体配置如下所示。

```
<filter>
<filter - name>struts2</filter - name>
<filter - class>org.apache.struts2.dispatcher.FilterDispatcher</filter - class>
</filter>
<filter - mapping>
<filter - name>struts2</filter - name>
<url - pattern>/ * </url - pattern>
</filter - mapping>
```

（3）创建 LoginAction. java 类，代码如下所示。

```
/**
 * 用来处理登录请求的类
 */
package com.bcpl.sshmodel.web.action;

import com.bcpl.sshmodel.service.UserService;
import com.opensymphony.xwork2.ActionSupport;
public class LoginAction extends ActionSupport {
     private String username ;
     private String password ;
```

```
    private UserService userService ;//添加服务层接口,setter 方法

    @Override
    public String execute() throws Exception {
        System.out.println("username:" + username);
        System.out.println("password" + password);

        boolean f = userService.validate(username,password); //调用服务层方法
        if(f){
            return this.SUCCESS ;
        }else{
            return this.LOGIN ;
        }

    }
    public void setUserService(UserService userService) {
        this.userService = userService;
    }
    public String getUsername() {
        return username;
    }

    public void setUsername(String username) {
        this.username = username;
    }
    public String getPassword() {
        return password;
    }

    public void setPassword(String password) {
        this.password = password;
    }
}
```

(4) 在 src 目录下配置 struts.xml(也可以根据需要在 classes 目录下)。

```
<?xml version = "1.0" encoding = "UTF-8" ?>
<!DOCTYPE struts PUBLIC
    "-//Apache Software Foundation//DTD Struts Configuration 2.0//EN"
    "http://struts.apache.org/dtds/struts-2.0.dtd">
<struts>
    <!--
    <constant name = "struts.i18n.encoding" value = "GBK"></constant>
    -->
    <package name = "sshmodel" extends = "struts-default">
        <action name = "login" class = "loginAction">
            <result name = "success">/welcome.jsp</result>
            <result name = "login">/login.jsp</result>
        </action>
    </package>
</struts>
```

（5）添加 login.jsp 页面文件和 welcome.jsp 文件，文件内容如下所示。

login.jsp 页面：

```
<%@ page language="java" import="java.util.*" pageEncoding="GB18030"%>
        <%@ taglib prefix="s" uri="/struts-tags" %>
        <html>
            <body>
                <s:form action="login">
                    <s:textfield name="username" label="username" />
                    <s:password name="password" label="password" />
                    <s:submit></s:submit>
                </s:form>
            </body>
        </html>
```

welcome.jsp 页面：

```
<%@ page language="java" import="java.util.*" pageEncoding="GB18030"%>
<%
String path=request.getContextPath();
String basePath=request.getScheme()+"://"+request.getServerName()+":"+request.getServerPort()+path+"/";
%>

<!DOCTYPE HTML PUBLIC "-//W3C//DTD HTML 4.01 Transitional//EN">
<html>
  <head>
    <base href="<%=basePath%>">

    <title>My JSP 'welcome.jsp' starting page</title>
    <!--
    <link rel="stylesheet" type="text/css" href="styles.css">
    -->

  </head>

  <body>
    欢迎,bcpl<br>
  </body>
</html>
```

（6）部署工程和启动 Tomcat，进行工程测试。

2. Struts 集成 Spring

（1）选中工程名，然后单击右键，选择“MyEclipse”，然后单击“Add Spring Capabilities…”，打开添加 Spring 3.0 支持配置界面，如图 10-2 所示。

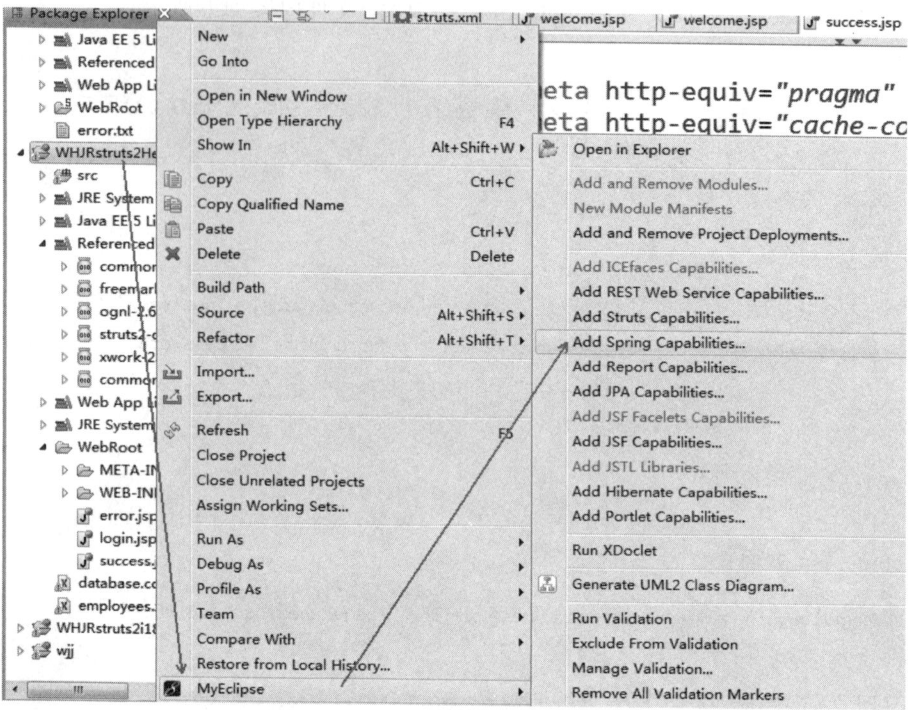

图 10-2　添加 Spring 支持配置界面

(2) 点开配置界面,选择配置 jar,如图 10-3 所示。

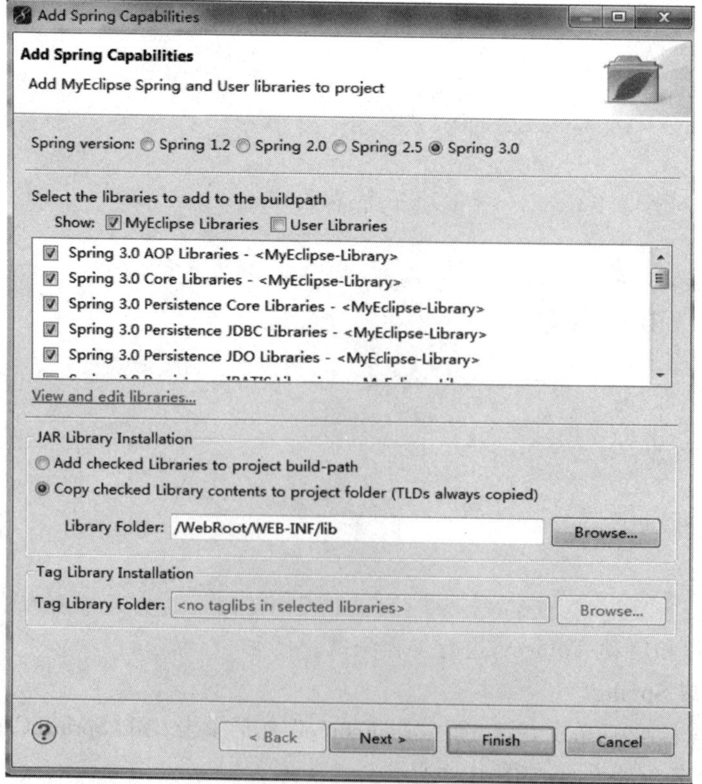

图 10-3　Spring 配置界面

(3) 进入 ApplicationContext. xml 生成界面，如图 10-4 所示。

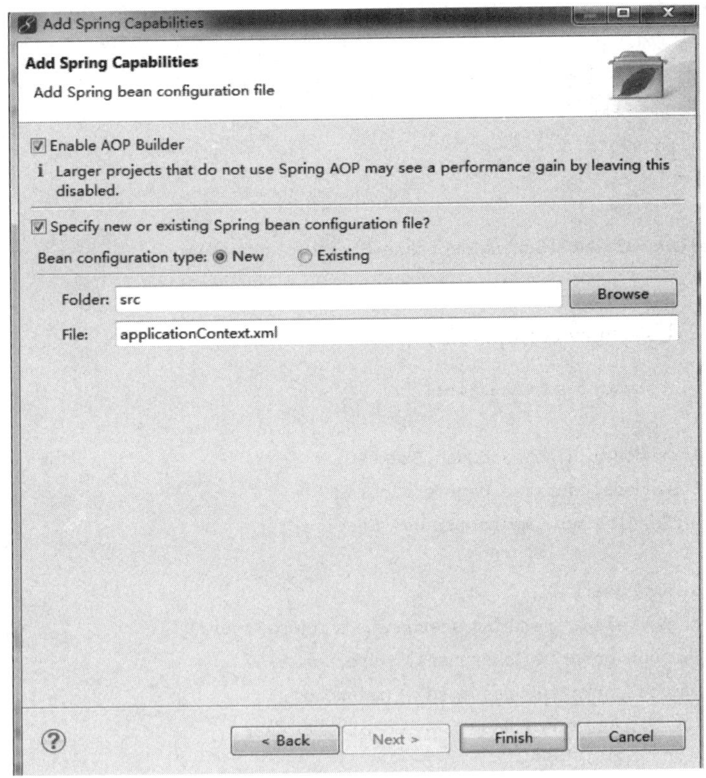

图 10-4 ApplicationContext. xml 生成界面

选择默认的“New”，Folder 选择工程下的“src”，File 为“ApplicationContext. xml”，最后单击“Finish”完成对 Spring 3. 0 的添加。

(4) Spring 3. 0 和 Struts 2. 3 整合在一起，需要添加 Struts 2-spring-plugin-2. 1. 2. jar 插件，添加到 lib 目录下，需要在 web. xml 中配置 Spring 初始化监听，如下代码所示。

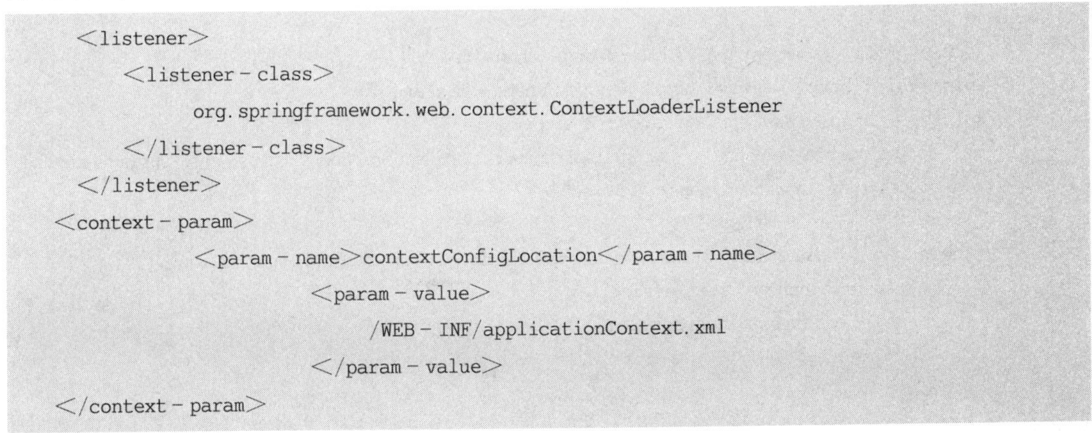

```
  <listener>
      <listener-class>
            org.springframework.web.context.ContextLoaderListener
      </listener-class>
  </listener>
<context-param>
            <param-name>contextConfigLocation</param-name>
                     <param-value>
                          /WEB-INF/applicationContext.xml
                     </param-value>
</context-param>
```

(5) 添加 Service 接口和实现类代码，并在 LoginAction 中调用(如上所述)，具体代码如下所示。

首先创建接口超级类 UserService：

```
package com.bcpl.sshmodel.service;
public interface UserService {
    /**
     *用来验证用户登录操作
     * @param username
     * @param password
     * @return
     */
    public boolean validate(String username,String password);
}
```

然后创建接口 UserServiceImpl，如下：

```
package com.bcpl.sshmodel.service.impl;

import com.bcpl.sshmodel.hibernate.dao.UserDAO;
import com.bcpl.sshmodel.service.UserService;
public class UserServiceImpl implements UserService {

    private UserDAO userDAO ;
    public boolean validate(String username,String password) {
        System.out.println("username:" + username);
        System.out.println("password" + password);
        return userDAO.checkUser(username,password) ;
    }
    public void setUserDAO(UserDAO userDAO) {
        this.userDAO = userDAO;
    }
}
```

(6) 在 ApplicationContext.xml 文件中配置 Bean，具体如下所示。

```
<?xml version = "1.0" encoding = "UTF-8"?>
<beans
    xmlns = "http://www.springframework.org/schema/beans"
    xmlns:xsi = "http://www.w3.org/2001/XMLSchema-instance"
    xmlns:p = "http://www.springframework.org/schema/p"
    xsi:schemaLocation = "http://www.springframework.org/schema/beans http://www.springframework.
    org/schema/beans/spring-beans-2.5.xsd">

    <bean id = "loginAction" class = "com.bcpl.sshmodel.web.action.LoginAction">
        <property name = "userService">
            <ref bean = "userService"/>
        </property>
    </bean>

    <bean id = "userService" class = "com.bcpl.sshmodel.service.impl.UserServiceImpl">
        <property name = "userDAO">
            <ref bean = "userDAOProxy"/>
        </property>
```

```xml
    </bean>

    <bean id = "sessionFactory" class =
    "org.springframework.orm.hibernate3.LocalSessionFactoryBean">
        <property name = "configLocation"
            value = "classpath:hibernate.cfg.xml">
        </property>
    </bean>


    <bean id = "userDAO" class = "com.bcpl.sshmodel.hibernate.dao.impl.UserDAOImpl">
        <property name = "sessionFactory">
            <ref bean = "sessionFactory"/>
        </property>
    </bean>

    <bean id = "transactionManager" class = "org.springframework.orm.hibernate3.HibernateTransactionManager">
        <property name = "sessionFactory">
            <ref local = "sessionFactory" />
        </property>
    </bean>


    <bean id = "userDAOProxy" class =
    "org.springframework.transaction.interceptor.TransactionProxyFactoryBean">
            <property name = "transactionManager">
            <ref bean = "transactionManager"/>
            </property>
            <property name = "target">
                <ref bean = "userDAO"/>
            </property>
            <property name = "transactionAttributes">
                <props>
                    <prop key = "check * ">PROPAGATION_REQUIRED</prop>
                </props>
            </property>
    </bean>

</beans>
```

同时修改 struts.xml 中<action name="login" class="loginAction">，并添加登录失

败返回页 error. jsp。

(7) 启动服务器,启动后运行登录得到成功效果, Struts 2 集成整合 Spring 3 完成。

3. 集成 Hibernate

(1) 添加工程的包结构,分别为 DAO、DAO. imp 和 PO 的类包,然后添加 Hibernate 3. 3 支持,如图 10-5 和 10-6 所示。

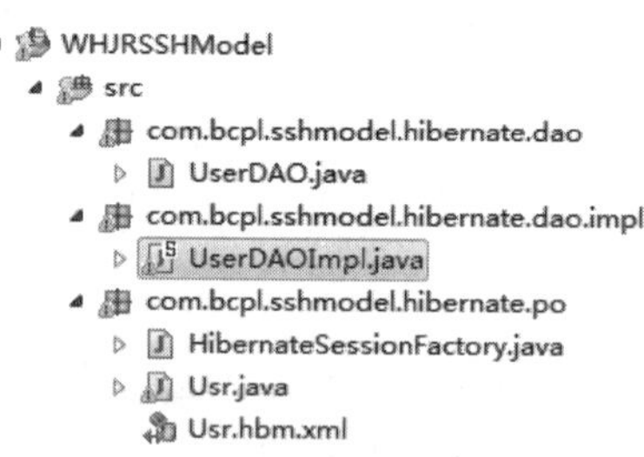

图 10-5　包结构

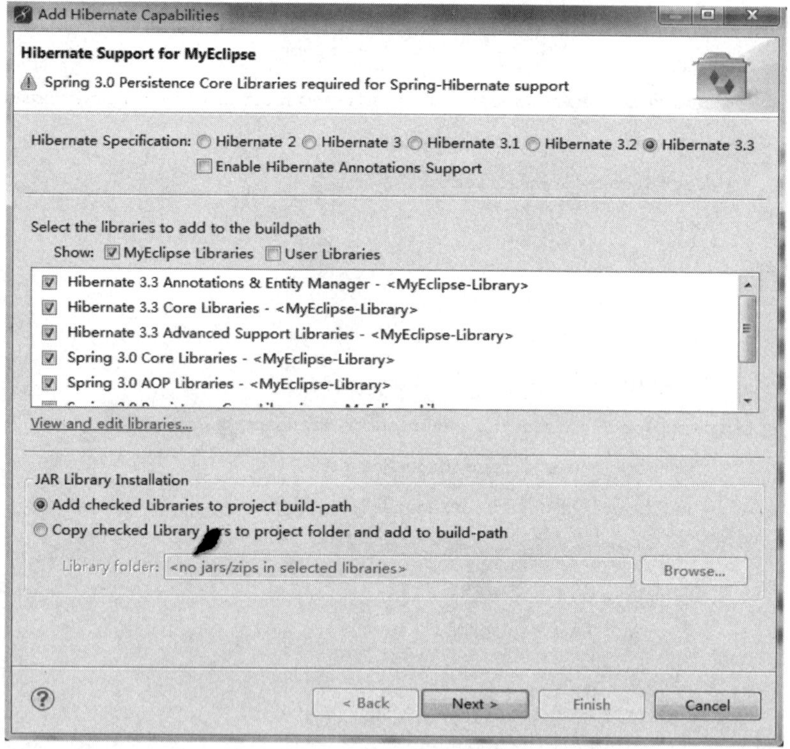

图 10-6　添加 Hibernate3. 3 支持

选择"Copy checked Library Jars to project folder and add to build-path",Library folder 路径为"WebRoot/WEB-INF/lib",选择"Next"。

(2) 选择"Spring configuration file (ApplicationContext. xml)",将 Hibernate 连接库的操作交给 Spring 来控制,然后单击"Next"。选择"Existing Spring configuration file",为前面配置好的 Spring 配置文件,SessionFaction ID 写为 SessionFactory,即为 Hibernate 产生连接的 Bean 的 id,如图 10-7 所示,然后单击"Next"。

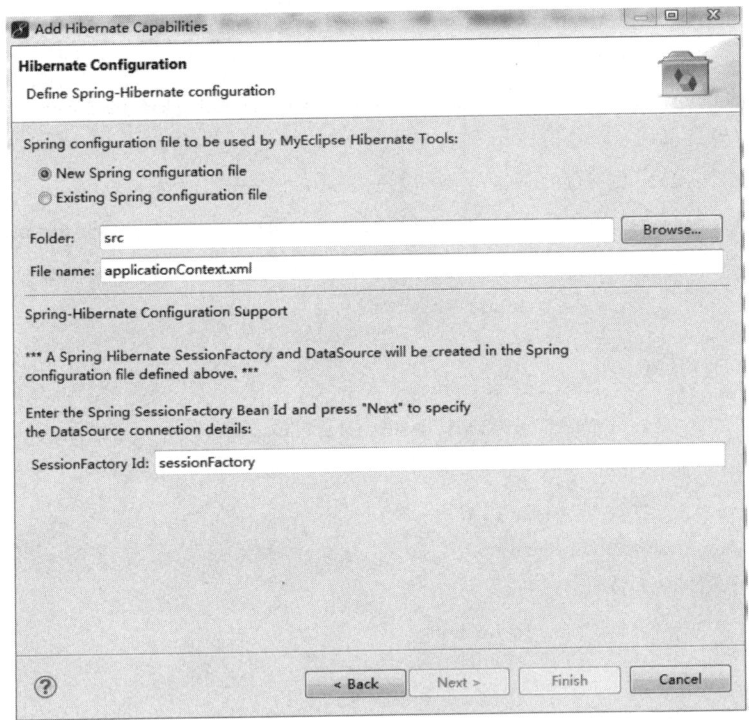

图 10-7　Hibernate 配置界面

(3) 创建数据库的表，创建语句如下所示。

```
CREATE TABLE "user" (
  "ID" int(11) NOT NULL auto_increment,
  "USERNAME" varchar(50) character set latin1 NOT NULL default '',
  "PASSWORD" varchar(50) character set latin1 NOT NULL default '',
  PRIMARY KEY  (´ID´)
) ENGINE = InnoDB DEFAULT CHARSET = utf8
```

然后在创建后的表中添加记录 insert into user (USERNAME, PASSWORD) values ('test','test')，在配置界面中配置数据库的连接部分，重要的是要将 JDBC 驱动复制到 lib 目录中，使用 MyEclipse 的数据 Database Explorer 工具创建 User. hmb. xml、AbstractUser. java 和 User. java 映射文件。

(4) 设置数据源，填写 Bean id 为 dataSource，选择 JDBC，DB Driver 为设置好的 mysql driver，如图 10-8 所示，然后选择"Next"。

(5) 因为上述已经将 SessionFactory 交由 Spring 来产生，所以取消创建 SessionFactory class，不勾选，然后单击"Finish"完成 Hibernate 支持的添加。

(6) 创建接口 UserDAO. java 和类 UserDAOImp. java，具体如下所示。

```
public interface UserDAO {
    public abstract boolean isValidUser(String username,String password);

}
```

```
import java.util.List;

import org.springframework.orm.hibernate.support.HibernateDaoSupport;

public class UserDAOImp extends HibernateDaoSupport implements UserDAO {
    private static String hql = "from User u where u.username = ? and password = ?";
        public boolean isValidUser(String username,String password) {
        String[] userlist = new String[2];
        userlist[0] = username;
        userlist[1] = password;
        List userList = this.getHibernateTemplate().find(hql,userlist);

        if (userList.size() > 0) {
            return true;
        }else
            return false;
    }
}
```

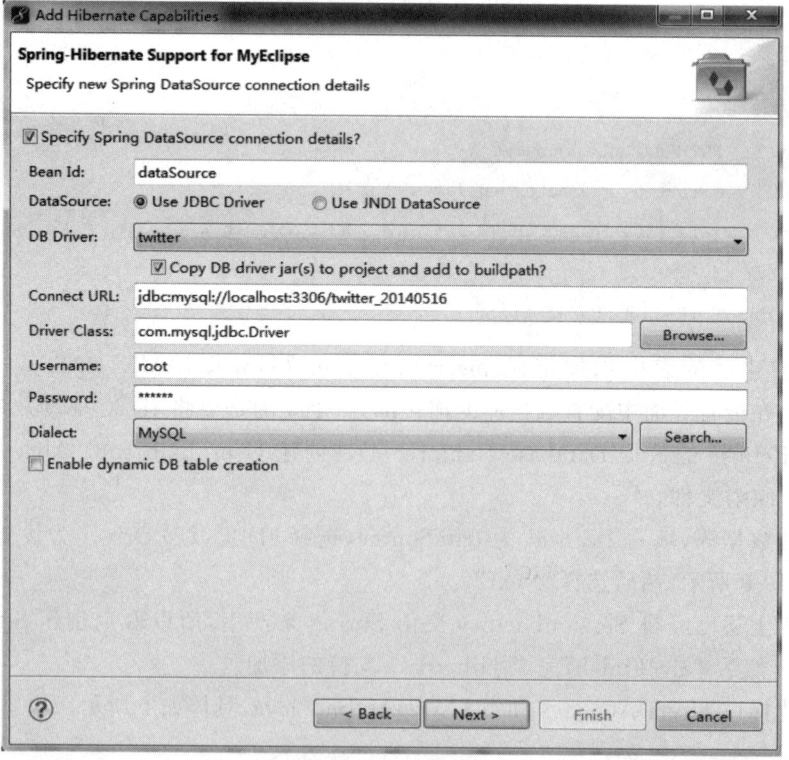

图 10-8　设置数据源

（7）修改 LoginAction.java 文件，使用 userDao 的方法来进行用户验证。

```java
package com.bcpl.struts.action;
import javax.servlet.http.HttpServletRequest;
import javax.servlet.http.HttpServletResponse;
import org.apache.struts.action.Action;
import org.apache.struts.action.ActionForm;
import org.apache.struts.action.ActionForward;
import org.apache.struts.action.ActionMapping;
import org.apache.struts.validator.DynaValidatorForm;
import com.bcpl.UserDAO;

public class LoginAction extends Action {
  private UserDAO userDAO;
  public UserDAO getUserDAO() {
      return userDAO;
  }

  public void setUserDAO(UserDAO userDAO) {
      this.userDAO = userDAO;
  }

  public ActionForward execute(ActionMapping mapping,ActionForm form,
                              HttpServletRequest request,HttpServletResponse response) {
    DynaValidatorForm loginForm = (DynaValidatorForm) form;
    String username = (String) loginForm.get("username");
    String password = (String) loginForm.get("password");
    loginForm.set("password",null);
    if (userDAO.isValidUser(username,password)) {
        return mapping.findForward("ok");
    } else {
        return mapping.getInputForward();
    }
  }
}
```

(8) 完成后，ApplicationContext.xml 中已经添加了数据源和 SessionFactory 的配置，如图 10-9 所示。

```xml
<?xml version = "1.0" encoding = "UTF-8"?>
    <!DOCTYPE beans PUBLIC "-//SPRING//DTD BEAN//EN"
    "http://www.springframework.org/dtd/spring-beans.dtd">

    <beans>
      <bean id = "dataSource" class =
      "org.apache.commons.dbcp.BasicDataSource" destroy-method = "close">
            <property name = "driverClassName">
              <value>com.mysql.jdbc.Driver</value>
            </property>
```

```
            <property name = "url">
              <value>jdbc:mysql://localhost:3306/test</value>
            </property>
            <property name = "username">
              <value>root</value>
            </property>
            <property name = "password">
              <value>123</value>
            </property>
      </bean>
```

```
<!-- 配置 sessionFactory,注意这里引入的包的不同 -->
      <bean id = "sessionFactory" class =
      "org.springframework.orm.hibernate.LocalSessionFactoryBean">
            <property name = "dataSource">
                  <ref local = "dataSource" />
            </property>
            <property name = "mappingResources">
                <list><!-- 注意这里 -->
                      <value>hibernate/po/St.hbm.xml</value>
                </list>
            </property>
            <property name = "hibernateProperties">
                <props>
                    <prop key = "hibernate.dialect">org.hibernate.dialect.MySQLDialect
                    </prop>
                      <prop key = "hibernate.show_sql">true</prop>
                </props>
            </property>
      </bean>

      <bean id = "transactionManager" class =
      "org.springframework.orm.hibernate.Hibernate TransactionManager">
            <property name = "sessionFactory">
                <ref local = "sessionFactory" />
            </property>
      </bean>

      <bean id = "userDAO" class = "com.bcpl.UserDAOImp">
            <property name = "sessionFactory">
                <ref local = "sessionFactory" />
            </property>
      </bean>

      <bean id = "userDAOProxy" class =
      "org.springframework.transaction.interceptor.TransactionProxyFactoryBean">
            <property name = "transactionManager">
```

```
                <ref bean = "transactionManager" />
            </property>
            <property name = "target">
                <ref local = "userDAO" />
            </property>
            <property name = "transactionAttributes">
                <props>
                    <prop key = "insert * ">PROPAGATION_REQUIRED</prop>
                    <prop key = "get * ">PROPAGATION_REQUIRED,readOnly</prop>
                    <prop key = "is * ">PROPAGATION_REQUIRED,readOnly</prop>
                </props>
            </property>
    </bean>

    <bean name = "/login" class = "com.bcpl.struts.action.LoginAction" singleton = "false">
        <property name = "userDAO">
            <ref bean = "userDAOProxy" />
        </property>
    </bean>
</beans>
```

配置 SessionFactory 也可以写成：

```
<bean id = "sessionFactory" class = "org.springframework.orm.hibernate.LocalSessionFactoryBean">
    <property name = "configLocation">
        <value>hibernate.cfg.xml</value>
    </property>
</bean>
```

(9) 完成了 Struts 2.3、Spring 3.0 和 Hibernate 3.3 的添加与整合，重新启动工程，进行测试。

10.7 小　　结

本章重点介绍了 Spring 的高级技术：Spring 持久层、AOP、AOP 代理及 Spring 事务编程，并就框架的分别集成进行了介绍，但应用层的框架集成 SSH 只是其中的一种方式，除此之外，还有 SSE(Struts＋Spring＋EJB)、SSJ(Spring＋Spring＋JDO)、SSJ(Struts＋Spring＋JDBC)和 FSH(Flex＋Spring＋Hibernate)等不同的组合方式，具体需要根据项目的实践进行选择，也可单独采用 Spring 框架，Spring 框架本身也是采用 MVC 设计。通常采用“客户端(浏览器)＋OS 框架＋数据库”的模式，进行项目的开发，OS 框架可以采用上述的框架集成中的任意一种，例如，本书采用的 SSH 模式，Struts 充当表示层角色，Spring 充当业务层角色，Hibernate 充当持久层角色。Struts 通过服务定位连接 Spring，Spring 通过数据访问连接 Hibernate，Hibernate 与数据库进行交互，客户端通过浏览器与 Struts 进行交互，整个过程形成完整的数据流操作、控制和访问流程。

参 考 文 献

[1] 邓子云.Java Web 轻量级开发全体验[M].北京:电子工业出版社,2012.

[2] 刘京华.Java Web 整合开发王者归来[M].北京:清华大学出版社,2010.

[3] 郑阿奇.JSP 编程教程[M].北京:电子工业出版社,2012.

[4] 毋建军,郑宝昆,郭锐.网页制作案例教程[M].北京:清华大学出版社,2011.

[5] 张浩军,张凤玲,毋建军,等.数据库设计开发技术案例教程[M].北京:清华大学出版社,2012.

[6] 刘志成.JSP 程序设计实例教程[M].北京:人民邮电出版社,2009.

[7] 陆荣幸,郁洲,阮永良,等.J2EE 平台上 MVC 设计模式的研究与实现[J].计算机应用研究,2003(3):144-146.